Studies in Classification, Data Analysis, and Knowledge Organization

More information about this series at http://www.springer.com/series/1564

Theodore Chadjipadelis · Berthold Lausen ·
Angelos Markos · Tae Rim Lee ·
Angela Montanari · Rebecca Nugent
Editors

Data Analysis and Rationality in a Complex World

Springer

Editors
Theodore Chadjipadelis
Department of Political Sciences
Aristotle University of Thessaloniki
Thessaloniki, Greece

Berthold Lausen
Department of Mathematical Sciences
University of Essex
Colchester, UK

Angelos Markos
School of Education
Democritus University of Thrace
Alexandroupolis, Greece

Tae Rim Lee
Department of Data Science and Statistics
Korea National Open University
Seoul, Korea (Republic of)

Angela Montanari
Department of Statistical Sciences
"Paolo Fortunati"
University of Bologna
Bologna, Italy

Rebecca Nugent
Department of Statistics and Data Science
Carnegie Mellon University
Pittsburgh, PA, USA

ISSN 1431-8814 ISSN 2198-3321 (electronic)
Studies in Classification, Data Analysis, and Knowledge Organization
ISBN 978-3-030-60103-4 ISBN 978-3-030-60104-1 (eBook)
https://doi.org/10.1007/978-3-030-60104-1

Mathematics Subject Classification: 62-XX, 62-06, 62-07, 62Hxx, 62H30, 62Pxx, 62Jxx

This Springer imprint is published by the registered company Springer Nature Switzerland AG
The registered company address is: Gewerbestrasse 11, 6330 Cham, Switzerland

Preface

This volume contains revised versions of the selected papers presented at the 16th Biennial Conference of the International Federation of Classification Societies (IFCS 2019) organized by the Greek Society of Data Analysis (GSDA), held in Thessaloniki, Greece on 26–29 August 2019. The theme of the conference was "Data Analysis and Rationality in a Complex World". Rationality is a critical issue, as we experience it today. The COVID-19 outbreak revealed also the complexity. Data Analysis is -not the only, but a critical tool for handling information and making decisions under uncertainty on many occasions and for many scientific areas. Rationality is about decision-making based on facts, political and social choice, and the Interest of the People. Authorities, universities, and institutions should take care in order to improve everyday life and solve major political and social problems bringing together Data Science [improve rationality], free and fair Elections [secure free choice and responsibility], and Governance [handling a complex World].

Theodore Chadjipadelis (Aristotle University of Thessaloniki) chaired the Local Organizing Committee and the Scientific Program Committee with Berthold Lausen (IFCS President) and Tae Rim Lee (Korea National Open University) as the vice-chairpersons. The conference encompassed 178 presentations in 56 sessions, including 8 plenary talks and 2 workshops. With 224 attendees from 29 countries, the conference provided a very attractive interdisciplinary international forum for discussion, mutual exchange of knowledge, and cross-disciplinary cooperation.

This volume presents 37 articles dealing with theoretical aspects, methodological advances, and practical applications in domains relating to classification and clustering. The contributions were selected in a second reviewing process after the conference. In addition to the fundamental areas of classification and clustering, the volume contains manuscripts concerning data analysis and statistical modelling in application areas such as economics and finance, computer science, political science, and education. The contributions are listed in alphabetical order with respect to the authors' names.

For the convenience of the reader, the content of this volume is briefly reviewed: *Bellanger et al.* present an agglomerative hierarchical clustering method with temporal ordering constraints. *Chadjipadelis & Teperoglou* employ hierarchical clustering and multiple correspondence analysis to analyze political competition in EU member states at the occasion of the 2019 European Parliament elections. *Champagne Gareau et al.* present a graph clustering technique to improve the efficiency of an electric vehicle planner. *Di Mari et al.* present an approach for computing the coefficient of determination for mixtures of regressions in the Gaussian framework. *Dziechciarz & Dziechciarz-Duda* present a procedure for survey data collection based on fuzzy coding. *Ferreira & Marques* study the relationships between performance measures in discrete supervised classification. *Ganczarek-Gamrot et al.* evaluate value-at-risk measures to assess the risk of price changes in the energy market. *Górecki et al.* define and evaluate measures of mutual dependence for multivariate functional data. *Iodice D'Enza et al.* present a chunk-wise version of iterative principal component analysis for single imputation of "tall" data sets. *Jimeno et al.* run a benchmarking study to evaluate the performance of different clustering methods for mixed-type data. *Kazana et al.* employ a joint dimension reduction and clustering approach to investigate entrepreneurs' attitudes toward a green infrastructure plan. Kitanishi et al. apply a topological data analysis mapper and a spatial perception method to systematically visualize the relationships among pharmaceutical data. *Koutsoupias & Mikelis* combine the use of text mining and multivariate data analysis methods to explore a set of textual documents. *Krężołek & Trzpiot* present an approach to estimate extreme risk using the Hill estimator and its modifications. *Lelu & Cadot* evaluate a series of clustering methods on text data. *Liang & Lee* present experimental results to obtain a rule-of-thumb for choosing the basis spacing for process convolution Gaussian process models. *McLachlan & Ahfock* review and present new results about using the Gaussian mixture model for partially classified data. *Menexes & Koutsos* combine correspondence analysis and ordinary kriging to display values of quantitative variables as supplementary onto factorial maps. *Moschidis & Thanopoulos* apply dimension reduction and clustering techniques to study heterogeneity in e-commerce data from official statistics. *Murugesan et al.* run a benchmarking study to highlight the advantages and drawbacks of spectral clustering, DBSCAN, and k-means on simulated and empirical data. Nakayama employs Bayesian network analysis to model trends in consumer web communication data of new products. *Nicolussi et al.* consider chain graph models for categorical variables to evaluate the level of perceived health in the EU. *Nienkemper-Swanepoel et al.* present a visualization approach to identify the missing data mechanism in incomplete multivariate categorical data. *Okada & Yokoyama* introduce a procedure for assembling one-mode three-way proximities from one-mode two-way proximities, and a method for hierarchical clustering of one-mode three-way proximities. *Panagiotidou & Chadjipadelis* explore the views and attitudes of first-time young voters about Europe and Democracy using multivariate data analysis techniques. *Pratsinakis et al.* compare hierarchical clustering approaches for binary data from molecular markers using external criteria for cluster validation. Smaga introduces permutation and bootstrap tests for the repeated measures analysis of variance for functional data. *Sokołowski & Markowska* present an algorithm for creating a robust distance matrix between observations with outliers. *Srakar & Vecco* present a clustering algorithm for polygonal data. *Stalidis et al.* evaluate the performance of multiple correspondence analysis and hierarchical clustering, as well as a two-layer shallow neural network for personalized supermarket offer recommendations. *Szilágyi & Lengyel* present the results of an empirical study on what motivates the participants of the sharing economy in Hungary using structural equation modeling. *Tai & Frisoli* run a benchmark comparison of minimax linkage to other hierarchical clustering methods using multiple performance metrics on data sets with known clustering structure. *Trejos-Zelaya et al.* implement and evaluate clustering algorithms based on combinatorial optimization metaheuristics. *Tsimperidis et al.* employ keystroke dynamics and machine learning models to classify unknown Internet users according to age, handedness, and educational level. *Varga & Fodor* use hierarchical clustering to derive a typology of critical raw materials with regard to technological innovation. *Vicente-Villardón et al.* extend redundancy analysis to binary data using logistic regression. *Warrens & Ebert* study the predictive power of cluster solutions based on normal mixture models when relevant outcomes are involved in the estimation procedure, using a real-world data set on school motivation.

We would like to express our gratitude to all members of the scientific program committee, for their ability in attracting interesting contributions. A special thanks is due to the local organizing committee for a well-organized conference. We also thank the session organizers for supporting the spread of information about the conference, and for inviting speakers, the reviewers for their timely reports, and Veronika Rosteck and Boopalan Renu of Springer Nature for their support and dedication to the production of this volume. Last but not least, we would like to thank all participants of the conference for their interest and various activities which made the IFCS 2019 conference and this volume an interdisciplinary possibility for scientific discussion.

Colchester, UK
Thessaloniki, Greece
Alexandroupolis, Greece
Seoul, Korea (Republic of)
Bologna, Italy
Pittsburgh, USA
June 2020

Berthold Lausen
Theodore Chadjipadelis
Angelos Markos
Tae Rim Lee
Angela Montanari
Rebecca Nugent

Contents

Contributors

Daniel Ahfock University of Queensland, Brisbane, QLD, Australia

Luis Eduardo Amaya-Briceño Guanacaste Campus, University of Costa Rica, Liberia, Costa Rica

Avi Arampatzis Democritus University of Thrace, Xanthi, Greece

Éric Beaudry Université du Québec á Montréal, QC, Montréal, Canada

Lise Bellanger Université de Nantes Laboratoire de Mathématiques Jean Leray UMR CNRS 6629, Nantes Cedex 03, France

Martine Cadot LORIA Nancy France, Vandoeuvre-lès-Nancy, France

Manuela Cazzaro University of Milano-Bicocca, Milano MI, Italy

Theodore Chadjipadelis School of Political Sciences, Aristotle University of Thessaloniki, Thessaloniki, Greece

Jaël Champagne Gareau Université du Québec á Montréal, QC, Montréal, Canada

Irene Cho Department of Mathematics and Statistics, San José State University, San José, CA, USA

Efthimia Chrisanthidou International Hellenic University, Drama, Greece

Arthur Coulon CNRS/Université de Tours, UMR 7324 CITERES, Laboratoire Archéologie et Territoires, Tours, France

Marina Delianidi International Hellenic University, Thermi, Greece

Agnese Maria Di Brisco University of Milano-Bicocca, Milano MI, Italy

Roberto Di Mari Department of Economics and Business, University of Catania, Catania, Italy

Konstantinos Diamantaras International Hellenic University, Thermi, Greece

Józef Dziechciarz Wrocław University of Economics, Wroclaw, Poland

Marta Dziechciarz-Duda Wrocław University of Economics, Wroclaw, Poland

W. Miro Ebert University of Groningen, Faculty of Behavioural and Social Sciences, Groningen, TS, The Netherlands

Ilias Eleftherohorinos Aristotle University of Thessaloniki, Thessaloniki, Greece

Ana Sousa Ferreira Faculdade de Psicologia, Universidade de Lisboa, Business Research Unit (BRU-IUL), Lisboa, Portugal

Kitti Fodor University of Miskolc, Institute of Economic Theory and Methodology, Miskolc-Egyetemváros, Hungary

Kayla Frisoli Carnegie Mellon University, Pittsburgh, PA, USA

Alicja Ganczarek-Gamrot University of Economics in Katowice, Katowice, Poland

Sugnet Gardner-Lubbe Stellenbosch University, Stellenbosch, South Africa

Tomasz Górecki Faculty of Mathematics and Computer Science, Adam Mickiewicz University, Poznań, Poland

Philippe Husi CNRS/Université de Tours, UMR 7324 CITERES, Laboratoire Archéologie et Territoires, Tours, France

Masaya Iizuka Okayama University, Okayama, Japan

Salvatore Ingrassia Department of Economics and Business, University of Catania, Catania, Italy

Alfonso Iodice D'Enza Universitá degli Studi di Napoli Federico II, Napoli, Italy

Fumio Ishioka Okayama University, Okayama, Japan

Jarrett Jimeno Department of Mathematics and Statistics, San José State University, San José, CA, USA

Alejandra Jiménez-Romero School of Mathematics, Costa Rica Institute of Technology, Cartago, Costa Rica

Pantelis I. Kaplanoglou International Hellenic University, Thermi, Greece

Lefkothea Karapetsi Centre for Research and Technology, Thessaloniki, Greece

Iphigenia Karaveli International Hellenic University, Thermi, Greece

Alkiviadis Katsalis International Hellenic University, Thermi, Greece

Stella Kazakli International Hellenic University, Drama, Greece

Angelos Kazaklis OLYMPOS Non Profit Integrated Centre for Environmental Management, Drama, Greece

Vassiliki Kazana Department of Forestry and Natural Environment 1st km Drama-Mikrohori, International Hellenic University, Drama, Greece

Yoshitake Kitanishi Okayama University, Okayama, Japan

Thomas Koutsos School of Agriculture Faculty of Agriculture Forestry and Natural Environment Hellas, Aristotle University of Thessaloniki, Thessaloniki, Greece

Nikos Koutsoupias University of Macedonia, Thessaloniki, Greece

Dominik Krężołek Department of Demographics and Economic Statistics, University of Economics in Katowice, Katowice, Poland

Mirosław Krzyśko Interfaculty Institute of Mathematics and Statistics, Calisia University, Kalisz, Poland

Koji Kurihara Okayama University, Okayama, Japan

Niël Le Roux Stellenbosch University, Stellenbosch, South Africa

Herbert K. H. Lee University of California, Santa Cruz, USA

Alain Lelu Université de Franche-Comté (rtd), Besançon, France

Levente Lengyel Institute of Economic Theory and Methodology, University of Miskolc, Miskolc-Egyetemváros, Hungary

Waley W. J. Liang University of California, Santa Cruz, USA

Panagiotis Madesis Centre for Research and Technology, Thessaloniki, Greece

Vladimir Makarenkov Université du Québec á Montréal, QC, Montréal, Canada

Angelos Markos Democritus University of Thrace, Alexandroupoli, Greece

Małgorzata Markowska Wrocław University of Economics and Business, Wrocław, Poland

Anabela Marques Escola Superior de Tecnologia do Barreiro, IPS, CIIAS, Barreiro, Portugal

Geoffrey McLachlan University of Queensland, Brisbane, QLD, Australia

George Menexes School of Agriculture Faculty of Agriculture Forestry and Natural Environment Hellas, Aristotle University of Thessaloniki, Thessaloniki, Greece

Kyriakos Mikelis University of Macedonia, Thessaloniki, Greece

Stratos Moschidis Hellenic Statistical Authority, Piraeus, Greece

Alex Murillo-Fernández Atlantic Campus, University of Costa Rica, Turrialba, Costa Rica

Nivedha Murugesan Department of Mathematics and Statistics, San José State University, San José, CA, USA

Atsuho Nakayama Tokyo Metropolitan University, Hachioji-shi, Japan

Federica Nicolussi University of Milan, Milano MI, Italy

Johané Nienkemper-Swanepoel Stellenbosch University, Stellenbosch, South Africa

Symela Ntoanidou Aristotle University of Thessaloniki, Thessaloniki, Greece

Akinori Okada Rikkyo University 3-18-1 Ozenji Higashi Asao-ku Kawasaki-shi, Kanagawa-ken, Japan

Francesco Palumbo Universitá degli Studi di Napoli Federico II, Napoli, Italy

Georgia Panagiotidou School of Political Sciences, Aristotle University of Thessaloniki, Thessaloniki, Greece

Georgios Peikos Democritus University of Thrace, Xanthi, Greece

Eduardo Piza-Volio CIMPA & School of Mathematics, Faculty of Science, University of Costa Rica, San José, Costa Rica

Emmanouil D. Pratsinakis Aristotle University of Thessaloniki, Thessaloniki, Greece

Antonio Punzo Department of Economics and Business, University of Catania, Catania, Italy

Dimitrios Raptis International Hellenic University, Drama, Greece

Madhumita Roy Department of Mathematics and Statistics, San José State University, San José, CA, USA

Theodosios Siomos International Hellenic University, Thermi, Greece

Łukasz Smaga Faculty of Mathematics and Computer Science, Adam Mickiewicz University, Poznań, Poland

Andrzej Sokołowski Cracow University of Economics, Cracow, Poland

Andrej Srakar Institute for Economic Research (IER), Ljubljana and Faculty of Economics, University of Ljubljana, Ljubljana, Slovenia

George Stalidis International Hellenic University, Thermi, Greece

Roland Szilágyi Institute of Economic Theory and Methodology, University of Miskolc, Miskolc-Egyetemváros, Hungary

Xiao Hui Tai University of California, Berkeley, CA, USA

Eftichia Teperoglou School of Political Sciences, Aristotle University of Thessaloniki, Thessaloniki, Greece

Athanasios Thanopoulos Hellenic Statistical Authority, Piraeus, Greece

Cristina Tortora Department of Mathematics and Statistics, San José State University, San José, CA, USA

Javier Trejos-Zelaya CIMPA & School of Mathematics, Faculty of Science, University of Costa Rica, San José, Costa Rica

Grażyna Trzpiot Department of Demographics and Economic Statistics, University of Economics in Katowice, Katowice, Poland

Ioannis Tsimperidis Democritus University of Thrace, Xanthi, Greece

Beatrix Varga University of Miskolc, Institute of Economic Theory and Methodology, Miskolc-Egyetemváros, Hungary

Marilena Vecco Burgundy School of Business—Universite Bourgogne Franche-Comte, Bourgogne Franche-Comte, France

Laura Vicente-Gonzalez Departamento de Estadística, Universidad de Salamanca, Salamanca, Spain

Jose L. Vicente-Villardon Departamento de Estadística, Universidad de Salamanca, Salamanca, Spain

Mario Villalobos-Arias CIMPA & School of Mathematics, Faculty of Science, University of Costa Rica, San José, Costa Rica

Matthijs J. Warrens University of Groningen, Faculty of Behavioural and Social Sciences, Groningen Institute for Educational Research, Groningen, TG, The Netherlands

Waldemar Wołyński Faculty of Mathematics and Computer Science, Adam Mickiewicz University, Poznań, Poland

Satoru Yokoyama Aoyama Gakuin University 4-4-25 Shibuya, Tokyo, Japan

Nefeli Zagourgini International Hellenic University, Drama, Greece

PerioClust: A Simple Hierarchical Agglomerative Clustering Approach Including Constraints

Lise Bellanger, Arthur Coulon, and Philippe Husi

Abstract PerioClust is a hierarchical agglomerative clustering (HAC) method including temporal (resp. spatial) ordering constraints. This new semi-supervised learning algorithm is designed to consider two potentially error-prone sources of information associated with the same observations. One reflects dissimilarities in the "feature space" and the other the temporal (resp. spatial) constraint structure between the observations. A distance-based approach is adopted to modify the distance measure in the classical HAC algorithm using a convex combination to take into account the two initial dissimilarity matrices. The choice of the mixing parameter is, therefore, the key point. We define a criterion based on cophenetic distances, as well as a resampling procedure to ensure the good robustness of the proposed clustering method. The dendrogram associated with this HAC can be interpreted as the result of a compromise between each source of information analysed separately. We illustrate our clustering method on two real data sets: (i) an archaeological one containing temporal information, (ii) a socio-economical one containing geographical information.

Keywords Semi-supervised learning algorithm · Non-strict constrained approach · Hierarchical agglomerative clustering · Weighted average distance · Cophenetic matrix

L. Bellanger (✉)
Université de Nantes Laboratoire de Mathématiques Jean Leray UMR CNRS 6629, 2 rue de la Houssinière BP 92208, 44322 Nantes Cedex 03, France
e-mail: lise.bellanger@univ-nantes.fr

A. Coulon · P. Husi
CNRS/Université de Tours, UMR 7324 CITERES, Laboratoire Archéologie et Territoires, 40 rue James Watt, ActiCampus 1, 37200 Tours, France

T. Chadjipadelis et al. (eds.), *Data Analysis and Rationality in a Complex World*, Studies in Classification, Data Analysis, and Knowledge Organization,
https://doi.org/10.1007/978-3-030-60104-1_1

1 Introduction and Motivation

Clustering problems can be addressed with a variety of methods, all requiring dedicated techniques for the data preprocessing phase of its own. There is an abundant literature on the clustering subject, see, for example, Aggarwal and Reddy (2014), Everitt et al. (2001), Kaufman and Rousseeuw (2005). The two most widely used clustering algorithms are partitional and hierarchical clustering. In this paper, we concentrate on hierarchical clustering whose approach consists in developing a binary-tree-based data structure called the dendrogram. More specifically we propose a new constrained Hierarchical Agglomerative Clustering (HAC) method named PerioClust, which belongs to the class of semi-supervised learning algorithms. It is a non-strict constrained procedure that has been originally developed to answer chronological problems in archaeology based on artefacts data. The method is designed to consider two potentially error-prone sources of information associated with the same observations. One reflects dissimilarities in the "feature space" and the other the temporal (resp. spatial) constraint structure between the observations. A distance-based approach is adopted to modify the distance measure in the classical HAC algorithm using a convex combination of the two initial dissimilarity matrices. The choice of the mixing parameter is, therefore, the key point. We define a selection criterion for this parameter based on cophenetic distances, as well as a resampling procedure to ensure the good robustness of the proposed clustering method.

This article is organized as follows. In Sect. 2, we describe the existing methods. In Sect. 3, we present the proposed procedure. In Sect. 4, we illustrate and compare our approach using two real datasets.

2 Existing Constrained HAC Methods

Constrained clustering is a class of semi-supervised learning algorithms that differ from its unconstrained counterpart in that the only admissible clusters are those that more or less strictly respect the relationship. User-specified constraints we are interested in here are those called instance-level constraints (Davidson and Basu 2007), specifying requirements on pairs of objects. Several researchers proposed extending classical algorithms for handling instance-level constraints. Two general approaches exist: (i) constrained-based ones where the clustering algorithm is modified to integrate pairwise constraints, (ii) distance-based ones where only the distance measure is modified in the existing clustering algorithm.

In constrained-based approach, the HAC methods included in the Lance and Williams general clustering model are easily modified to incorporate the constraint of continuity. Clustering algorithms with temporal (or spatial) constraint need to state unambiguously which objects are neighbours. The most common constrained clustering solution is to use simple connecting schemes, proceeding as described in Legendre and Legendre (2012) in the case of temporal constraints. In this approach called the chronological clustering method, the constraint is imposed on the clustering activity. Another constrained-based approach, proposed by Chavent et al. (2018) and called hclustgeo, consists of proposing a Ward-like hierarchical algorithm including spatial constraints through two dissimilarity matrices and a mixing parameter. It has the potential advantage to be a non-strict constrained procedure. However, it imposes the underlying aggregation measure which leads to a Ward-like hierarchical clustering process. In general terms, HAC with constraint-based approaches have some disadvantages: (i) it can occasionally produce reversals in the dendrogram, except with complete linkage (Ferligoj and Batagelj 1982), (ii) it usually considers only the dissimilarities between linked units (e.g. chronological clustering method), which may be too restrictive in some fields such as archaeology, as we will see below, (iii) the choice and interpretation of the mixing parameter could be sensitive points (e.g. hclustgeo method).

Finally, HAC also has a long history of using spatial constraints to find specific types of clusters with the distance-based approach: the dissimilarity matrix is modified differently as, for example, a combination of geographical dissimilarities and dissimilarities on non-geographical variables. But a problem arises on how to define weight given to the geographical dissimilarities in the combination. In this work, we propose a non-strict constrained HAC approach that takes up this idea by (i) adapting it to the temporal or spatial constraints and (ii) defining weights in an objective and interpretable way.

3 The Proposed Clustering Method: PerioClust

3.1 A Distance-Based Approach

Let us consider a set of n observations. Let $\mathbf{D}_1$ be a $n \times n$ normalized[1] dissimilarity matrix (not necessary an Euclidean distance) giving dissimilarity values in the "feature space". Let $\mathbf{D}_2$ be a $n \times n$ matrix containing the normalized dissimilarities in the "constraint space". In this clustering work, we apply the HAC method to the following convex combination:

$$\mathbf{D}_\alpha = \alpha \mathbf{D}_1 + (1 - \alpha)\mathbf{D}_2 \tag{1}$$

where $\alpha \in [0; 1]$ is a fixed parameter given the importance of each dissimilarity matrix in the clustering procedure. Formula (1) defines a weighted average distance and as such makes it possible to weight each of the two sources of information calculated on the data set. When $\alpha = 0$ (resp. $\alpha = 1$), the dissimilarities obtained from dissimilarity matrix $\mathbf{D}_1$ (resp. $\mathbf{D}_2$) are not taken into account in the hierarchical clustering process. An agglomerative strategy could be chosen among those satisfying the Lance and Williams formulation. Thus, the key point here is the choice of α.

[1]Dissimilarity values are between 0 and 1.

Sokal and Rohlf (1962) developed a simple criteria, the cophenetic correlation, which provides a simple and effective method for comparing dendrograms of various sorts. The starting point is the so-called cophenetic matrix whose elements are the dissimilarity levels at which objects become members of the same cluster in the dendrogram. The correlation between the original dissimilarities and the cophenetic dissimilarities (called cophenetic correlation) is a "suitability index" of the clustering. It judges the extent to which the hierarchical structure produced by a dendrogram actually represents the data itself (see Everitt et al. 2001; Sokal and Rohlf 1962). We will base the determination of α on the optimization of an objective function based on the spirit of the cophenetic correlation.

We defined the following criterion that "balances" the weight of $\mathbf{D}_1$ and $\mathbf{D}_2$ in the final clustering:

$$CorCrit_\alpha = |Cor(\mathbf{D}_\alpha^{coph}, \mathbf{D}_1) - Cor(\mathbf{D}_\alpha^{coph}, \mathbf{D}_2)| \tag{2}$$

where $\mathbf{D}_\alpha^{coph}$ is the cophenetic matrix obtained from the HAC dendrogram with α fixed in (1). The $CorCrit_\alpha$ criterium in (2), therefore, represents the difference in absolute value between two correlations, each correlation measuring how faithfully the dendrogram with $\mathbf{D}_\alpha^{coph}$ preserves the pairwise distances between the original data points. The first correlation is associated with the comparison between the original dissimilarity matrix $\mathbf{D}_1$ and $\mathbf{D}_\alpha^{coph}$ with α fixed; while the second one compares the original dissimilarity matrix $\mathbf{D}_2$ with $\mathbf{D}_\alpha^{coph}$, α fixed. Then, in order to compromise between the information provided by $\mathbf{D}_1$ and $\mathbf{D}_2$, we determine α with $\hat{\alpha}$ such that

$$\hat{\alpha} = \text{argmin}_\alpha \, CorCrit_\alpha \tag{3}$$

In practice, we use a one-dimensional optimization procedure, combination of golden section search and successive parabolic interpolation, to obtain the value $\hat{\alpha}$ that minimizes (3). However, since this choice does not consider the potential errors in the data corpus, we also decide to create a resampling procedure adapted to obtain a percentile confidence interval for α and study its variability.

3.2 Resampling Strategy

To do this, a set of "clones" is created for each observation. A clone c of observation $i \in \{1, ..., n\}$ is defined as a copy of observation i for which dissimilarities in the "feature space" have the same values as observation i but those in the "constraint space" have been modified taking into account for a fixed i all possible profiles $i' \neq i$. Let $\mathbf{D}_1^{(c)}$ and $\mathbf{D}_2^{(c)}$ be the two $(n+1) \times (n+1)$ dissimilarity augmented matrices for clone $c \in \{1, ..., n(n-1)\}$. A HAC is then carried out using the combination defined in (1) with $\mathbf{D}_1^{(i)}$ and $\mathbf{D}_2^{(i(i'))}$. Let $CorCrit_\alpha^{(c)}$ define the same criterion as in (2) in which $\mathbf{D}_1$ and $\mathbf{D}_2$ are replaced, respectively, by $\mathbf{D}_1^{(c)}$ and $\mathbf{D}_2^{(c)}$. The adaptation of the previous reasoning to estimate α with (3) using our resampling procedure leads us to define

$$\hat{\alpha}^{(c)} = \text{argmin}_{\alpha}\, CorCrit_{\alpha}^{(c)};\, c \in \{1, ..., n(n-1)\} \tag{4}$$

Based on replications, the same spirit as the Bootstrap method Efron and Tibshirani (1993), we obtain an estimated percentile confidence interval using the empirical percentiles of the distribution of $\hat{\alpha}^{(c)}$, average and standard error. A HAC can then be made using $\mathbf{D}_{\hat{\alpha}}$ or $\mathbf{D}_{\tilde{\alpha}}$, where $\tilde{\alpha}$ is in the vicinity of $\hat{\alpha}$ with respect to the confidence interval $CI_{95\%}\ (\alpha)$.

4 Results and Comparison on Two Real Datasets

In this section, we present the results obtained with our distance-based approach PerioClust on two real datasets: (i) Angkor data with temporal constraint, (ii) Estuary data with geographical constraint. We also compare them with those obtained with the constrained-based approach hclustgeo (Chavent et al. 2018), a comparable non-strict constrained HAC with a mixing parameter. All statistical analyses were performed using R.[2]

4.1 Archaeological Dataset: Temporal Constraints

The archaeological data come from excavations carried out in Angkor Thom (Cambodia), capital of the Khmer empire between the ninth and fifteenth centuries (Gaucher 2004). One of the major objectives here is to specify the periodization of the city, particularly from the seriation diagram (Fig. 1) otherwise called "Harris matrix" (Harris 1989) with connected sets coming from 3 disconnected archaeological sites and assemblages of pottery (quantities of different types of sherds of pottery contained in sets).

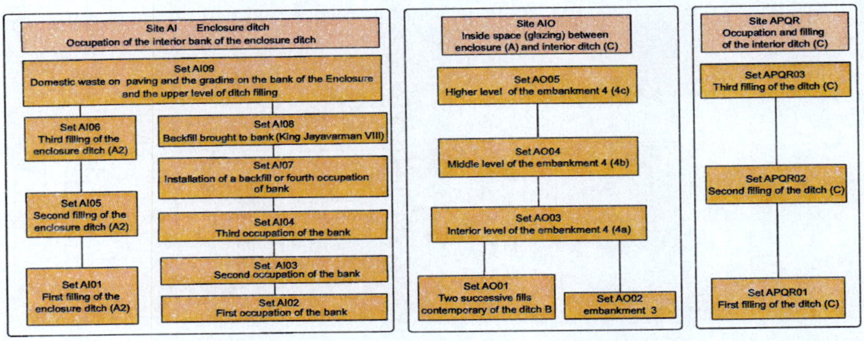

Fig. 1 Angkor: seriation diagram for three archaeological sites in relation to enclosure system

[2]http://www.r-project.org.

From the seriation diagram, it is therefore possible to construct $\mathbf{S}_2$, the symmetric adjacency matrix defined as a binary matrix of connectivity and then $\mathbf{D}_2 = \mathbf{1}_{17\times 17} - \mathbf{S}_2$ associated with the 17 archaeological sets (see Sect. 3.1). Information on pottery is contained in a contingency table $\mathbf{N}$ of size 17×12 where the rows correspond to the sets and the columns to the pottery categories. As very often on this type of data Bellanger and Husi (2012), Correspondence Analysis[3] Greenacre (2016) on $\mathbf{N}$ allows to observe an arch pattern of row and column points. The first factor is related to the best chronological seriation order obtained with pottery only. Hence, the construction and overall interpretation of the chronology of the 3 sites can be enriched by combining these two sources of information using an adapted clustering method such as PerioClust. Euclidean distances between sets are calculated from all CA row components on $\mathbf{N}$. HAC on $\mathbf{D}_1$ representing pottery and $\mathbf{D}_2$ representing stratigraphy lead to the highest values of the Agglomerative Coefficient (Kaufman and Rousseeuw 2005) for Ward's criterion. It can be considered as the best aggregation strategy to adopt for this data. As a lower entanglement coefficient corresponds to a good alignment, the entanglement value of 0.39 between trees from Ward HAC with $\mathbf{D}_1$ and $\mathbf{D}_2$ indicates that they are similar, but not identical. This confirms that the information provided by pottery and stratigraphy must be considered simultaneously to solve the clustering problem.

To apply PerioClust, we define $\mathbf{D}_\alpha$ from (1) and determine an optimal α using (3) with a confidence interval from the resampling strategy (see Sect. 3.2). We obtain $CI_{95\%}$ (α)= [0.55; 0.80] and choose $\hat{\alpha} = 0.7$. This value indicates that for Angkor data the weight of each information source is distributed as follows: 70% for pottery and 30% for stratigraphy. This imbalance may result from the difficulty of fine interpretation of a disturbed stratigraphy, often without well-defined limits.

A Ward HAC is performed with $\mathbf{D}_{0.7}$ as defined in (1). The number of clusters to be retained was selected based on the examination of the scale of aggregation indices associated with the dendrogram (Fig. 2a) but also on the archaeologist's knowledge of the site. Indeed, the choice of 4 clusters with cluster D divided into 3 sub-clusters (Fig. 2a) seems better adapted to the chronological rhythms of the city. The fact that some clusters only include one set is archaeologically explained by the chronology:

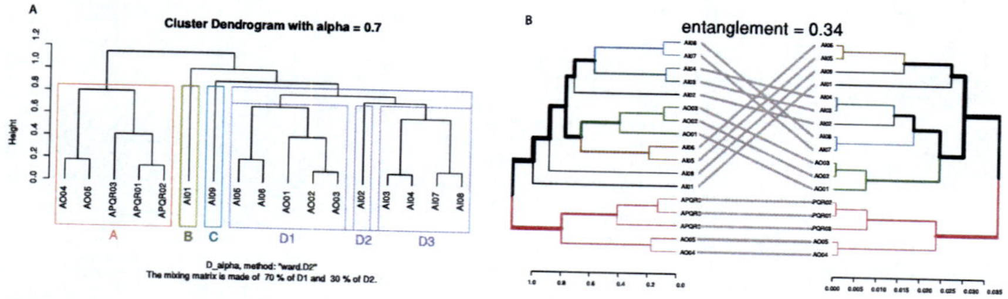

Fig. 2 **a** PerioClust with $\alpha = 0.7$, 4 clusters and 3 sub-clusters for D cluster. **b** Tanglegram between PerioClust tree (left) and hclustgeo tree (right), $\alpha = 0.7$

---

[3]In abbreviated form CA.

AI01 and AI02 are anterior and AI09 posterior to the most intense activity around the enclosure shown by a more rapid succession of the many other sets.

By applying hclustgeo with $\alpha = 0.7$ [4] and making 4 groups, we obtain a partition different from that of PerioClust (Fig. 2b). $\alpha = 0.7$ is also the value to choose using the quality criterion described in Chavent et al. (2018). An entanglement value of 0.34, a correlation between cophenetic matrices of 0.94 and an ARI of 0.71 indicate that PerioClust and hclustgeo give slightly different results.

From an archaeological point of view, hclustgeo's results with a very rapid grouping of the AI09, AI05 and AI06 sets are not very satisfactory (see Figs. 1 and 3). Indeed, AI09 is much more recent than the other sets and should therefore remain isolated as is the case with PerioClust. In the same way, AI01 and AI09 are very quickly grouped in the same class with hclustgeo, which poses a real problem because the oldest and most recent set are then grouped together. PerioClust presents a grouping reading more adapted to this archaeological dataset.

4.2 Estuary Dataset: Geographical Constraints

Estuary dataset is available in the ClustGeo package (Chavent et al. 2018). It is an extraction of 4 quantitative socio-economic variables for a subsample of 303 French municipalities located on the Atlantic Coast between Royan and Mimizan. The two considered dissimilarity matrices are $\mathbf{D}_1$ the Euclidean distance matrix between the municipalities performed with the 4 socio-economic variables and $\mathbf{D}_2$ the geographical distances. We set the number of classes to 5 as in Chavent et al. (2018) and compare the α values obtained with each method. For hclustgeo, $\alpha = 0.8$ was retained by the authors. If we set $\alpha = 0.8$ in PerioClust, we obtain an entanglement value of 0.84, indicating that the trees are very different.

For PerioClust, using the resampling strategy, we retain $\alpha = 0.45$ ($CI_{95\%}(\alpha)$= [0.33; 0.50]). An entanglement value of 0.8 indicates that the trees obtained with the

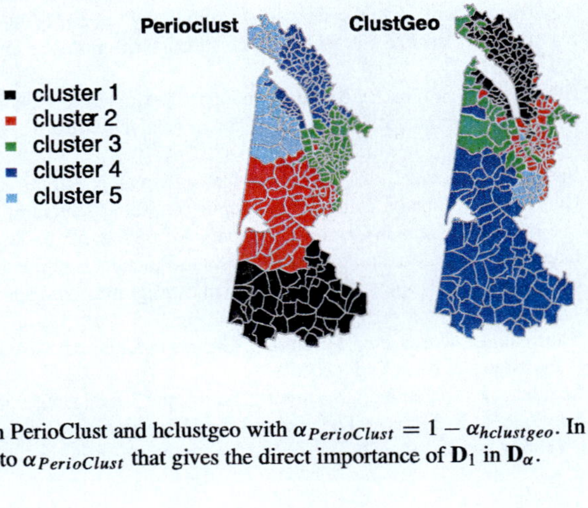

Fig. 3 Estuary: Map of the partition in 5 clusters: $\alpha = 0.45$ for PerioClust (left), $\alpha = 0.8$ hclustgeo (right)

[4]$\mathbf{D}_\alpha$ is differently defined between PerioClust and hclustgeo with $\alpha_{PerioClust} = 1 - \alpha_{hclustgeo}$. In the following, α will always refer to $\alpha_{PerioClust}$ that gives the direct importance of $\mathbf{D}_1$ in $\mathbf{D}_\alpha$.

alpha values adapted to each method are very different. Figure 3 shows that the 5 clusters are more spatially compact than those obtained for hclustgeo with $\alpha = 0.8$.

In summary, the two methods applied to these two datasets result in different dendrograms and then different partitions even if the mix parameter choice is the same.

5 Conclusions

Here, with PerioClust, we proposed a new HAC approach using temporal or spatial constraints, designed to take into consideration two sources of information. This distance-based and non-strict constrained approach is simple to implement. The modified dissimilarity matrix in the HAC is a combination of two dissimilarity matrices, so by construction all existing linkage criteria can be used. The problems of the choice and the interpretation of the mixing parameter, key points for this type of clustering methods, are solved. The mixing parameter α sets the importance of the constraint in the clustering procedure. Although PerioClust was firstly designed for archaeology, it may have a great interest in many other fields, including, for example, ecology or health (e.g. in Genome-Wide Association Studies (GWAS)).

PerioClust will be soon implemented in an R package with a shiny version also to facilitate its use.

Acknowledgements This research was supported in part by the ANR project ModAThom coordinated by Philippe Husi and Jacques Gaucher (EFEO). The authors wish to thank Jacques Gaucher for comments and suggestion.

References

Aggarwal, C., Reddy, C.: Data Clustering: Algorithms and Applications. Chapman and Hall/CRC, Boca Raton (2014)
Bellanger, L., Husi, P.: Statistical tool for dating and interpreting archaeological contexts using pottery. J. Archaeol. Sci. **39**, 777–790 (2012)
Chavent, M., Kuentz-Simonet, V., Labenne, A., Saracco, J.: ClustGeo: an R package for hierarchical clustering with spatial constraints. Computat. Stat. **33**, 1799–1822 (2018)
Davidson, I., Basu, S.: A survey of clustering with instance level. ACM T Knowl. Discov. D. **77**, 1–41 (2007)
Efron, B., Tibshirani, R.: An Introduction to the Bootstrap. Chapman & Hall/CRC, New York (1993)
Everitt, B., Landau, S., Morven, L.: Cluster Analysis, 4th edn. Oxford University Press Inc., Oxford (2001)
Ferligoj, A., Batagelj, V.: Clustering with relational constraint. Psychometrika **47**, 413–426 (1982)
Gaucher, J.: Angkor Thom, une utopie réalisée?: structuration de l'espace et modèle indien d'urbanisme dans le Cambodge ancien. Arts Asiat. **59**, 58–86 (2004)
Greenacre, M.: Correspondence Analysis in Practice. Chapman & Hall/CRC, Boca Raton (2016)
Harris, E.C.: Principles of Archaeological Stratigraphy, 2nd edn. Academic Press, London and San Diego (1989)
Kaufman, L., Rousseeuw, P.: Finding Groups in Data: An Introduction to Cluster Analysis. Wiley-Interscience, New York (2005)
Legendre, P., Legendre, L.: Numerical Ecology, 3rd edn. Elsevier Science BV, Amsterdam (2012)
Sokal, R.R., Rohlf, F.J.: The comparison of dendrograms by objective methods. Taxon **11**, 33–40 (1962)

What Was Really the Case? Party Competition in Europe at the Occasion of the 2019 European Parliament Elections

Theodore Chadjipadelis and Eftichia Teperoglou

Abstract The main aim of the paper is to analyse political competition in EU member states at the occasion of the 2019 European Parliament elections. At the core of our analysis are both the priorities of the national parties campaigning for the 2019 European elections and the manifestos of the transnational party groups, each consisting of national member parties from the 28 member states of the European Union. By comparing the major priorities of national actors/parties and those of the European political groups, we will be able to gauge out whether they share different or same dimensions of policy. More broadly, we will depict whether the dynamism in policy competition at the national level affects EP political groups or vice versa. The analysis is implemented through the use of correspondence analysis. Through this approach, the axes of political competition are realized.

Keywords European elections 2019 · European parliament · Policy positions · Party cohesion · Party competition

1 Introduction

Since 1979 when the first European Parliament election (EP elections) took place, national politics have dominated the electoral campaigns, party strategies and as a result, political competition at the EU level as well. Since then, in most member states of the European Union (EU), European issues have not been at stake and the EP elections have been mainly regarded as national second-order national contests (Reif and Schmitt 1980). Given the fact that the role of the European parties and transnational party groups in the EP is less clear for European citizens, at the EU level it could be argued that the so-called 'electoral connection' between the politi-

T. Chadjipadelis (✉) · E. Teperoglou
School of Political Sciences, Aristotle University of Thessaloniki, Thessaloniki, Greece
e-mail: chadji@polsci.auth.gr

E. Teperoglou
e-mail: efteperoglou@polsci.auth.gr

T. Chadjipadelis et al. (eds.), *Data Analysis and Rationality in a Complex World*, Studies in Classification, Data Analysis, and Knowledge Organization,
https://doi.org/10.1007/978-3-030-60104-1_2

cal elites and the voters is weak. This aspect is considered as an important cause of legitimacy problems (e.g. Hix and Lord 1997) at the European level. In addition, in the EP elections national parties have a predominant role. This feature has somehow changed with the nomination of the Spitzenkandidaten back in the 2014 EP elections.[1] Nevertheless, despite the fact that the "nationalization" of the EU politics is a distinguishing characteristic, some scholars argue for the opposite, namely that over the past years is observed a growing policy authority of the EU level of European governance. The European Parliament has become an important actor in the EU-level policymaking process. In other words, we can observe a possible "Europeanization" of the domestic political arena in each EU member state (see among others Schmitt and Teperoglou 2018). It could be argued that the politicization of European integration has changed the content, as well as the process of decision-making (Hooghe and Marks 2009). In this framework of analysis, a central question regards the axes of political competition in EP elections.

The main objective of this article is to depict some insights into the position of national parties and European political parties/ groups competing for the 2019 EP elections on different issues. The main question of our study is whether we can identify the same or different patterns across different party families. Or in other words, to what extent is contestation at the EU level in 2019 related to the classic left-right dimension or other dimensions?

Before presenting our method and data, in the next section, we will briefly refer to the context of the 2019 EP election.

2 The Context of the 2019 European Parliament Elections

The 2019 EP election was the first one in the shadow of the Brexit referendum (for an overview about the context see Russo et al. 2019). Furthermore, the election took place in a period of an ongoing debate about the challenges for EU especially related to the immigration and refugee crises and the rise of populist parties. However, EU membership remained at quite high levels (59 % of the Europeans considered the EU membership as "something good"), but however with remarkable country variations (see the results of the post-electoral survey Eurobarometer 91.5). An important characteristic of the electoral campaign was that different initiatives have been taken in order to increase the participation especially among the younger voters.[2] Contrary to the 2014 EP elections in which the economic crisis was the main issue at stake, in the last EP election with the exception of Southern European countries, the economy was

[1]In 2014, for the first time in the history of the EP elections, there was an explicit attempt to link the results of the election with the appointment of the European Commission President. The Lisbon Treaty specified that the President of the European Commission will be elected by the European Parliament based on a proposal by the European Council, taking into account the results of the European elections (Article 17(7) TEU).

[2]Another striking feature of the last EP election was that overall turnout increased and reached 50.6%.

stabilized.[3] Finally, like in previous EP elections, the contest had in most EU member states a referendum character for some incumbent parties. The electorate wanted to send a "message" to the governmental parties and protest vote (or voting with the boot) is documented (see for the term among others Schmitt and Teperoglou 2018).

3 Operationalization: Data and Method

In our analysis, we use data from the "Your Vote Matters" platform.[4] This platform includes information about the positions of MEPs, national parties and EP political groups on 25 key issues (with three possible answers: against, in favour or abstain/ no vote for each issue]. In total, we have collected the answers of 148 national-level parties across EU and 7 groups of the European Parliament.[5] Given the fact that there was no other available data at the time of presenting and writing this article (e.g. data from Euromanifestos studies), this source is considered as a valuable one. It covers all national- and EU-level parties and a variety of issues.

The first step in our method was to select and group the issues/ questions in specific categories. In total, we have selected items that are grouped into five different categories. These are: economic issues, law and order issues, immigration and refugee issues, environmental issues and finally, institutional issues[6] (for further information about the issues see Figs. 4, 5, 6, 7 and 8 in the Appendix). The second main step was the use of hierarchical cluster analysis for grouping the issues to categories based on the agreement, disagreement or no vote-no opinion of the party on a specific issue (see Fig. 1). The third step was the creation of our dataset with the answers of the parties to these questions per category. Finally, 18 out of the 25 items were used in the analysis.

Data analysis was based on Hierarchical Cluster Analysis (HCA) and Multiple Correspondence Analysis (MCA) in two steps (Chatzipadelis 2017). In the first step, HCA was used to assign subjects to distinct groups according to their response patterns. The main output of HCA was a group or cluster membership variable, which

---

[3]For example the EU28 unemployment rate was at 6.3% in June 2019 (Eurostat). It remained high in Greece (17.6% in April 2019), Spain (14.0%) and Italy (9.7%).

[4]Data from: https://yourvotematters.eu. YourVoteMatters.eu is a multilingual digital platform designed as an innovative communication tool between the 2019 European elections' candidates and their electorate. The platform is developed by a consortium of five European organizations: Riparte il Futuro (Italy), VoteWatch Europe (Belgium), European Citizen Action Service (Belgium), Vouliwatch (Greece) and Collegium Civitas (Poland) with the aim of enhancing the dialogue between all the actors involved in the elections (politicians, political parties, citizens, organizations and stakeholders).

[5]These are: European People's Party (EPP), Progressive Alliance of Socialists and Democrats (S&D), Renew Europe (Renew), Greens-European Free Alliance (Greens–EFA), Identity and Democracy (ID), European Conservatives and Reformists (ECR) and European United Left–Nordic Green Left (GUE-NGL).

[6]We have excluded from the analysis in total seven issues related mostly to external relations of the EU with US, China, Russia, etc. or specific economic measures.

reflects the partitioning of the subjects into groups. Furthermore, for each group, the contribution of each question (variable) to the group formation was investigated, in order to reveal a typology of behavioural patterns.

In the second step, the group membership variable, obtained from the first step, was jointly analysed with the existing variables via Multiple Correspondence Analysis (MCA) on the so-called Burt table (Greenacre 2017). The Burt table is a symmetric, generalized contingency table, which cross-tabulates all variables against each other. The main MCA output is a set of orthogonal axes or dimensions, which summarize the associations between variable categories into a space of lower dimensionality, with the least possible loss of the original information contained in the Burt table. HCA is then applied on the coordinates of variable categories on the factorial axes. Note that this is now a clustering of the variables, instead of the subjects. The groups of variable categories can reveal complex discourses. Bringing the two analyses together, behavioural patterns and complex discourses are used to construct a semantic map for the variables and the subjects.

ISSUES	1st GROUP	2nd GROUP	3rd GROUP	4th GROUP	5th GROUP
Economy A1		1	2		0
Economy A2		1		2	0
Law & order A4	1		0		2
Immigr A5	1		0	2	
Immigr A6	0		1	2	
Law & order A7		1		2	0
Environ.A8		1		2	0
Environ A10	1		0	2	
Economy A11		1		2	0
Economy A12		1/2			0
Economy A13	1		0	2	
Economy A14	1		0	2	
Economy A15		1		2	0
Economy A17		1		2	0
Economy A18		1		2	0
Institutional A19	1		0	2	
Institutional A20		1		2	0
Institutional A21		0		2	1

Fig. 1 Operationalization

From the hierarchical cluster analysis, as presented in Fig. 1, the main result is that all the issues are split into five subgroups. More specifically, the first and second group represents mainly an agreement with most of the issues, regardless if the issue is about economy, institutional issues, etc. On the contrary, the last group (5th) is in contrast with the second one, while the third one mainly with the first one. Finally, the fourth group is the one that includes parties that do not express in most of the cases any opinion on the various issues. In the next section of the presentation of the main findings, we try to answer the question of whether the party family plays any role for this variation.

4 Findings

In order to include party family in our analysis, we have grouped the total 148 national parties of the dataset in the respective EU party groups.

As we can see in Fig. 2, a main finding from our analysis is that the parties belonging to the centre-left and left constitute one group (node 303) which is distinguished from right-wing parties (mainly node 302) on the issues presented in Figs. 4, 5, 6, 7 and 8 in the Appendix. Moreover, the parties belonging to the EPP and RenewEurope (node 302) do not share the same views on these issues with parties that are classified to the groups of ECR and ID (node 304). On the other hand, the last group (node 305) is more difficult to interpret since it includes a mixture of left-wing, right-wing parties and those parties which do not belong to any political group. Finally, the first group is mainly composed of parties that belong to GUE/NGL and those which do not belong to any group (node 300). Overall, from this cluster analysis, we might conclude that the left-right dimension is a salient characteristic of the political competition at the occasion of the 2019 EP elections. However, as aforementioned, we have also identified two clusters that include parties belonging to different ideological party families. This aspect underlines the importance of the left-right divide at least regarding some issues. Therefore, a next step in our analysis is pivotal to link parties and items.

CLUSTER PARTIES (per party family)	**300**	**302**	**303**	**304**	**305**	**Total**
EPP		**69.2%**	5.1%	5.1%	**20.5%**	**39**
S&D	3.2%	3.2%	**87.1%**		6.5%	**31**
ECR	7.7%	7.7%	0.0%	**76.9%**	7.7%	**13**
RENEW EUROPE	7.7%	**69.2%**	7.7%	3.8%	11.5%	**26**
GUE/NGL	**16.7%**		**66.7%**		**16.7%**	**12**
GREENS/EFA	10.0%	5.0%	**80.0%**		5.0%	**20**
ID	12.5%	12.5%		**75.0%**	0.0%	**8**
NI	**33.3%**		16.7%	**33.3%**	**16.7%**	**6**
Total	7.1%	31.6%	36.1%	13.5%	11.6%	**155**

Fig. 2 Parties clustering

ITEMS	113	88	100	111	113	104	110
A1	NV			AGAINST	In Favour		
A2			NV	AGAINST	In Favour		
A4		In Favour		NV		AGAINST	
A5		In Favour				AGAINST	NV
A6		AGAINST				AGAINST	NV
A7				AGAINST	In Favour		NV
A8				AGAINST	In Favour		NV
A10		In Favour				AGAINST	NV
A11				AGAINST	In Favour		NV
A12	NV			AGAINST	In Favour		
A13		In Favour				AGAINST	NV
A14		In Favour				AGAINST	NV
A15			NV	AGAINST	In Favour		
A17			NV	AGAINST	In Favour		
A18				AGAINST	In Favour		NV
A19		In Favour				AGAINST	NV
A20	In Favour			AGAINST			NV
A21	AGAINST			In Favour			NV

Fig. 3 Parties and issues clustering

In Fig. 3 the results of this analysis are presented. The analysis, using as cases the items and as variables the parties, is implemented through the use of two-way cross-tabulation, contingency tables, and correspondence analysis by using the software MAD [Méthodes de l'Analyse des Données] Karapistolis (2002, 2011). The central conclusion is that the parties belonging to the centre-left and left, as well as the green parties, are linked mainly with the second group of items, whereas the parties belonging to the right and centre-right are mainly opposed on various issues with the group of parties belonging to the left and centre-left. The latter group supports protectionism on different economic indicators, while the former is in favour of economic liberal positions. In addition, the group of parties belonging to the right are in favour of economic growth, while those on the left and centre-left in favour of measures to protect the environment. Interestingly, this division is less clear regarding socio-cultural issues and opinions regarding the immigration crisis.

5 Concluding Remarks

The main conclusion from our analysis is that at the occasion of the 2019 EP elections there is strong evidence of the contestation around the left-right divide both for national and European parties, thus party family as a variable in our analysis matters at the EU level. Parties that belong either to the left-wing group or to the greens share some same views on a variety of issues with the socialist group. In particular, it is observed a significant divide among the conservative group versus the socialist one mainly on economic issues and to a lesser extent on socio-cultural issues. Overall, in this contribution, we have attempted to provide some first insights on the axes of political competition in 2019. Further research is needed in order to shed light on political contestation at the EU level in a period of an increasing role of the European issues in each domestic political arena.

6 Appendix

See Figs. 4, 5, 6, 7 and 8

A1	Should there be a tax on companies that use robots, as a way to support the social security system?
A2	Should the EU introduce stricter rules on the privacy of online communication (with potential implications for the growth of digital businesses)?
A11	Should a common minimum corporate tax rate of at least 20% be introduced in the EU?
A12	Should there be a European authority empowered to enforce fiscal compliance?
A13	Should a tax on financial activities fund the EU budget?
A14	Should there be a bigger increase in the EU's multi-annual budget for the period 2021-2027?
A15	Should the EU spend more money to support small farmers against volatility of agricultural prices?
A17	Should the EU introduce common rules on minimum income, which would likely force some Member States to increase their minimum income levels?
A18	Should policymakers take stronger measures to restrict the use of temporary work contracts?

Fig. 4 Economic issues

A4	Should verbal abuse related to gender identity and sexual orientation be punished by means of criminal law (with potential implications for freedom of expression)?
A7	Should the European Union introduce penalties for the EU Member States that do not follow EU rules on the treatment of animals?

Fig. 5 Law and order issues

A5	Should asylum seekers be redistributed across EU countries on the basis of a quota system?
A6	Should EU countries be allowed to reintroduce border controls within the Schengen area?

Fig. 6 Immigration/refugee crises issues

A8	Should EU policymakers support a quicker phase-out of fossil fuels subsidies?
A10	Should the EU levy a tax on plastic and single-use items to fund the EU budget, as a way to restrict their usage?

Fig. 7 Environmental issues

A19	Should the EU financially sanction countries found to be violating EU principles on rule of law?
A20	Should a majority of national parliaments get the power to veto EU legislation?
A21	Should the European Defence Union ultimately lead to the establishment of a European Armed Forces?

Fig. 8 Institutional issues

References

Chatzipadelis, T.: What really happened: party competition in the January and September 2015 parliamentary elections. Eur. Quart. Polit. Attitudes Ment. **6**, 8–39 (2017)

Greenacre, M.: Correspondence Analysis in Practice. CRC Press (2017)

Hix, S., Lord, C: Party groups in the European parliament: election and formation. In: Hix, S., Lord, C. (eds.), Political Parties in the European Union, pp. 77–110, Palgrave, London (1997)

Hooghe, L., Marks, G.: A postfunctionalist theory of European integration: from permissive consensus to constraining dissensus. Br. J. Polit. Sci. **39**, 1–23 (2009)

Karapistolis, D.: Multidimensional Statistical Analysis. Altintzis Publications (2011)

Karapistolis, D.: The software MAD. Dat. Anal. Bull. **2**, 133–147 (2002)

Reif, K., Schmitt, H.: Nine second-order national elections-a conceptual framework for the analysis of European Election results. Eur. J. Political Res. **8**, 3–44 (1980)

Russo, L., Franklin, M.N., De Sio, L.: Understanding the European Parliament elections of 2019. In: De Sio, L., Franklin, M.N., Russo, L. (eds.) The European Parliament Elections of 2019, pp. 9–12. Luiss University Press (2019)

Schmitt, H., Teperoglou, E.: Voting behavior in multi-level electoral systems. In: Fisher, J., Fieldhouse, E., Franklin, M.N., Gibson, R., Cantijoch, M., Wlezien, C. (eds.) Handbook of Elections, Voting Behavior and Public Opinion, pp. 232–243, Routledge, Abingdon (2018)

A Fast Electric Vehicle Planner Using Clustering

Jaël Champagne Gareau, Éric Beaudry, and Vladimir Makarenkov

Abstract Over the past few years, several studies have considered the problem of Electric Vehicle Path Planning with intermediate recharge (EVPP-R) that consists of finding the shortest path between two given points by traveling through one or many charging stations, without exceeding the vehicle's range. Unfortunately, the exact solution to this problem has a high computational cost. Therefore, speedup techniques are generally necessary (e.g., contraction hierarchies). In this paper, we propose and evaluate a new fast and intuitive graph clustering technique, which is applied on a real map with charging station data. We show that by grouping nearby stations, we can reduce the number of stations considered by a factor of 13 and increase the speed of computation by a factor of 35, while having a very limited trade-off increase, of less than 1%, on the average journey duration time.

Keywords Electric vehicles · Charging stations · Planning · Clustering · Graphs

1 Introduction

Electric vehicles (EVs) are an attractive alternative to fossil-fuel vehicles to reduce air pollution. However, their limited range and their high-charging time represent a major obstacle to their massive adoption Smart and Schey (2012). Moreover, long journeys require careful planning to determine the charging stations to be used in order to avoid running out of energy. For example, an average EV currently has a range of around 250 km and needs to make many recharging stops on long journeys.

J. Champagne Gareau (✉) · É. Beaudry · V. Makarenkov
Université du Québec á Montréal, P.O. Box 8888, H3C 3P8, QC, Montréal, Canada
e-mail: champagne_gareau.jael@courrier.uqam.ca

É. Beaudry
e-mail: beaudry.eric@uqam.ca

V. Makarenkov
e-mail: makarenkov.vladimir@uqam.ca

T. Chadjipadelis et al. (eds.), *Data Analysis and Rationality in a Complex World*,
Studies in Classification, Data Analysis, and Knowledge Organization,
https://doi.org/10.1007/978-3-030-60104-1_3

EV path planning with intermediate recharges (EVPP-R) is a complex problem which cannot be effectively solved by conventional approaches because one needs to consider not only the different variables applicable to conventional vehicles (the wind, the energy needed to fight the air resistance relative to the speed, the traffic, eventual detours, etc.), but also aspects specific to EVs, such as the level of charging stations (which influences the charging speed), the non-linearity of the charging curve of the battery, the topography of the map (EVs can recover some energy when moving downhill) as well as the expected waiting time at the charging station.

In addition to these considerations, a non-negligible characteristic of a good EV planner is its running time. Many well-known techniques are commonly used to accelerate the graph search, including graph contraction hierarchies (Geisberger et al. 2012) and various search heuristics. However, no one has yet tried to decrease the journey computation time by using clustering techniques in order to decrease the number of nodes considered in the graph. *The main contributions of our paper are as follows*:

- a fast and intuitive clustering technique to solve the EVPP-R problem;
- evaluation of the proposed technique on real data to assess its performance.

An EV planning framework that takes into account the fact that the EV recharges itself during a journey (by using regenerative braking) already exists (Sachenbacher 2011). However, it does not take into account the possibility of using charging stations to charge the EV battery midway. The algorithm used in Sachenbacher (2011) is a variant of A* (Hart et al. 1968), called *Energy-A**. This method allows one to find a solution in $\mathcal{O}(n \log n)$, where n is the number of nodes of the graph. Techniques that tackle the problem of midway charging have also been proposed. Some of them consider the problem of a single EV path planning (what we call EVPP-R), while others consider the problem of an EV fleet routing (EVRP). In this article, we focus on the former. The two main approaches which have been used to solve the EVPP-R problem are Dijkstra/A*-based (Baouche et al. 2014, Champagne Gareau 2018), or dynamic programming/MDP-based (Sweda and Klabjan 2012) algorithms. While these techniques try to minimize the sum of the travel time and charging time, some new techniques also consider the expected waiting time (Champagne Gareau 2019). The majority of Dijkstra/A*-based techniques use heuristic graph searches as well as contraction hierarchies (Geisberger et al. 2012) in order to accelerate the computations. In this paper, we show that clustering is another way to speedup the computation (that can be combined with the other speedup techniques).

2 Problem Formulation and Base Planner

We begin by presenting how we model the problem, including the road network representation, the problem definition, and what we consider an optimal solution.

Definition 3.1 A *road network* M is modeled by a tuple $(V, E, \lambda, \sigma, S)$, where (V, E) is a digraph and λ, σ are two labelings of the edges. More specifically:

- V is the set of nodes (latitude, longitude) on the map;
- E is the set of road segments (edges);
- $\lambda : E \to \mathbb{R}^+$ gives the length (in m) at every edge;
- $\sigma : E \to \mathbb{R}^+$ gives the expected speed (in m/s) at every edge;
- S is the set of all charging stations.

Every charging station $s \in S$ is associated to the nearest vertex $v_s \in V$. The σ labeling can be based on empirical data of the mean speed on every edge, or can simply be the maximum allowed speed of each road segment.

Definition 3.2 An *EVPP-R* problem is defined by the tuple $(M, \rho, \alpha, \omega)$, where

- M is the road network;
- $\rho \in \mathbb{R}^+$ is the range of the EV;
- α and ω are the points of departure and arrival.

Remark 3.1 We assume that $\alpha, \omega \in V$. If it is not the case, a KD-Tree can simply be used to find the nearest corresponding nodes in the graph.

Definition 3.3 A *solution* of an EVPP-R problem $(M, \rho, \alpha, \omega)$ is a tuple (P, Q), where

- P is a finite sequence $(P_i)_{i=0}^{k}$ (where $P_i \in V$);
- Q is a subsequence $(P_{i_j})_{j=0}^{b+1}$ of P containing the P_i's where a station is used;
- $P_{i_0} = P_0 = \alpha$ and $P_{i_{b+1}} = P_k = \omega$;
- $\forall j \in \{0, 1, \ldots, b\}, \mathrm{d}(P_{i_j}, P_{i_{j+1}})$[1] $\leq \rho$.

In other words, P is the sequence of nodes that the EV needs to travel by according to the solution, and Q is a subsequence of P containing the charging stations that need to be used in the journey (as well as α and ω). Our objective is to find a solution that minimizes the total time of the journey, including the travel time and the charging time (we don't consider the waiting time, but the planner can easily be modified to consider it as well (Champagne Gareau 2019; Sweda et al. 2017)). This is formalized in the next definition.

Definition 3.4 An *optimal solution* to an EVPP-R is a solution (P, Q), as stated in Definition 3.3, which minimizes the objective function $\mathrm{Z}(P, Q) = \mathrm{TT}(P) + \mathrm{CT}(Q)$, where TT is the expected travel time and CT is the expected charging time:

$$\mathrm{TT}(P) = \sum_{i=0}^{k-1} \frac{\lambda(P_i, P_{i+1})}{\sigma(P_i, P_{i+1})} \quad \text{and} \quad \mathrm{CT}(Q) = \sum_{i=1}^{b} \mathrm{ST}(Q_i).$$

$\mathrm{ST}(Q_i)$ is the charging time when considering Q_i's state of charge and level.

[1] $\mathrm{d}(A, B)$ is the distance in the graph between A and B.

We now present our baseline planner. First, the distance between each pair of stations is pre-computed and stored in a matrix $D = (D_{ij})$, where $D_{ij} = \mathrm{d}(S_i, S_j)$. We also store the optimal path between each of them. We then build a simplified graph (s-graph) (V', E') containing the nodes associated with the stations and the edges with weights corresponding to the pre-computed distances between every pair of stations. For every request (α, ω, ρ), we temporarily add α and ω to the s-graph and use Dijkstra's algorithm twice (from α, and from ω on the reversed graph) to add edges from α to every station and from each station to ω. The computations for this step can be accelerated by using contraction hierarchies. The new graph obtained is a complete graph, from which we remove every edge whose length is larger than ρ. We then execute the A* algorithm on this new graph (using the great-circle distance as heuristic) from α to ω, which is enough to get a sequence Q, as specified in Definition 3.3. This sequence satisfies the last part of Definition 3.3 insofar as every intermediary node is a charging station. Consequently, there is no need to consider the range of the vehicle at this point. The sequence P can then be found from Q using the previously computed paths.

The base planner has a time complexity of $\mathcal{O}(|V|\log|V| + |E|)$, equivalent to Dijkstra's algorithm's complexity. In a real road network, the maximum number of intersections at a given node is bounded by a small constant (i.e., $|E| \in \mathcal{O}(|V|)$). This implies that we can simplify the time complexity of the algorithm to $\mathcal{O}(|V|\log|V|)$. In the next section, we show how to improve the aforementioned algorithm by clustering the charging stations before creating the s-graph.

3 Clustering

To implement a faster EV planning algorithm, we propose to cluster the stations. We introduce the parameter $d_{\max} \in \mathbb{R}^+$ corresponding to the maximal distance between the center of a cluster and the stations it contains. We then create the clusters as described in Algorithm 1, which has the time complexity of $\mathcal{O}(K(|S|^2 + |V|))$, where K, $|S|$ and $|V|$ are, respectively, the number of clusters, stations and nodes in the graph. $|S|^2$ is currently small compared to $|V|$, but is expected to grow in the future.

Algorithm 1 Fast charging stations clustering

```
1: Find the two nearest charging stations s1, s2 ∈ S          ▷ using the pre-computed matrix
2: while d(s1, s2) ≤ dmax do
3:     Find the midway node m ∈ V between s1 and s2            ▷ using Dijkstra from s1 to s2
4:     Find C = {s ∈ S | dist(s, m) ≤ dmax}                    ▷ using Dijkstra from m
5:     S ← (S \ C) ∪ {m}                    ▷ m is the representative of the new cluster
6:     Find the two nearest charging stations s1, s2 ∈ S
```

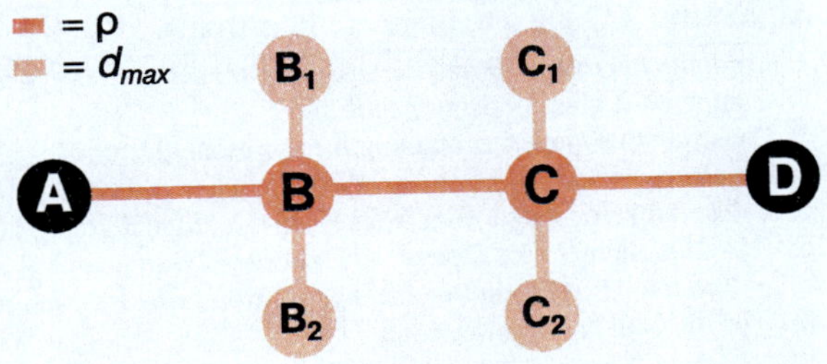

Fig. 1 The distance between clusters B and C is equal to the range ρ. To reach C_1 or C_2 from B_1 or B_2, one needs to travel $\rho + 2d_{\max}$ km. Alas, the planner will consider only the distance between the centers B and C

When we use the base planner described previously along with the clustering, some previously feasible paths may become unfeasible because the new itinerary will always pass through a cluster's center before reaching one of its stations. To solve this problem, the range considered by the algorithm needs to be modified. Figure 1 gives an instance of this problem, and Proposition 3.1 gives the solution.

Proposition 3.1 *Let ρ be the true range of the EV and ρ' be the range considered by the planner. To always have a feasible plan, ρ' needs to be set to $\rho - d_{\max}$ between α and the first traversed cluster, and between the last traversed cluster and ω. Between pairs of clusters, it needs to be set to $\rho - 2d_{\max}$.*

Proof Let $\{c_1, \ldots, c_k\}$ be the traversed clusters in the path from α to ω returned by the planner. The path from α to c_1 is less than or equal to ρ'. Since c_1 is the center of a cluster[2] and the stations have at most a distance of $d_{\max}$ from the center, the range of the EV needs to be $\rho = \rho' + d_{\max}$, so the considered range must be $\rho' = \rho - d_{\max}$. The same argument applies to the segment of the plan between the last cluster used and the arrival point. Similarly, for the path between $s_i \in c_i$ and $s_{i+1} \in c_{i+1}$, we have

$$d(s_i, s_{i+1}) \leq d(s_i, c_i) + d(c_i, c_{i+1}) + d(c_{i+1}, s_{i+1}) \leq d_{\max} + \rho' + d_{\max}$$

In some cases (e.g., Fig. 1) we have equality, so the bound is tight. □

Proposition 3.1 presents the theoretical worst case. However, in the implementation, it is possible to consider the *radius* r_i of every cluster c_i (the maximal distance between the center and a station inside it). The range ρ' that needs to be considered between pairs of clusters (c_i, c_j) can then be set to $\rho' = \rho - r_i - r_j \leq \rho - 2d_{\max}$.

The new range ρ' can make some previously feasible paths not found by the planner. While at first it may seem to be an unacceptable compromise, a simple solution is to compute the journey using clustering, and if the journey is not found because of the new range, recompute the journey without clustering. In the evaluation of our technique, we call this strategy the *amortized version* of our planner.

Algorithm 2 shows how the different steps fit together in the complete planner. Note that another version of the problem starts with the s-graph given as input, in which case lines 3, 5, and 10 can be omitted. Nevertheless, even when the s-graph

---

[2]For the sake of simplicity, c_i denotes both the set of stations in the cluster and the center node.

Algorithm 2 Complete planner algorithm

```
1: Compute the matrix D and the optimal path between every pair of stations
2: Compute the clusters using Algorithm 1
3: Construct the s-graph containing the representative of each cluster
4: for all request (α, ω, ρ) do
5:     Run Dijkstra's algorithm from α (on the original graph) and ω (on the reversed graph)
6:     Add α and ω to the s-graph and add edges with length ≤ ρ
7:     Run the A* algorithm on the s-graph from α to ω to find the sequence Q
8:     if the path is infeasible then              ▷ When we use the amortized strategy
9:         Run A* on the s-graph (without clustering)
10:    Find the sequence P from Q using the already computed paths
```

needs to be constructed, line 5 is done in the original graph, so the time complexity of this step is not affected by the clustering (but can be accelerated using contraction hierarchies). Furthermore, lines 6 and 10 have a negligible time compared to the others. Therefore, we focus on the time complexity of lines 7–9 in the evaluation.

When we use clustering ($d_{\max} > 0$), we can compute P (Line 10) as in the previous section. Note, however, that Q now contains the clusters that the EV will pass through on its journey (instead of containing raw stations), but gives no information about which station to select inside the cluster. This choice is made at runtime (i.e., a short time before the vehicle arrives at a decision point before the cluster) rather than during the initial planning. Therefore, clustering gives the opportunity of considering real-time data (e.g., road conditions, charging stations occupancy state, etc.) to select the most suitable station inside the next cluster on the path.

4 Empirical Evaluation

Our algorithm was evaluated using real data from the Province of Québec, Canada. The map data (i.e., the nodes and the road segments) were extracted from the OpenStreetMap project. We chose this map in our tests because the territory is vast, the journeys between pairs of cities can be very long and the network of stations is relatively well developed. The graph we generated from these data had 2 923 013 vertices and 5 907 672 edges. The stations considered in our tests were real stations from this territory. Our dataset included 1162 charging stations (Level 2 and 3). We considered all stations in the data as if they were Level 3 because EV planners are primarily used for a long itinerary where fast charging is a must and there were not enough Level 3 stations in the dataset to test our algorithm adequately (only 122).

To evaluate our method, we generated 1000 requests consisting of the departure α, arrival ω (both chosen at random in the graph) and EV range ρ (generated uniformly between 90 and 550 km, based on the most common EV ranges available on the market). The optimal solution length ranged from 150 to 1000 km. Every generated request required at least one stop at a station for the EV to be able to reach the destination. Our tests consisted of running these 1000 requests by our planner on

Table 1 Empirical results obtained for 1162 real stations in Québec (Canada)

Problem Param.	Clust. Param.		Base version		Amortized version	
d_{max} km	C #	JDIR %	FR %	CPU ms	FR %	CPU ms
0.0	1162	0.0	0.0	26.56	0.0	26.56
2.5	487	0.0	0.0	3.385	0.0	3.385
5.0	342	0.2	0.4	1.541	0.0	1.647
10.0	236	0.2	0.8	0.588	0.0	0.801
15.0	188	0.6	0.9	0.523	0.0	0.762
20.0	150	1.0	1.4	0.382	0.0	0.754
30.0	111	2.3	2.0	0.265	0.0	0.796
40.0	87	2.8	8.2	0.226	0.0	2.404

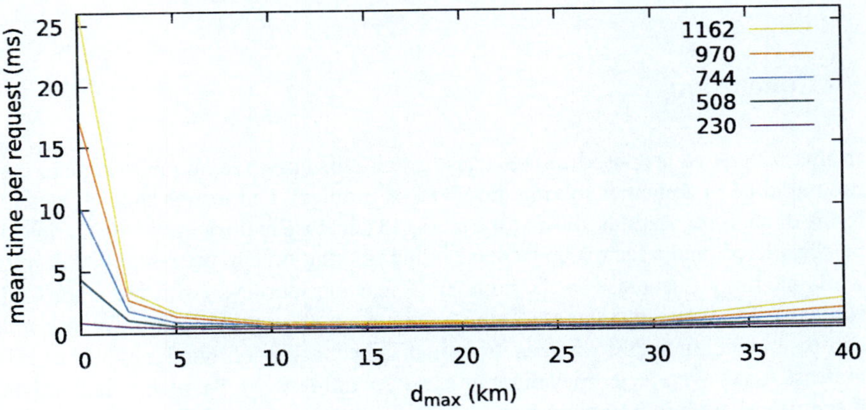

Fig. 2 Mean CPU time for different density of stations using amortized strategy

an Intel Core i5 7600k processor. We compared the results obtained with different combinations of parameters: different subsets of stations (20, 40, 60, 80, or 100% of the total) to measure the effect of stations density on clustering, and different values of the parameter d_{max} (0, 2.5, 5, 10, 15, 20, 30, and 40 km). For the sake of simplicity, it was assumed in the tests that $\sigma(e) = 90$km/h for all $e \in E$ and that $ST(s) = 30$ min for all $s \in S$ since it doesn't influence the clustering efficiency.

Table 1 presents the results obtained by running the tests described previously. The columns denote, respectively, the parameter d_{max}, the number of clusters (**C**), the Journey Duration Increase Rate (**JDIR**), the journey Failure Rate (**FR**), and the average computation time (**CPU**) (with and without amortized strategy). When considering the columns d_{max} and C, we observe that even a small value of d_{max} reduces drastically the number of clusters (since they are merged). The percentage of remaining stations decreased from 58.1% (with the smallest d_{max}) to 92.5%.

When using the amortized strategy, the majority of the performance improvement is preserved, while eliminating the infeasible paths (as can be seen by comparing the two pairs of columns (FR, CPU) and by looking at Fig. 2). Column JDIR shows that the use of our clustering technique can slightly increase the duration of the journeys returned by the planner. The running time increase is proportional to d_{max}.

Finally, by comparing the different curves in Fig. 2, we can conclude that when the number of stations is larger (i.e., the density of stations on the map is higher), the decrease in the running time due to clustering is also larger. Thus the usefulness of our clustering technique will increase while more stations are getting installed. Our results suggest (when considering the columns JDIR and CPU (Amortized)) that the optimal d_{max} value for our dataset is between 15 and 20 km. A larger value of d_{max} results in an increase of more than 1% of the average journey duration and an increase in the failure rate which is large enough to mitigate the time saving (i.e., the CPU time starts to increase when d_{max} becomes larger than 20).

5 Conclusion

In this paper, we proposed to use a fast graph clustering technique to reduce the running time of a planner solving the EVPP-R problem. Our results show that clustering of charging stations allows for a factor 13 decrease in the number of stations to be considered, and a factor 35 decrease in the running time in the simplified graph, while having no decrease in the number of feasible journeys (with the amortized strategy). The main advantages of our technique are its simplicity, intuitiveness and computational efficiency, whereas its main disadvantage is a limited trade-off of 1% increase to the average journey duration time. As future work, we plan to investigate if the use of traditional graph clustering techniques such as MCL and k-medoids on graph spectral embeddings, which are slower than the proposed algorithm, could improve some of our results.

Acknowledgements This work was supported by the *Natural Sciences and Engineering Research Council of Canada* (NSERC) and by the *Fonds de recherche du Québec—Nature et technologies* (FRQNT).

References

Baouche, F., Billot, R., Trigui, R., El Faouzi, N.E.: Electric vehicle green routing with possible en-route recharging. In: International Conference on Intelligent Transportation Systems (ITSC), pp. 2787–2792 (2014)

Champagne Gareau, J., Beaudry, É., Makarenkov, V.: Planification d'itinéraires quasi-optimaux pour un véhicule électrique en considérant le regroupement de bornes de recharge et leur probabilité d'occupation. In: XXV-émes Rencontres de la Société Francophone de Classification (SFC2018), pp. 5–8. Paris, France (2018)

Champagne Gareau, J., Beaudry, É., Makarenkov, V.: An efficient electric vehicle path-planner that considers the waiting time. In: Proceedings of the 27th ACM SIGSPATIAL International Conference on Advances in Geographic Information Systems. ACM, Chicago, United States (2019)

Geisberger, R., Sanders, P., Schultes, D., Vetter, C.: Exact routing in large road networks using contraction hierarchies. Transport. Sci. **46**(3), 388–404 (2012)

Hart, P., Nilsson, N., Bertram, R.: A formal basis for the heuristic determination of minimum cost paths. IEEE Trans. Syst. Man Cybern. A. Syst. Humans - TSMCA **4**(2), 100–107 (1968)

Sachenbacher, M., Leucker, M., Artmeier, A., Haselmayr, J.: Efficient energy-optimal routing for electric vehicles. In: Proceedings of the Association for the Advancement of Artificial Intelligence (AAAI), pp. 1402–1407 (2011)

Smart, J., Schey, S.: Battery electric vehicle driving and charging behavior observed early in the EV project. In: SAE Technical Papers, pp. 27–33 (2012)

Sweda, T.M., Dolinskaya, I.S., Klabjan, D.: Adaptive routing and recharging policies for electric vehicles. Transport. Sci. **51**(4), 1326–1348 (2017)

Sweda, T.M., Klabjan, D.: Finding minimum-cost paths for electric vehicles. In: IEEE International Electric Vehicle Conference (IEVC), pp. 1–4 (2012)

A Generalized Coefficient of Determination for Mixtures of Regressions

Roberto Di Mari, Salvatore Ingrassia, and Antonio Punzo

Abstract One of the challenges in cluster analysis is the evaluation of the obtained clustering results without using auxiliary information. To this end, a common approach is to use internal validity criteria. For mixtures of linear regressions whose parameters are estimated via the maximum likelihood approach, we propose a three-term decomposition of the total sum of squares as a starting point to define some internal validity criteria. Exploiting this decomposition, local and overall coefficients of determination are, respectively, defined to judge how well the model fits the data group-by-group but also taken as a whole. An application to real data illustrates the use and the usefulness of these proposals.

Keywords Cluster validation · Em algorithm · Maximum likelihood · Mixtures of regressions · Model-based clustering

1 Introduction and Background Results

The concept of R^2 measures is well understood in regression analysis for checking model fitting to data. Quite recently, in Ingrassia and Punzo (2019) similar statistics have been introduced for clusterwise linear regression. In particular, model fit can be typically evaluated both in terms of within-class fit of the regression hyperplanes and class separation, but the formulation of an R^2 is not as straightforward as the two concepts overlap to a certain degree. Anyway, this index does not take into account the variability within groups. To this end, a generalized index is proposed here based

R. Di Mari · S. Ingrassia (✉) · A. Punzo
Department of Economics and Business, University of Catania, Corso Italia 55, 95129 Catania, Italy
e-mail: salvatore.ingrassia@unict.it

R. Di Mari
e-mail: roberto.dimari@unict.it

A. Punzo
e-mail: antonio.punzo@unict.it

T. Chadjipadelis et al. (eds.), *Data Analysis and Rationality in a Complex World*, Studies in Classification, Data Analysis, and Knowledge Organization,
https://doi.org/10.1007/978-3-030-60104-1_4

on a definition of the coefficient of determination given in Cameron and Windmeijer (1996). The proposed pseudo R^2 is illustrated for finite mixtures of linear regressions and is evaluated by means of a simulation study.

Let $\mathbf{X}$ be a vector of covariates with values in $\mathbb{R}^d$, and let Y be a dependent variable taking values in $\mathbb{R}$; assume that the regression of Y on $\mathbf{X}$ varies across the k levels (groups or clusters) of a categorical latent variable G. Finite mixtures of linear regressions (FMLR) McLachlan and Peel (2000) are characterized by the following conditional density function:

$$p\,(y|\mathbf{x};\boldsymbol{\psi}) = \sum_{g=1}^{k} \pi_g \phi(y;\mu(\mathbf{x};\boldsymbol{\beta}_g),\sigma_g^2), \tag{1}$$

where π_g are the mixing weights, with $\pi_g > 0$ and $\sum_{g=1}^{k}\pi_g = 1$, and $\phi(y;\mu(\mathbf{x};\boldsymbol{\beta}_g))$ denotes group conditional Gaussian distributions with conditional mean $\mu(\mathbf{x};\boldsymbol{\beta}_g) = \beta_{0g} + \boldsymbol{\beta}'_{1g}\mathbf{x}$, $\boldsymbol{\beta}_g$ is a $(d+1)$-dimensional vector, and σ_g^2 is the group variance.

Given a random sample $(\mathbf{x}'_1, y_1)', \ldots, (\mathbf{x}'_n, y_n)'$ of $\left(\mathbf{X}', Y\right)'$, for a fixed number k of groups, we can specify the following log-likelihood function

$$\ell(\boldsymbol{\psi}) = \sum_{i=1}^{n} \ln \sum_{g=1}^{k} \pi_g \phi(y;\mu(\mathbf{x};\boldsymbol{\beta}_g),\sigma_g^2), \tag{2}$$

which is maximized with respect to the model's parameters in order to get maximum likelihood (ML) estimates. This is usually accomplished via iterative procedures, like the expectation-maximization (EM) algorithm. The EM algorithm considers the following complete-data log-likelihood:

$$\begin{aligned}\mathcal{L}(\boldsymbol{\psi}) &= \sum_{g=1}^{k}\sum_{i=1}^{n} z_{ig}\ln\pi_g + \sum_{g=1}^{k}\sum_{i=1}^{n} z_{ig}\ln\left\{\phi\left[y_i;\mu(\mathbf{x}_i;\boldsymbol{\beta}_g),\sigma_g^2\right]\right\} \\ &= \sum_{g=1}^{k}\sum_{i=1}^{n} z_{ig}\ln\pi_g + \ell(\boldsymbol{\beta},\boldsymbol{\sigma}^2)\end{aligned} \tag{3}$$

where $z_{ig} = 1$ if $(\mathbf{x}'_i, y_i)'$ comes from component g and $z_{ig} = 0$ otherwise, $\ell(\boldsymbol{\beta},\boldsymbol{\sigma}^2) = \sum_{g=1}^{k}\sum_{i=1}^{n} z_{ig}\ln\left\{\phi\left[y_i;\mu(\mathbf{x}_i;\boldsymbol{\beta}_g),\sigma_g^2\right]\right\}$, $\boldsymbol{\beta} = (\boldsymbol{\beta}'_1,\ldots,\boldsymbol{\beta}'_k)'$ and $\boldsymbol{\sigma}^2 = (\sigma_1^2,\ldots,\sigma_k^2)'$.

Total sum of squares decomposition for FMLRs. Once the model has been fitted to given data and parameters have been estimated, in Ingrassia and Punzo (2019) the classical total sum of squares decomposition for regression models is generalized when data come from a heterogeneous population and are modeled through a mixture of regressions with Gaussian components. Therefore, the following three-term decomposition of the Total Sum of Squares (TSS) is proposed:

$$\mathrm{TSS} = \sum_{i=1}^{n}(y_i - \bar{y})^2 = \sum_{i=1}^{n}\sum_{g=1}^{k}\widehat{z}_{ig}\left(\bar{y}_g - \bar{y}\right)^2 + \sum_{i=1}^{n}\sum_{g=1}^{k}\widehat{z}_{ig}\left[\mu(\mathbf{x}_i;\widehat{\boldsymbol{\beta}}_g) - \bar{y}_g\right]^2 +$$

$$+\sum_{i=1}^{n}\sum_{g=1}^{k}\widehat{z}_{ig}\left[y_i - \mu(\mathbf{x}_i;\widehat{\boldsymbol{\beta}}_g)\right]^2 = \mathrm{BSS} + \mathrm{RWSS} + \mathrm{EWSS} \tag{4}$$

where BSS is the (soft) between-group sum of squares (i.e., the variability of Y explained by the latent group variable G), EWSS is the (soft) within-group sum of squares explained by the model (i.e., due to the covariates) and RWSS is the (soft) residual within-group sum of squares. In terms of clustering validation BSS can be seen as a separation measure along the y-axis, i.e., as a measure of how well-separated clusters are along the y-axis, and RWSS can be seen as a compactness measure. Finally, we remark that when $k = 1$, the BSS term vanishes and then the decomposition reduces to the classical decomposition for the standard linear regression model whose parameters are estimated by least squares. We remark also that local and overall coefficient of determinations is also proposed and discussed in Ingrassia and Punzo (2019) starting from (4).

Actually, the decomposition (4) does not take into account the variability within the groups. To overcome this problem, we build from the definition of the R^2 coefficient proposed by Cameron and Windmeijer (1996) in the framework of Poisson regression. This proposal is based on the decomposition of the deviance residuals.

2 A Deviance-Based Coefficient of Determination R^2

In Cameron and Windmeijer (1996), some R^2 measures are discussed in the framework of Poisson regression model, where the dependent variable y_i $(i = 1, \ldots, n)$ is independent Poisson with log-density

$$l_i(\mu_i) = \mu_i + y_i \ln \mu_i - \ln y_i! \tag{5}$$

where $\mu_i = \mathbb{E}(y_i|\mathbf{x}_i) = \mu(\mathbf{x}_i, \boldsymbol{\beta})$ and $\boldsymbol{\beta}$ is a $d+1$ parameter vector. In particular, in Cameron and Windmeijer (1996) is first considered the *scaled deviance* based on the log-likelihood computed in the data $\mathbf{y}$ and in the fitted data $\widehat{\mu}$, say $\mathcal{L}(\mathbf{y};\mathbf{y})$ and $\mathcal{L}(\widehat{\mu};\mathbf{y})$, respectively,

$$D(\mathbf{y};\widehat{\mu}) = 2\left\{\mathcal{L}(\mathbf{y};\mathbf{y}) - \mathcal{L}(\widehat{\mu};\mathbf{y})\right\}. \tag{6}$$

Then from the decomposition

$$\mathcal{L}(\mathbf{y};\mathbf{y}) - \mathcal{L}(\bar{\mathbf{y}};\mathbf{y}) = \left\{\mathcal{L}(\mathbf{y};\mathbf{y}) - \mathcal{L}(\widehat{\mu};\mathbf{y})\right\} + \left\{\mathcal{L}(\mathbf{y};\mathbf{y}) - \mathcal{L}(\bar{\mathbf{y}};\mathbf{y})\right\}, \tag{7}$$

the coefficient of determination for the model (5) can be defined as follows:

$$R^2 = 1 - \frac{2\{\mathcal{L}(\mathbf{y};\mathbf{y}) - \mathcal{L}(\widehat{\mu};\mathbf{y})\}}{2\{\mathcal{L}(\mathbf{y};\mathbf{y}) - \mathcal{L}(\bar{\mathbf{y}};\mathbf{y})\}} = \frac{\mathcal{L}(\widehat{\mu};\mathbf{y}) - \mathcal{L}(\bar{\mathbf{y}};\mathbf{y})}{\mathcal{L}(\mathbf{y};\mathbf{y}) - \mathcal{L}(\bar{\mathbf{y}};\mathbf{y})}$$

It can be shown that this measure satisfies the following 5 criteria, see Cameron and Windmeijer (1996):

1. $0 \leq R^2 \leq 1$;
2. R^2 does not decrease as regressors are added (without degree-of-freedom correction);
3. R^2 based on residual sum of squares coincides with R^2 based on explained sum of squares;
4. R^2 has an interpretation in terms of content of data;
5. There is a correspondence between R^2 and a significance test on all slope parameters and between changes in R^2 as regressors are added and significance tests.

Generalized decompositions of deviance for FMLRs. When we move from a regression model to a mixture of regressions, only criteria 1–4 presented earlier can be retained, because the criterion #5 concerns inference about a homogeneous population. To begin with, in (3) let us focus on the model term in the log-likelihood function and set

$$\ell(\mu, \boldsymbol{\sigma}^2; \mathbf{y}) = \sum_{g=1}^{k}\sum_{i=1}^{n} z_{ig}\left[-\frac{1}{2}\ln\sigma_g^2 - \frac{(y_i - \mu_{i,g})^2}{2\sigma_g^2}\right] \tag{8}$$

where $\mu_{i,g} = \beta_{0g} + \boldsymbol{\beta}'_{1g}\mathbf{x}_i$ and the term $-(\ln 2\pi)/2$ has been omitted from $\ell(\boldsymbol{\beta}, \boldsymbol{\sigma}^2)$. After the parameters have been estimated, let us introduce the following quantities:

$$\ell(\widehat{\mu}, \boldsymbol{\sigma}^2; \mathbf{y}) = \sum_{g=1}^{k}\sum_{i=1}^{n} z_{ig}\left[-\frac{1}{2}\ln\widehat{\sigma}_g^2 - \frac{(y_i - \widehat{\mu}_{i,g})^2}{2\widehat{\sigma}_g^2}\right] \tag{9}$$

$$\ell(\mathbf{y}, \widehat{\boldsymbol{\sigma}}^2; \mathbf{y}) = \sum_{g=1}^{k}\sum_{i=1}^{n} z_{ig}\left[-\frac{1}{2}\ln\widehat{\sigma}_g^2 - \frac{(y_i - y_i)^2}{2\widehat{\sigma}_g^2}\right] = \sum_{g=1}^{k}\sum_{i=1}^{n} z_{ig}\left[-\frac{1}{2}\ln\widehat{\sigma}_g^2\right] \tag{10}$$

$$\ell(\bar{\mathbf{y}}, \widehat{\boldsymbol{\sigma}}^2; \mathbf{y}) = \sum_{g=1}^{k}\sum_{i=1}^{n} z_{ig}\left[-\frac{1}{2}\ln\widehat{\sigma}_g^2 - \frac{(y_i - \bar{y})^2}{2\widehat{\sigma}_g^2}\right] \tag{11}$$

and consider:

$$\ell(\mathbf{y}, \widehat{\boldsymbol{\sigma}}^2; \mathbf{y}) - \ell(\bar{\mathbf{y}}, \widehat{\boldsymbol{\sigma}}^2; \mathbf{y}) = \left\{\ell(\mathbf{y}, \widehat{\boldsymbol{\sigma}}^2; \mathbf{y}) - \ell(\widehat{\mu}, \boldsymbol{\sigma}^2; \mathbf{y})\right\} + \left\{\ell(\widehat{\mu}, \widehat{\boldsymbol{\sigma}}^2; \mathbf{y}) - \ell(\bar{\mathbf{y}}, \widehat{\boldsymbol{\sigma}}^2; \mathbf{y})\right\}. \tag{12}$$

Therefore once parameters have been estimated, based on (9)-(11), in the decomposition (12), we have

$$\ell(\mathbf{y}, \widehat{\boldsymbol{\sigma}}^2; \mathbf{y}) - \ell(\bar{\mathbf{y}}, \widehat{\boldsymbol{\sigma}}^2; \mathbf{y}) = \sum_{g=1}^{k}\sum_{i=1}^{n} \widehat{z}_{ig} \frac{(y_i - \bar{y})^2}{2\widehat{\sigma}_g^2} \tag{13}$$

$$\ell(\mathbf{y}, \widehat{\boldsymbol{\sigma}}^2; \mathbf{y}) - \ell(\widehat{\mu}, \widehat{\boldsymbol{\sigma}}^2; \mathbf{y}) = \sum_{g=1}^{k}\sum_{i=1}^{n} \widehat{z}_{ig} \frac{(y_i - \widehat{\mu}_{i,g})^2}{2\widehat{\sigma}_g^2} \tag{14}$$

$$\ell(\widehat{\mu}, \widehat{\boldsymbol{\sigma}}^2; \mathbf{y}) - \ell(\bar{\mathbf{y}}, \widehat{\boldsymbol{\sigma}}^2; \mathbf{y}) = \sum_{g=1}^{k}\sum_{i=1}^{n} \widehat{z}_{ig} \frac{(y_i - \bar{y})^2}{2\widehat{\sigma}_g^2} - \sum_{g=1}^{k}\sum_{i=1}^{n} \widehat{z}_{ig} \frac{(y_i - \widehat{\mu}_{i,g})^2}{2\widehat{\sigma}_g^2} \tag{15}$$

According to (Ingrassia and Punzo, 2019, Sect. 4), in (15) we have

$$\sum_{g=1}^{k}\sum_{i=1}^{n} \widehat{z}_{ig} \frac{(y_i - \bar{y})^2}{2\widehat{\sigma}_g^2} = \sum_{g=1}^{k}\sum_{i=1}^{n} \hat{z}_{ig} \frac{(y_i - \bar{y}_g)^2}{2\widehat{\sigma}_g^2} + \sum_{g=1}^{k} \hat{n}_g \frac{(\bar{y}_g - \bar{y})^2}{2\widehat{\sigma}_g^2}$$

where the first term can be further decomposed as

$$\sum_{g=1}^{k}\sum_{i=1}^{n} \hat{z}_{ig} \frac{(y_i - \bar{y}_g)^2}{2\widehat{\sigma}_g^2} = \sum_{g=1}^{k}\sum_{i=1}^{n} \widehat{z}_{ig} \frac{(y_i - \widehat{\mu}_{i,g})^2}{2\widehat{\sigma}_g^2} + \sum_{g=1}^{k}\sum_{i=1}^{n} \hat{z}_{ig} \frac{(\widehat{\mu}_{i,g} - \bar{y}_g)^2}{2\widehat{\sigma}_g^2}.$$

Thus finally (15) yields

$$\begin{aligned}
\ell(\widehat{\mu}, \widehat{\boldsymbol{\sigma}}^2; \mathbf{y}) - \ell(\bar{\mathbf{y}}, \widehat{\boldsymbol{\sigma}}^2; \mathbf{y}) &= \sum_{g=1}^{k}\sum_{i=1}^{n} \widehat{z}_{ig} \frac{(y_i - \widehat{\mu}_{i,g})^2}{2\widehat{\sigma}_g^2} + \sum_{g=1}^{k}\sum_{i=1}^{n} \hat{z}_{ig} \frac{(\widehat{\mu}_{i,g} - \bar{y}_g)^2}{2\widehat{\sigma}_g^2} \\
&\quad + \sum_{g=1}^{k} \hat{n}_g \frac{(\bar{y}_g - \bar{y})^2}{2\widehat{\sigma}_g^2} - \sum_{g=1}^{k}\sum_{i=1}^{n} \widehat{z}_{ig} \frac{(y_i - \widehat{\mu}_{i,g})^2}{2\widehat{\sigma}_g^2} \\
&= \sum_{g=1}^{k}\sum_{i=1}^{n} \hat{z}_{ig} \frac{(\widehat{\mu}_{i,g} - \bar{y}_g)^2}{2\widehat{\sigma}_g^2} + \sum_{g=1}^{k} \hat{n}_g \frac{(\bar{y}_g - \bar{y})^2}{2\widehat{\sigma}_g^2}
\end{aligned} \tag{16}$$

Therefore, putting together (14) and (16), from (12), we get

$$\begin{aligned}
\ell(\mathbf{y}, \widehat{\boldsymbol{\sigma}}^2; \mathbf{y}) - \ell(\bar{\mathbf{y}}, \widehat{\boldsymbol{\sigma}}^2; \mathbf{y}) &= \{\ell(\mathbf{y}, \widehat{\boldsymbol{\sigma}}^2; \mathbf{y}) - \ell(\bar{\mathbf{y}}, \widehat{\boldsymbol{\sigma}}^2; \mathbf{y})\} + \{\ell(\widehat{\mu}, \widehat{\boldsymbol{\sigma}}^2; \mathbf{y}) - \ell(\bar{\mathbf{y}}, \widehat{\boldsymbol{\sigma}}^2; \mathbf{y})\} \\
&= \sum_{g=1}^{k}\sum_{i=1}^{n} \widehat{z}_{ig} \frac{(y_i - \widehat{\mu}_{i,g})^2}{2\widehat{\sigma}_g^2} + \sum_{g=1}^{k}\sum_{i=1}^{n} \hat{z}_{ig} \frac{(\widehat{\mu}_{i,g} - \bar{y}_g)^2}{2\widehat{\sigma}_g^2} \\
&\quad + \sum_{g=1}^{k} \hat{n}_g \frac{(\bar{y}_g - \bar{y})^2}{2\widehat{\sigma}_g^2}.
\end{aligned} \tag{17}$$

Let us set

$$\Delta\ell_t(\mathbf{y}, \bar{\mathbf{y}}) = \ell(\mathbf{y}, \widehat{\boldsymbol{\sigma}}^2; \mathbf{y}) - \ell(\bar{\mathbf{y}}, \widehat{\boldsymbol{\sigma}}^2; \mathbf{y}) = \sum_{g=1}^{k}\sum_{i=1}^{n} \widehat{z}_{ig} \frac{(y_i - \bar{y})^2}{2\widehat{\sigma}_g^2}$$

$$\Delta\ell_r(\mathbf{y}, \widehat{\mu}) = \sum_{g=1}^{k}\sum_{i=1}^{n} \widehat{z}_{ig} \frac{(y_i - \widehat{\mu}_{i,g})^2}{2\widehat{\sigma}_g^2} \tag{18}$$

$$\Delta\ell_f(\widehat{\mu}, \bar{\mathbf{y}}_G) = \sum_{g=1}^{k}\sum_{i=1}^{n} \hat{z}_{ig} \frac{(\widehat{\mu}_{i,g} - \bar{y}_g)^2}{2\widehat{\sigma}_g^2} \tag{19}$$

$$\Delta\ell_b(\bar{\mathbf{y}}_G, \bar{\mathbf{y}}) = \sum_{g=1}^{k} \hat{n}_g \frac{(\bar{y}_g - \bar{y})^2}{2\widehat{\sigma}_g^2}$$

the decomposition (12) can be written as

$$\Delta\ell(\mathbf{y}, \bar{\mathbf{y}}) = \Delta\ell_r(\mathbf{y}, \widehat{\mu}) + \Delta\ell_f(\widehat{\mu}, \bar{\mathbf{y}}_G) + \Delta\ell_b(\bar{\mathbf{y}}_G, \bar{\mathbf{y}}) \tag{20}$$

which generalizes the relation (32) in Ingrassia and Punzo (2019) because here the component variances are included. In particular, the decomposition (20) reduces to earlier result in the homoschedastic case, i.e. when $\sigma_1^2 = \cdots = \sigma_k^2 = \sigma^2$.

Some special cases:

- $\Delta\ell_b(\bar{\mathbf{y}}_G, \bar{\mathbf{y}}) = 0$ when $\bar{y}_1 = \cdots = \bar{y}_k = \bar{y}$, regardless of the group sizes $\hat{n}_1, \ldots, \hat{n}_k$;
- $\Delta\ell_f(\widehat{\mu}, \bar{\mathbf{y}}_G) = 0$ when when $\hat{\boldsymbol{\beta}}_{11} = \cdots = \hat{\boldsymbol{\beta}}_{k1} = \mathbf{0}$ so that $\hat{\beta}_{g0} = \bar{y}_g, g = 1, \ldots, k$, regardless of the values of $\hat{z}_{ig}$;
- $\Delta\ell_r(\mathbf{y}, \widehat{\mu}) = 0$ when either $(y_i - \widehat{\mu}_{i,g})^2 = 0$ or $\widehat{z}_{ig} = 0$ for $i = 1, \ldots, n$ and $g = 1, \ldots, k$, i.e. the n data points lie in either regression line (plane or hyperplane).

The generalized coefficient of determination. According to (Cameron and Windmeijer, 1996, (1.16)), we can define the local coefficient of determination for the gth group ($g = 1, \ldots, k$) as

$$R_g^2 = \frac{\Delta\ell_{f,g}(\widehat{\mu}, \bar{\mathbf{y}}_G)}{\Delta\ell_{r,g}(\mathbf{y}, \widehat{\mu}) + \Delta\ell_{f,g}(\widehat{\mu}, \bar{\mathbf{y}}_G)}. \tag{21}$$

We can see immediately that R_g^2 satisfies conditions 1–4 of Cameron and Windmeijer (1996), see Sect. 2.

With the same principle, it is natural to define the overall coefficient of determination as

$$R^2 = \frac{\Delta\ell_f(\widehat{\mu}, \bar{\mathbf{y}}_G)}{\Delta\ell_r(\mathbf{y}, \widehat{\mu}) + \Delta\ell_f(\widehat{\mu}, \bar{\mathbf{y}}_G)}, \tag{22}$$

which can be interpreted as the proportion of the weighted within-group response variation explained (accounted for) by the fitted mixture of regression. Based on

(21), R^2 is related to $R_1^2, \ldots, R_k^2$ by the following relation:

$$R^2 = \frac{\sum_{g=1}^{k} \Delta\ell_{f,g}(\widehat{\mu}, \bar{\mathbf{y}}_G)}{\Delta\ell_r(\mathbf{y}, \widehat{\mu}) + \Delta\ell_f(\widehat{\mu}, \bar{\mathbf{y}}_G)} = \sum_{g=1}^{k} \frac{\Delta\ell_{r,g}(\mathbf{y}, \widehat{\mu}) + \Delta\ell_{f,g}(\widehat{\mu}, \bar{\mathbf{y}}_G)}{\Delta\ell_r(\mathbf{y}, \widehat{\mu}) + \Delta\ell_f(\widehat{\mu}, \bar{\mathbf{y}}_G)} R_g^2 \tag{23}$$

which is a more general version of Ingrassia and Punzo (2019)'s Eq. (37)—here the terms in (23) for the Gaussian case depend on the variances $\widehat{\sigma}_1^2, \ldots, \widehat{\sigma}_k^2$. According to (23), R^2 can be seen as a weighted average of the local coefficients of determination $R_1^2, \ldots, R_k^2$ weighted on the proportion of the within-group of $\Delta\ell_r(\mathbf{y}, \widehat{\mu}) + \Delta\ell_f(\widehat{\mu}, \bar{\mathbf{y}}_G)$.

The local coefficient of determination for the gth group ($g = 1, \ldots, k$) with conditional Gaussian components can be defined as

$$R_g^2 = \frac{\Delta\ell_{f,g}(\widehat{\mu}, \bar{\mathbf{y}}_G)}{\Delta\ell_{r,g}(\mathbf{y}, \widehat{\mu}) + \Delta\ell_{f,g}(\widehat{\mu}, \bar{\mathbf{y}}_G)} = \frac{\sum_{i=1}^{n} \hat{z}_{ig}(\widehat{\mu}_{i,g} - \bar{y}_g)^2}{\sum_{i=1}^{n} \hat{z}_{ig}(\widehat{\mu}_{i,g} - \bar{y}_g)^2 + \sum_{i=1}^{n} \hat{z}_{ig}(y_i - \widehat{\mu}_{i,g})^2}, \tag{24}$$

where we note that R_g^2 does not depend on the variances $\widehat{\sigma}_g^2$, $g = 1, \ldots, k$; similarly, the overall coefficient of determination is defined as

$$R^2 = \frac{\Delta\ell_f(\widehat{\mu}, \bar{\mathbf{y}}_G)}{\Delta\ell_r(\mathbf{y}, \widehat{\mu}) + \Delta\ell_f(\widehat{\mu}, \bar{\mathbf{y}}_G)} = \frac{\displaystyle\sum_{g=1}^{k}\sum_{i=1}^{n} \hat{z}_{ig} \frac{(\widehat{\mu}_{i,g} - \bar{y}_g)^2}{2\widehat{\sigma}_g^2}}{\displaystyle\sum_{g=1}^{k}\sum_{i=1}^{n} \hat{z}_{ig} \frac{(\widehat{\mu}_{i,g} - \bar{y}_g)^2}{2\widehat{\sigma}_g^2} + \sum_{g=1}^{k}\sum_{i=1}^{n} \hat{z}_{ig} \frac{(y_i - \widehat{\mu}_{i,g})^2}{2\widehat{\sigma}_g^2}}. \tag{25}$$

3 A Real Data Example

The dataset we use for this illustration contains $n = 480$ observations coming from a system for monitoring the progress of new sales force put in place by a firm (Stine and Foster 2014). The main predictor is the log of the number of accounts opened by new sale agents, which we use to predict the log of the profit from sales of new agents over the past two years. In Fig. 1a) the plot of the raw data is given, to which a simple regression model fit yields $R^2 = 0.1762$.

Actually, data concern two groups, depending on if new sales agent worked in existing offices (group 1) or new offices (group 2). Thus, data were fitted using a mixture of regression (1) with $k = 2$ and the two regression lines are plotted in Fig. 1b).

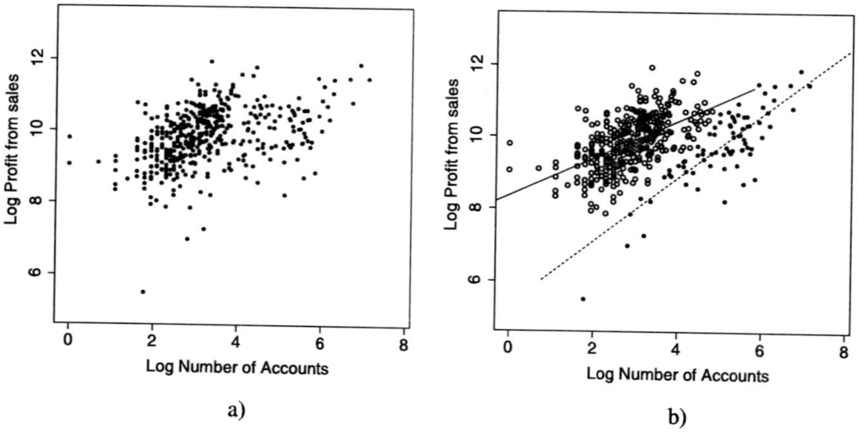

Fig. 1 Hiring dataset: **a** raw data and **b** data and lines based on mixture of regression model

Table 1 Coefficient of determinations for the fitted mixture of regression model

R^2	R_1^2	R_2^2
0.4187	0.3127	0.6664

In Table 1, we list the values of the overall coefficient of determination R^2 and the two local coefficients of determination R_1^2 and R_2^2. We see that the overall coefficient of determination $R^2 = 0.4187$ is now quite larger than the value obtained through a simple regression model. In detail, as for the agent worked in new offices (group 2), we get $R_2^2 = 0.6664$ showing a good fit to the model, while for sales agent worked in existing offices (group 1) the local coefficient of determination yields $R_1^2 = 0.3127$.

4 Conclusions

In this work, we have presented a new approach for computing the coefficient of determination for mixtures of regressions in the Gaussian framework. This approach is based on the decomposition of the scaled deviance and is general enough to be extended also to mixtures of regression with a non-continuous response and other link functions. Although this index has an intuitive appeal in terms of model interpretation, it need not be used for model selection.

References

Cameron, A.C., Windmeijer, F.A.: R-squared measures for count data regression models with applications to health-care utilization. J. Bus. Econ. Stat. **14**(2), 209–220 (1996)

Ingrassia, S., Punzo, A.: Cluster validation for mixtures of regressions via the total sum of squares decomposition. J. Classif. (2019)

McLachlan, G.J., Peel, D.: Finite Mixture Models. Wiley, New York (2000)

Stine, R., Foster, D.: Statistics for Business. Decision Making and Analysis, Pearson, Upper Saddle River (2014)

Distance Measurement When Fuzzy Numbers Are Used. Survey of Selected Problems and Procedures

Józef Dziechciarz and Marta Dziechciarz-Duda

Abstract The goal is to identify and to discuss distance and dissimilarity measures calculated with fuzzy numbers. It is crucial to define the distance and dissimilarity measures for unconventional fuzzy numbers, i.e. asymmetric, overlapping triangular, with unequal width. Resulting distance measures are to be used for clustering and linear ordering of objects. The method applied consists of an attempt to identify and to discuss the applicability of specialised techniques for unconventional fuzzy measurement. The emphasis is put on distance (similarity and dissimilarity) of measurement concepts when unconventional fuzzy numbers are used. The use of conventional fuzzy numbers, i.e. symmetric, not overlapping triangular, with equal width is limited when Computer-Aided Web Interviewing is applied. Respondents tend to use asymmetric fuzzy numbers with overlapping shape and unequal width. Several problems arise in the multivariate statistical analysis of measurement results. Proposals from pattern recognition literature are not applicable and new methods based on directed fuzzy numbers are involved.

Keywords Ordered fuzzy numbers · Unconventional fuzzy numbers · Fuzzy measurement · Fuzzy data · Fuzzy distance

1 Introduction

The literature widely recognises the importance of fuzzy numbers. Public statistics, quantitatively oriented socio-economic researchers, marketing and capital market analytics tend to use metric data. Since well-being measurement is subjective, opinion-based judgement and subjective opinions are hardly metric, it seems natural to extend the methodology on fuzzy tools. Although the use of fuzzy numbers

J. Dziechciarz (✉) · M. Dziechciarz-Duda
Wrocław University of Economics, Wroclaw, Poland
e-mail: jozef.dziechciarz@ue.wroc.pl

M. Dziechciarz-Duda
e-mail: marta.dziechciarz@ue.wroc.pl

T. Chadjipadelis et al. (eds.), *Data Analysis and Rationality in a Complex World*, Studies in Classification, Data Analysis, and Knowledge Organization,
https://doi.org/10.1007/978-3-030-60104-1_5

is widely discussed in the literature, the discussion is limited to the conventional understanding of fuzzy numbers, i.e. symmetric, not overlapping triangular, with equal width. In this paper, the term "unconventional fuzzy numbers" will stand for fuzzy numbers that lack one or all characteristics of conventional fuzzy numbers, meaning they may be asymmetric, overlapping, or with unequal width. The public opinion poll and market survey specialists, especially those using Computer-Aided Web Interviewing, report that respondents tend to use unconventional fuzzy numbers, such as (triangle) numbers which are asymmetric, overlapping and have unequal width. The scientific routine and the methodical discipline of socio-economic science require tools which are adequate for unconventional fuzzy measurement, for statistical analysis of collected data, for its econometric modelling and for formulating policy recommendations. Using fuzzy measurement results for multivariate statistical analysis is not easy. There are many problems, especially when it comes to unconventional fuzzy numbers. First of all, the issue of measurement result representation (coding) is completely omitted when they are in the form of unconventional fuzzy numbers. The next step requires a specialized concept of distance or dissimilarity. Some suggestions can be found in the literature on pattern recognition, but the socio-economic phenomena can seldom be considered as patterns.

2 Distance Measurement for Results in the Form of Fuzzy Numbers

Conventional fuzzy numbers. Data analysis when measurement results are in the form of unconventional fuzzy numbers is hardly discussed in the literature. Analogously with other issues in the fuzzy framework, distance measurement tools are also usually referring to conventional fuzzy numbers. The most widespread definition of fuzzy distance is known as the Hausdorff measure, named after a German mathematician Hausdorff (1914, 1918), HuttenLocher et al. (1993), Im (2019). The source of this idea may be seen in a doctoral thesis of a French mathematician Frechet (1906). The intuitive explanation of the of Hausdorff distance when used to measure the difference between two subsets is that it represents the greatest of all the distances from a point in one set to the closest point in the other set Gerla and Volpe (1986), Kaleva and Seikkala (1984), Osman (1983), Szmidt and Kacprzyk (2005, 2009). The specialised fuzzy distance measures, where classic metric space is generalized for the fuzzy environment, may be found, among others, in Li et al. (2011), Zhang et al. (2009). Extensive discussion on measurement theory, including distance measurement, may be found in five volumes of a fundamental work by Fremlin (2015). The frequently used solutions in specialised fuzzy distance measures may be attributed to three approaches, including a generalization of the classic metric space distance measure for use in fuzzy subsets, distance based on the difference between membership functions of the fuzzy set or the use of distance based on differences in membership functions of a fuzzy set and, finally, a fuzzy metric introduced by generalizing a metric space (Luo and Cheng 2015). Dudek and Pełka (2015) reviewed, evaluated and compared five clustering algorithms using Hausdorff-type distance measures. The list of tested procedures includes self-organizing maps; Fuzzy Learning Vector Quantization; Fuzzy Adaptive Resonance Theory; Growing Neural Gas and Self-Organizing Simplified Adaptive Resonance Theory. Because Hausdorff-type distance measures, known as an α-cut concept (Fig. 1, right panel), is designed for pattern recognition, it is not the best choice for socio-economic phenomena, where evaluative interpretation is crucial.

Unconventional fuzzy numbers. Unconventionality, in this context, means that the fuzzy numbers may have uneven length, can be asymmetric and overlapping (Delgado et al. Delgado et al. 1998a, b). In socio-economic problems, the usefulness of the traditional Hausdorff-like distance is limited. In a situation using linguistic variables, the construction is implicit, of evaluative character, leading to ordinal, hierarchical assessment. To address such issues, specifications coming from new developments in fuzzy theory are more appropriate. The first milestone was laid by K. Atanassov, who introduced the concept of intuitionistic fuzzy sets (Atanassov 1986; Szmidt and Kacprzyk 2000). The second was the definition of ambiguity and fuzziness (Delgado et al. 1998a, b). They were defined as follows: let A be a fuzzy number with α-cut representation (see Fig. 1, right panel, for explanation), then the ambiguity is an integral value on α-cut of a fuzzy number. Ambiguity may be seen as a kind of global spread of the membership function, whereas the fuzziness involves a comparison between the fuzzy set and its complement. Human intuition suggests that vagueness definitely exists in the distances between any two fuzzy numbers, but a less vague and less ambiguous distance is always acceptable from the stability point of view. Supposing there are two fuzzy numbers, A and B, with the same central value but with different spreads, then A is expected to be better than B in the sense

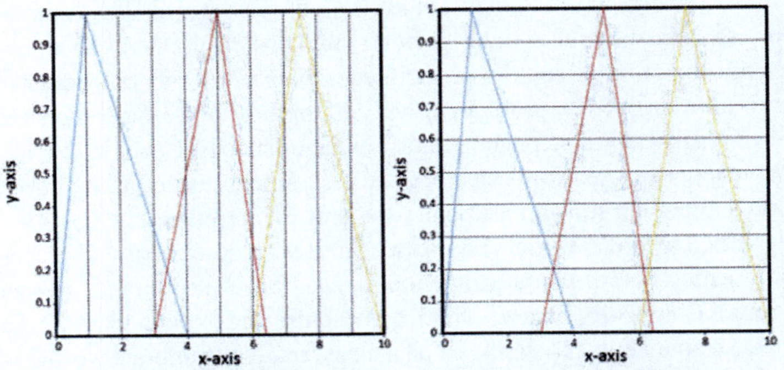

Left panel: positively oriented fuzzy numbers

$$d(A;B) = \frac{1}{n}\sum_{i=1}^{n} |\mu_A(x_i) - \mu_B(x_i)|$$

n – number of cuts of x-axis

Right panel: *alfa cuts* (classic fuzzy numbers)

$$d(A;B) = \frac{1}{m}\sum_{i=1}^{m} max\left\{\left|A_{\alpha_{il}} - B_{\alpha_{il}}\right|; \left|A_{\alpha_{ip}} - B_{\alpha_{ip}}\right|\right\}$$

n – number of cuts of y-axis

α_{il}, α_{ip} – left and right cut points

Fig. 1 Distance measurement

of stability or preciseness, provided a spread of A is smaller than B. Introduction of generalized fuzzy numbers may be seen as the third milestone (Zimmermann 2001). The generalized fuzzy number represented by the left value, the right value, middle point; left spread (distance between left value and middle point), the right spread (distance between middle point and right value). H. Zimmermann defined interval arithmetic for generalized fuzzy numbers with addition, subtraction, multiplication and division as admissible arithmetic operations.

The crucial, revolutionary milestone was laid by Kosinski et al. (2003), who introduced ordered fuzzy numbers. Ordered fuzzy numbers (sometimes referred to as oriented or directed), address the problem of assessment with evaluative character, leading to ordinal, hierarchical inference. McCulloch et al. (1983) described the directional distance measurement between fuzzy sets. The graphical illustration is given in Fig. 1, left panel.

3 Measurement. Linguistic Form of Determining the Values of Characteristics

Linguistic scale. Socio-economic phenomena are multidimensional in nature. Additionally, a large portion of those dimensions—socio-economic phenomenon characteristics—are inherently qualitative. When measuring, one attempts to quantify characteristics which have hidden values, immeasurable on metric scales (Liao et al. 2004; Walesiak and Dziechciarz 1999). In other words, the sole procedure used for data collection is a questionnaire interview, asking respondents to provide a subjective assessment of the value. Most of the routinely used quantitative measurement tools for determining attitudes or perception, at least partly, ignore the problem of subjectivity of attitudes and perception toward socio-economic phenomena. Nowadays, the most popular data acquisition tool is Computer-Aided Web Interviewing. It is obvious that the researcher (interviewer) cannot ask respondents to give numerical values of subjective perception of socio-economic phenomenon characteristics. The natural solution for the problem, when attempting to measure the subjective perception of socio-economic phenomenon characteristics is to use linguistic scales, with verbal categories used for describing the assessment results. It may be seen as an alternative approach for numeric measurement. In a routine survey situation, a common practice is to use verbal (linguistic) phrases to describe the characteristics (attitudes or perception) of the phenomenon in question. Based on that, one can recommend verbal (linguistic) phrases which can capture and explain the differences in the individual respondents' assessments as a measurement technique worth endorsement (Linguistic Variable and Fuzzy Inference System 2019; Qualitative Analysis 2017; Schnorr-Bäcker 2018; Zamri and Abdullah 2014). Practical use of linguistic scale is not straightforward. The problem with applying the verbal categories is the need to indicate quantitative (numerical) equivalents for verbal expressions used by respondents. Expecting that a person (respondent) will be able to code verbal categories, i.e. produce results on a metric scale, leads to the conclusion that a researcher is ready to accept highly subjective statements (Benoit and Foulloy 2013; Ryjov 2003). On the other hand, when the researcher takes on the responsibility for the difficult task of adequate coding of verbal statements with numeric equivalents, the subjectivity shifts onto the researcher. One should not unambiguously and arbitrarily define the way of interpreting the differences in assessments that have been expressed in linguistic terms and translate them into numeric values. In the coding procedure, verbal statements may be coded based on the concept of (triangular) fuzzy numbers (Lalla et al. 2005).

Measurement. The most common approach for data collection when fuzzy numbers are involved consists of several steps. After preparing the analysis plan, choosing multivariate data analysis tools and designing (not) fuzzy measurement scale, comes the data collection process (conducting measurement). Collected data undergoes the process of fuzzification (transformation into fuzzy numbers). As a rule, researchers' arbitrary decisions about the form of fuzziness are involved. In other words, each verbal statement is coded with numerical equivalent (numerical code), usually in such a way that the researcher and not the respondent determines values assigned to the verbal statement. Therefore, the researcher's action constitutes the shape of a fuzzy number. This potential coding procedure has severe disadvantages. It is fully arbitrary, with the researcher alone deciding how to translate the verbal term (low, medium, high, very high, etc.) into the numeric values of a fuzzy number.

It is much more advisable to design a fuzzy measurement scale disguised as a linguistic scale. To do that, the researcher has to decide on the form of the fuzzy number: triangle, trapeze, etc., and the permissible variants: asymmetric, overlapping, with unequal width, etc. During the data collection process, respondents are asked to define numeric equivalents of verbal statements. However, respondents should not be asked to define fuzzy numbers. Instead, respondents may be asked to specify, where on an

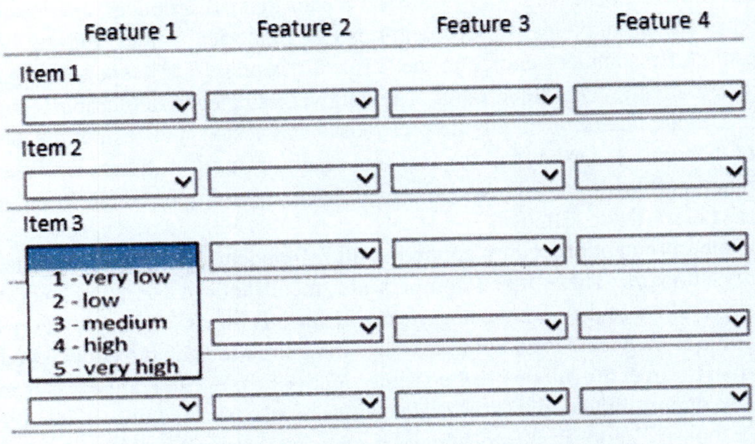

Fig. 2 Respondents' choice of verbal categories attached to assessed items

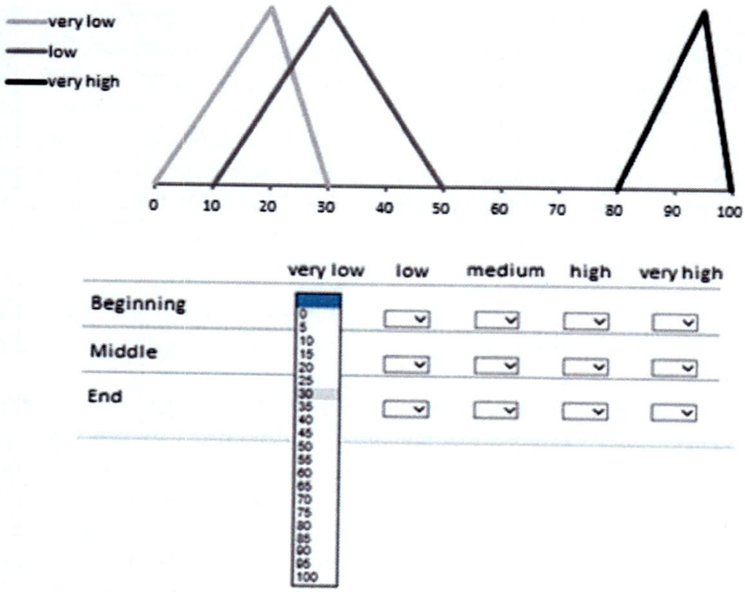

Fig. 3 Respondents' choice. Coding verbal categories with numerical values

axis (e.g. 1–100) begins the numeric value of a specific verbal term, where it has a middle point, and where does it end (Dziechciarz et al. 2010; Dziechciarz 2015). On Figs. 2 and 3, the illustration shows how is the data collection performed. In the first step, respondents are asked to choose verbal categories corresponding to their assessment of the phenomenon in question. Then, in the second step, the respondent is provided with instructions on how to define the beginning, middle and upper value of the chosen response (Fig. 3). In reality, it means that a respondent, unconsciously, without realizing it, codes verbal categories into numerical values. Figure 3 shows the results of respondent choice in the form of triangles. The respondent may see the immediate graphical illustration of coding verbal categories (computer monitor view). The applicability of this new proposal was tested, the results are promising and benefits go far beyond the fact that resulting fuzzy numbers are free from researcher's arbitrary interference. Detailed discussion may be found in the article by Dziechciarz-Duda (2020).

The measurement procedure applied to all respondents gives results in the form of fuzzy numbers. These fuzzy numbers are unconventional. Respondents define numerical values of the verbal categories with various forms of triangle fuzzy numbers. Some respondents tried not to overlap, some tried to keep the equal length and some tried to cover the full range of possible values (the proposed range was from 0 to 100). The opportunity to assess the scale adequacy may be an additional benefit. The Item Response Theory framework may be used for the purpose (Dziechciarz-Duda 2020). Figure 4 illustrates that.

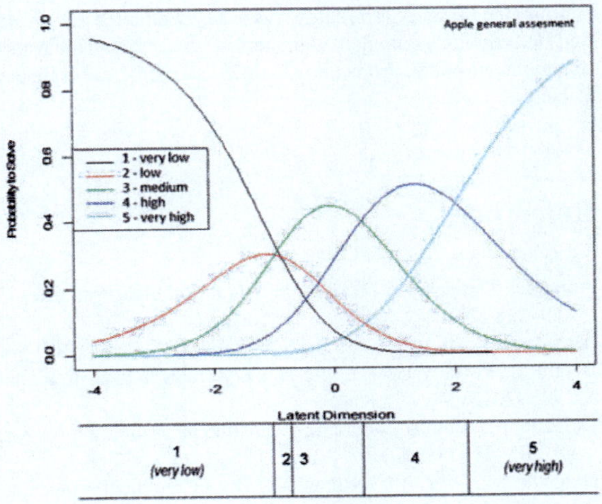

Fig. 4 Characteristic curves for subscale items (Item Characteristic Curves)

4 Conclusions and Future Work

This is a new proposal for data collection and data preliminary preparation for use in multivariate statistical analysis. The core of the proposal is the suggestion on how to measure the subjective perception of socio-economic phenomena. The proposed distance measurement using fuzzy numbers includes the use of linguistic variables for subjective opinions measurement. The novelty of this procedure lies in how are the unconventional fuzzy number values determined. The respondents alone define the type, shape and numeric characteristics of fuzzy numbers representing their subjective (verbal) assessment. The applicability of this new proposal was tested, the results are promising and benefits go far beyond freeing the resulting fuzzy numbers from the researcher's arbitrary interference. There is also a new tool for Computer-Aided Web Interviewing. Unconventional fuzzy numbers are considered for calculating distance and dissimilarity measures. The unconventional concept of ordered (directed) fuzzy numbers proved to be the most appropriate specification for subjective perception assessment of socio-economic phenomena. The discussion goal is to identify and to assess suitability and examine the applicability of specialised techniques for multidimensional statistical analysis. The stress is put on distance (similarity and dissimilarity) measurement concepts when using unconventional fuzzy numbers. The future work on the idea of fuzziness in socio-economic sciences will concentrate on the effort to directly exploit fuzzy measurement results. An equally important problem is discretization (defuzzification) of measurement results, i.e. the way to swap (encode) fuzzy values with numerical numbers, in order to apply them to traditional multidimensional statistical analysis techniques.

Acknowledgements The study was partly conducted in the framework of the research project entitled Households' equipment with durable goods in statistical analysis and econometric modelling of material well-being. Project no. 2018/29/B/HS4/01420 is financed by the National Science Centre, Poland.

References

Arguelles Mendez, L.: From fuzzy sets to linguistic variables. Stud. Fuzziness Soft Comput. **327**, 169–228 (2016)
Atanassov, K.: Intuitionistic fuzzy sets. Fuzzy Set Syst. **20**(1), 87–96 (1986)
Benoit, E., Foulloy, L.: The role of fuzzy scales in measurement theory. Measurement **46**(8), 2921–2926 (2013)
Delgado, M., Vila, A., Voxman, W.: A fuzziness measure for fuzzy numbers. Appl. Fuzzy Set Syst. **94**, 205–216 (1998a)
Delgado, M., Vila, A., Voxman, W.: On a Canonical representation of fuzzy numbers. Fuzzy Set Syst. **93**, 125–135 (1998b)
Dudek, A., Pełka, M.: The comparison of fuzzy clustering methods for symbolic interval-valued data. Przegląd. Polish Stat. **LXII/3**, 301–319 (2015)
Dziechciarz J., Dziechciarz M., Przybysz K.: Household possession of consumer durables on background of some poverty lines. In: Locarek-Junge H., Weihs C. (eds.) Classification as a Tool for Research. Studies in Classification, Data Analysis, and Knowledge Organization, pp. 735–745. Springer, Berlin (2010)
Dziechciarz, J., Dziechciarz-Duda, M.: Non-metric data in household durable goods analysis. Selected Aspects. Acta Universitatis Lodziensis. **4**(330), 111–128 (2017)
Dziechciarz J.: Pomiar i wycena wiedzy, umiejętności i kompetencji nabytych w formalnych i nieformalnych formach kształcenia. In: Wdowiński, P. (ed.) Nauczyciel akademicki wobec nowych wyzwań edukacyjnych, Uniwersytet Łódzki, Łódź, pp. 25–42 (2015)
Dziechciarz-Duda, M.: A proposal for linguistic scale adequacy assessment using the IRT tools. Przeglad Statystyczny. Polish Statistician. Submitted (2020)
Frechet, M.: Sur quelques points de calcul functionnel. Publisher not identified. These Faculte des sciences de Paris, Paris (1906)
Fremlin, D.: Measure Teory. Vol. 1, The Irreducible Minimum. Vol. 2, Broad Foundations. Vol. 3, Measure Algebras. Vol. 4, Topological Measure Spaces. Vol. 5, Set-theoretic Measure Theory. Torres Fremlin, Colchester (2015)
Gerla, G., Volpe, R.: The definition of distance and diameter in fuzzy set theory. Studia of the University of Babes-Bolyai. Mathematics. **31**, 21–26 (1986)
Hausdorff, F.: Grundzüge der Mengenlehre. Veit, Leipzig (1914)
Hausdorff, F.: Dimension und Äußeres Maß. Math. Ann. **79**(1–2), 157–179 (1918)
HuttenLocher, C., Klanderman, G., Rucklidge, W.: Comparing images using the Hausdorff distance. IEEE Trans. Pattern Anal. Mach. Intell. **15**(9), 850–863 (1993)
Im, M.: Nonstandard approach to Hausdorff outer measure. In: arXiv.org (2019). https://arxiv.org/abs/1912.00297v1
Kaleva, O., Seikkala, S.: On fuzzy metric spaces. Fuzzy Set Syst. **12**, 215–229 (1984)
Kosiński, W., Prokopowicz, P., Ślęzak, D.: Ordered fuzzy numbers. Bull. Polish Acad. Sci. Math. **51**(3), 327–339 (2003)
Lalla, M., Facchinetti, G., Mastroleo, G.: Ordinal scales and fuzzy set systems to measure agreement. An application to the evaluation of teaching activity. Qual Quant. **38**, 577–601 (2005)
Li, D., Li, Y., Xie, Y.: Robustness of interval - valued fuzzy inference. Inf Sci. **181**, 321–354 (2011)
Liao, S., Tang, T., Liu, W.: Finding relevant sequences in time series containing crisp, interval and fuzzy interval data. IEEE Trans. Syst. Man Cybern. B Cybern. **34**(5), 2071–2079 (2004)

Linguistic variable and fuzzy inference system. In: FisPro: An Open Source Portable Software for Fuzzy Inference Systems (2019). http://fispro.org
Luo, M., Cheng, Z.: The Distance between Fuzzy Sets in Fuzzy Metric Spaces (2015)
McCulloch, J., Wagner, C., Aickelin, U.: Measuring the Directional Distance between Fuzzy Sets. In: arXiv.org (2013). https://arxiv.org/pdf/1308.5137.pdf
Osman, A.: Fuzzy metric spaces and fixed fuzzy set theorem. B Malays Math. Sci. So. **6**(1), 1–4 (1983)
Qualitative analysis. Verticals and environments. In: Identification and Quantification of Key Socioeconomic Data to Support Strategic Planning for the Introduction of 5G in Europe, Publications Office of the EU, Brussels (2017)
Roubens, M., Vincke, P.: Fuzzy possibility graphs and their application to ranking fuzzy numbers. Lect. Notes Econ. Math. Syst. **301**, 119–128 (1988)
Ryjov, A.: Fuzzy linguistic scales: definition, properties and applications. Stud. Fuzziness Soft Comput. **127**, 23–39 (2003)
Schnorr-Bäcker, S.: The possibilities and limitations of measuring prosperity and wellbeing in official statistics. In: Essays by the Members of the Scientific Advisory Board Government Strategy on Wellbeing in Germany. German Government (2018). https://www.gut-leben-in-deutschland.de
Szmidt, E., Kacprzyk, J.: Distances between Intuitionistic fuzzy sets. Fuzzy Set Syst. **114**, 505–518 (2000)
Szmidt, E., Kacprzyk, J.: A New Concept of a Similarity Measure for Intuitionistic Fuzzy Sets and its Use in Group Decision Making (2005). https://doi.org/10.1007/11526018_27
Szmidt, E., Kacprzyk, J.: A note on the Hausdorff distance between Atanassov's Intuitionistic fuzzy sets. Notes Intuit. Fuzzy Sets **15**(1), 1–12 (2009)
Walesiak, M., Dziechciarz, J., Bak, A.: An application of conjoint analysis for preference measurement. Argumenta Oeconomica **1/7**, 169–179 (1999)
Zamri, N., Abdullah, L.: A new positive and negative linguistic variable of interval triangular type-2 fuzzy sets for MCDM. Adv. Intell. Syst. Comput. **287**, 69–78 (2014)
Zhang, H., Zhang, W., Mei, C.: Entropy of interval-valued fuzzy sets based on distance and its relationship with similarity Measure. Knowl Based Syst. **22**, 449–454 (2009)
Zimmermann, H.: Fuzzy Sets. Theory and its applications, Kluwer, Dordrecht (2001)

Performance Measures in Discrete Supervised Classification

Ana Sousa Ferreira and Anabela Marques

Abstract The evaluation of results in Cluster Analysis frequently appears in the literature, and a variety of evaluation measures have been proposed. On the contrary, in supervised classification, particularly in the discrete case, the subject of results' evaluation is relatively scarce in this field of the literature. This is the motto underlying this study. The evaluation of the performance of any model of supervised classification is, generally, based on the number of cases correctly or incorrectly predicted by the model. However, these measures can lead to a misleading evaluation when the data is not balanced. More recently, other types of measures have been studied as association or agreement coefficients, the Huberty index, Mutual information, and even ROC curves. Exploratory studies were conducted in this study to understand the relationship between each measure and data characteristics, namely, sample size, balance, and separability of classes. To this end, simulated data and a Beta regression model in the performance of the models were used.

Keywords Balanced classes · Performance measures · Separability of classes · Supervised classification

1 Introduction

In Statistics, a supervised classification problem exists when the aim is to identify to which, of a set of classes defined a priori, a new observation belongs, on the basis of a training set of data containing subjects whose class membership is known. For example, consider a breast cancer dataset that contains nine variables describing 300 females who have suffered from breast cancer with or without its recurrence within 5

A. S. Ferreira (✉)
Faculdade de Psicologia, Universidade de Lisboa, Business Research Unit (BRU-IUL), Lisboa, Portugal
e-mail: asferreira@psicologia.ulisboa.pt

A. Marques
Escola Superior de Tecnologia do Barreiro, IPS, CIIAS, Barreiro, Portugal
e-mail: anabela.marques@estbarreiro.ips.pt

T. Chadjipadelis et al. (eds.), *Data Analysis and Rationality in a Complex World*, Studies in Classification, Data Analysis, and Knowledge Organization,
https://doi.org/10.1007/978-3-030-60104-1_6

years. So, we are facing a binary classification problem: of the 300 observed females, how many will or will not suffer a recurrence of breast cancer within 5 years? Performance evaluation makes it possible both to evaluate the quality of a new classification model and to choose the most appropriate technique to solve a specific supervised classification problem. In fact, performance evaluation is fundamental in supervised classification: "it is almost unthinkable to carry out any research work without an experimental section where the performance of the new proposed algorithm is tested and compared with other already proposed methods" (Santafe et al. 2015, p. 1).

In the breast cancer example, *False Negatives* (females wrongly diagnosed with no breast cancer recurrence) are likely to be worse than *False Positives* (females wrongly diagnosed with recurrence) in this problem? In fact, the more detailed screening will certainly clarify the *Positives*, but in the case of *False Negatives* females will be sent home and will probably miss out on follow-up evaluations.

The results of a supervised classification problem can be resumed in a contingency table referred to as the confusion matrix. In Table 1, the confusion matrix for the breast cancer data is presented.

In the field of Medicine, the 25, 75, 18, and 182 values are habitually referred to as *True Positives (TP)*, *False Negatives (FN)*, *False Positives (FP)*, and *True Negatives (TN)*, respectively. This terminology, which has been extended to many other fields of application, stems, for example, from the fact that a diagnostic exam indicates that a given female is undergoing a recurrence while in reality, the woman is not. Therefore, here we are dealing with a *False Positive* case. Some evaluation measures in classification are associated with this type of classification problem.

In supervised classification, global accuracy (or misclassification error) is widely used in classification problems since it is easy to compute and understand. Sometimes, accuracy is selected without considering in depth whether it is the most appropriate score to measure the quality of a classifier for the specific classification problem at hand. For Table 1, the global accuracy value is 0.69 (and the misclassification error 0.31).

In the discrete field, there is often a problem of dimensionality, due to the fact that the number of parameters to be estimated in each model is too large, samples are frequently small and owing to sparseness. So, most of the discrete models perform poorly, especially when classes are unbalanced and there is also a class separability problem. Thus, in discrete supervised classification, the evaluation of results gains even more relevance, when comparing with newly proposed models with other

Table 1 Breast cancer confusion matrix

		Predicted classes		
		Recurrence	No recurrence	
True Classes	Recurrence	25 (TP)	75 (FN)	100
	No recurrence	18 (FP)	182 (TN)	200
		43	257	300

important models of the supervised classification literature. The aim of this paper is to explore the evaluation of results in supervised classification, by comparing the correct classification rate with other types of measures (Ferreira and Cardoso 2013; Paik 1998).

2 Performance Measures

In the statistical literature, the most reported measure is accuracy which evaluates the overall efficiency of an algorithm. However, accuracy can be a misleading evaluation measure when data is not balanced (Santafe et al. 2015; Ferreira and Cardoso 2013; Ho and Basu 2002; Paik 1998). Thus, several measures have been defined in order to correctly evaluate the performance of each algorithm.

The *Accuracy* (*Acc*) rate is the most commonly used measure, and quantifies the overall efficiency of the model. In fact, *Accuracy* seeks to respond to the question: "Overall, how frequently does the classification model decide correctly?"

The *Correctly classified rate* of cases in class 1 is also referred to as *Sensitivity*, and measures efficiency in class 1. The *Correctly classified rate* of cases in class 2 is also referred to as *Specificity*, and measures efficiency in class 2. In Table 2, some of the evaluation measures based on the confusion matrix are presented.

Clearly, a good classification model should be capable of identifying both *True Positive* and *True Negative* cases. In precise terms, *Sensitivity* is the rate of *True Positive* cases, while *Specificity* is the rate of *True Negative* cases. Finally, *Precision*, also referred to as the positive predictive value, measures the precision of the model, providing the answer to another question: "Among the cases classified by the model as *Positive*, that is, belonging to Class 1, how many effectively are?" Thus, a high *Precision* value shows a model that it is a good predictor.

In general, the performance measures used do not provide a balance between the *False Positive* and *False Negative* cases. The combined performance measures, presented in Table 3, seek to obtain improved parity between them.

Balanced accuracy is the arithmetic mean between *Sensitivity* and *Specificity* and, when compared with *Overall accuracy*, will tend to be lower when the model is unable to classify both classes equally and correctly. The *Geometric mean* measures the balance between the classification in the two classes. A low *Geometric mean*

Table 2 Performance measures based on the confusion matrix

Measures	Definition
Correctly classified rate or *Accuracy (Acc)*	$\frac{TP+TN}{TP+TN+FP+FN}$
Accuracy of class 1 (Acc1) or *Sensitivity*	$\frac{TP}{TP+FN}$
Accuracy of class 2 (Acc2) or *Specificity*	$\frac{TN}{TN+FP}$
Precision (Pre)	$\frac{TP}{TP+FP}$

Table 3 Combined performance measures

Measures	Definition
Balanced accuracy (B_Acc)	$\frac{Sensitivity+Specificity}{2}$
Geometric mean (G_mean)	$\sqrt{Sensitivity \times Specificity}$
F measure (F)	$\frac{2 \times Sensitivity \times Precision}{Sensitivity+Precision}$

Table 4 Less traditional performance measures

Measures	Definition
Phi coefficient (ϕ)	$\frac{TP \times TN - FP \times FN}{\sqrt{(TP+FP)(FN+TN)(TP+FN)(FP+TN)}}$
Cohen's Kappa statistic (K)	$\frac{A_{cc}-P_{random}}{1-P_{random}}$, where
	$P_{random} = (\frac{TP+FN}{N} \times \frac{TP+FP}{N}) + (\frac{TN+FP}{N} \times \frac{TN+FN}{N})$
	and $N = TP + TN + FP + FN$
Huberty index (H)	$\frac{P_{cc}-P_m}{1-P_m}$, where
	P_{cc} – % correctly classified cases and
	P_m – % correctly classified cases in accordance with the majority rule

value indicates a weak performance in the class considered to be positive (usually deemed the class of most interest). Finally, the *F measure* combines the *Sensitivity* and *Precision* measures, even when the classes of data are really balanced. The aforementioned evaluation measures, which are generally simple or combined rates, naturally vary in the [0, 1] interval.

Evaluation measures of a different type (Santafe et al. 2015; Ferreira and Cardoso 2013; Ho and Basu 2002; Paik 1998), which indicate association or agreement between real and predicted classes (Cohen 1960), have been referred to by several authors. On the other hand, an evaluation of the effective improvement the model brings to the majority rule appears to be of relevance (Huberty 2006). These less traditional measures in supervised classification are presented in Table 4.

The *Phi coefficient* (ϕ) is a known measure of association between two binary variables, and take values in the interval [−1, 1]. The positive sign of this coefficient indicates a higher number of cases where the classification model has decided correctly. The negative sign, on the contrary, points to the existence of more incorrectly decided cases. *Cohen's Kappa statistic* (Cohen 1960) may be defined as the proportion of agreement between two classifications after removal of the agreement proportion owing to the random, and may also take values in the interval [−1, 1]. Finally, the *Huberty index* (Huberty 2006) evaluates the performance of a model as the degree of classification correction achieved, in comparison to a percentage of correctly classified cases by the majority rule, defined as the ratio between effective improvement and possible improvement in the classification. This index is the only evaluation measure presented that take values outside the interval [−1, 1].

Table 5 Parameters of the Multinomial distribution used in data simulation

Separability	C_1	C_2
Low	(0,5;0,5;0,5;0,5; 0,5;0,5;0,5;0,5)	(0,5;0,5;0,5;0,5; 0,5;0,5;0,5;0,5)
High	(0,1;0,9;0,7;0,3; 0,2;0,8;0,6;0,4)	(0,9;0,1;0,3;0,7; 0,8;0,2;0,1;0,9)

3 Experimental Results

In order to understand the relationship between each performance measure and data characteristics, Beta regression (Cribari-Neto and Zeileis 2010) or Multiple linear regression models were used, according to the variation in intervals of the performance measures. For this purpose, we resorted to simulated data to predict performance measures based on data characteristics so as to understand the relative impact of each experimental complexity factor on performance. For the sake of simplicity, this study focuses on a problem of two classes and four binary predictors.

The data were simulated considering two levels of separability of the classes (Low and High) and according to the Multinomial distribution, with the occurrence probabilities of the four predicting binary variables, presented in Table 5.

Two other characteristics were considered for simulated data: Sample size (small ($n = 60$), moderate ($n = 120$), and large ($n = 400$); Balance (classes with equal size), classes with moderate unbalanced and with severe unbalanced) and 30 classification runs in each scenario were considered. To implement the regression models, the three complexity factors were used in order to study the impact of each one in the performance measure:

- Separability of classes: The Affinity coefficient (Bacelar-Nicolau 1985), defined in the interval [0, 1], is used to measure the separability of classes;
- Balance: The ratio between the minority and the majority class sizes is used to measure balance;
- Sample size: The ratio between the "number of degrees of freedom" and sample size is used to measure sample size importance.

Based on the 270 sets of generated data, the aforementioned performance measures referred previously were obtained by means of a reference model in discrete supervised classification, namely the First-Order Independence Model (FOIM) (Goldstein and Dillon 1978), and were estimated by twofold cross-validation. Forthe

Table 6 Estimated coefficients for performance measures based on the confusion matrix

	Accuracy – Pseudo $R^2 = 0.80$				*Sensitivity* – Pseudo $R^2 = 0.35$			
	Estimate	St. Error	z	Sig.	Estimate	St. Error	z	Sig.
Intercept	1.55	0.23	6.77	∗∗∗	2.90	0.53	5.49	∗∗∗
Separability	**–2.99**	0.37	**–8.04**	∗∗∗	**–4.86**	0.83	**–5.82**	∗∗∗
Balance	1.14	0.23	4.87	∗∗∗	0.19	0.54	0.36	0.72
Sample size	**1.74**	0.33	**5.31**	∗∗∗	0.53	0.73	0.73	0.47
Sep × Bal	**–0.49**	0.22	**–2.21**	∗	**0.97**	0.47	**2.08**	∗
Sep × S. size	−0.13	0.46	−0.28	0.78	1.47	1.01	1.46	0.15
Bal × S. size	−0.61	0.32	−1.88	0.06	−0.85	0.75	−1.13	0.26
	Specificity– Pseudo $R^2 = 0.60$				*Precision* – Pseudo $R^2 = 0.69$			
	Estimate	St. Error	z	Sig.	Estimate	St. Error	z	Sig.
Intercept	1.93	0.32	6.13	∗∗∗	−0.51	0.34	−1.51	0.13
Separability	**–3.45**	0.51	**–6.75**	∗∗∗	**–3.52**	0.63	**–5.57**	∗∗∗
Balance	**1.40**	0.32	**4.35**	∗∗∗	**4.66**	0.44	**10.65**	∗∗∗
Sample size	1.22	0.44	2.75	∗∗	1.45	0.47	3.09	∗∗
Sep × Bal	**–0.94**	0.30	**–3.11**	∗∗	**–1.53**	0.37	**–4.14**	∗∗∗
Sep × S. size	0.49	0.63	0.78	0.43	0.87	0.77	1.12	0.26
Bal × S. size	−0.49	0.44	−1.12	0.26	−1.26	0.60	−2.10	∗

$*p < 0.05$; $**p < 0.01$; $***p < 0.001$

performance measures that assume values in the standard unit interval (0,1) Beta regression models were used and the estimated coefficients were obtained using the Betareg R package (Cribari-Neto and Zeileis 2010); For the performance measures that assume values elsewhere, Linear regression models were used. In both cases, factors interaction was considered: Separability of classes × Balance, Separability of classes × Sample Size, and Balance × Sample Size. The estimated regression models are presented in Tables 6, 7 and 8.

The estimated regression models exhibited an adequate to good fit to the data and the three complexity measures impacted significantly on almost all evaluation measures.

Table 7 Estimated regression coefficients for combined performance measures

	Balanced accuracy – Pseudo $R^2 = 0.76$				*Geometric mean* – Pseudo $R^2 = 0.73$			
	Estimate	St. Error	z	Sig.	Estimate	St. Error	z	Sig.
Intercept	1.15	0.26	4.37	* * *	0.75	0.31	2.44	*
Separability	**–2.79**	0.43	**–6.51**	* * *	**–2.87**	0.50	**–5.69**	* * *
Balance	1.48	0.27	5.39	* * *	**1.91**	0.33	**5.85**	* * *
Sample size	**2.12**	0.38	**5.61**	* * *	2.50	0.44	5.63	* * *
Sep × Bal	**–0.52**	0.26	**–2.01**	*	–0.44	0.30	–1.45	0.15
Sep × S. size	–0.30	0.53	–0.55	0.58	–0.19	0.63	–0.30	0.77
Bal × S. size	–0.94	0.38	–2.48	*	**–1.46**	0.45	**–3.22**	**
	F measure – Pseudo $R^2 = 0.80$							
	Estimate	St. Error	z	Sig.				
Intercept	–0.25	0.29	–0.85	0.39				
Separability	**–2.35**	0.52	**–4.51**	* * *				
Balance	**3.11**	0.36	**8.65**	* * *				
Sample size	1.74	0.40	4.33	* * *				
Sep × Bal	**–1.15**	0.31	**–3.73**	* * *				
Sep × S. size	–0.39	0.64	–0.61	0.54				
Bal × S. size	–0.50	0.50	–1.00	0.31				

$*p < 0.05$; $**\ p < 0.01$; $***p < 0.001$

Separability, measured by the Affinity coefficient, emerged as one of the most important experimental factors with a negative impact on performance. Balance was the most important factor for *Precision, Geometric mean, F measure* and the *Phi coefficient*, with a positive impact on performance. Balance was also the second most important factor for *Specificity* and the *Cohen's Kappa statistic* with a positive impact on performance; Sample size was also the second most important factor for *Accuracy, Balanced Accuracy*, and the *Huberty index*, with a positive impact on performance. The Separability of classes × Balance interaction seems to be, also, one important factor of complexity.

Table 8 Estimated regression coefficients for less traditional performance measures

	Phi[1] – Pseudo $R^2 = 0.55$				*Kappa*[1] – Pseudo $R^2 = 0.84$			
	Estimate	St. Error	t	Sig.	Estimate	St. Error	t	Sig.
Intercept	0.35	0.13	2.69	**	0.31	0.08	3.92	* * *
Separability	−0.05	0.22	−0.23	0.82	**–0.76**	0.13	**–5.84**	* * *
Balance	**0.61**	0.15	**4.18**	* * *	**0.53**	0.09	**6.05**	* * *
Sample size	−0.02	0.19	−0.11	0.92	**0.67**	0.11	**6.10**	* * *
Sep × Bal	**–1.14**	0.12	**–9.48**	* * *	−0.12	0.07	−1.59	0.11
Sep × S. size	0.09	0.26	0.34	0.74	−0.20	0.16	−1.28	0.20
Bal × S. size	0.32	0.21	1.56	0.12	**–0.27**	0.12	**–2.20**	*
	Huberty index[1] Pseudo $R^2 = 0.71$							
	Estimate	St. Error	t	Sig.				
Intercept	−0.79	0.43	−1.83	0.07				
Separability	**–3.47**	0.72	**–4.85**	* * *				
Balance	1.65	0.48	3.43	**				
Sample size	**2.32**	0.61	**3.82**	* * *				
Sep × Bal	**3.30**	0.40	**8.30**	* * *				
Sep × S. size	−0.91	0.87	−1.06	0.29				
Bal × S. size	−1.83	0.68	−2.70	**				

[1]– Multiple linear regression

$*p < 0.05$; $**\,p < 0.01$; $***p < 0.001$

4 Conclusions

This study has proven to be an interesting contribution to the goal of understanding how to choose an evaluation measure that really takes into account the classification problem at hand. The separability of classes emerges as the factor with the most influence on classifier performance: the more weakly separated the classes are, the higher the affinity coefficient and the weaker the classification performance is. The size of the samples and the balance between them also has an important impact on the quality of the classifier performance. Sample size is the second most important factor for (*Accuracy*, *Balanced Accuracy*, and the *Huberty index*): the larger the sample size the better the classification performance is. Balance is the most important factor for *Precision*, *Geometric mean*, *F measure*, and the *Phi coefficient*: the more balanced the classes are the stronger the classification performance is. Naturally, classification results improve as the classification problem becomes easier (better separability, larger samples, and more balanced classes).

Although not presented here due to space constraints, exploratory analysis with real data showed that with balanced classes, all the performance measures yielded similar results; conversely, with low separability, considerable differences between the results of association or agreement measures and all the others were observed. In the unbalanced case with a high separability of classes, the measures tended to be similar, however, with low separability in the measured values were discrepant. Interaction terms seem to have an important role in performance measures and must continue to be further explored. Finally, it should be noted that the *Huberty index* is a highly demanding but interesting measure, barely reaching high values in real-life problems.

The evaluation of results in Discrete Supervised Classification will continue to be further explored, using both simulated and real data, particularly in the case of unbalanced classes, with a view to better understanding the interest of other performance measures almost always absent in the literature of the area.

References

Bacelar-Nicolau, H.: The affinity coefficient in cluster analysis. Methods Oper. Res. **53**, 507–512 (1985)

Cribari-Neto, F., Zeileis, A.: Beta regression in R. J. Stat. Softw. **34**, 1–24 (2010)

Cohen, J.: A coefficient of agreement for nominal scales. Educ. Psychol. Measur. **20**, 37–46 (1960)

Ferreira, A.S., Cardoso, M.G.: Evaluating discriminant analysis results. In: Lita da Silva, J., Caeiro, F., Natário, I., Braumann C. (eds.) Advances in Regression, Survival Analysis, Extreme Values, Markov Processes and and Other Statistical Applications. Studies in Theoretical and Applied Statistics, pp. 155–162. Springer, Heidelberg (2013)

Goldstein, M., Dillon, W.R.: Discrete Discriminant Analysis. Wiley, New York (1978)

Ho, T.K., Basu, M.: Complexity measures of supervised classification problems. IEEE Trans. Pattern Anal. Mach. Intell. **24**, 289–300 (2002)

Huberty, C.J., Olejnik, S.: Applied MANOVA and Discriminant Analysis. Wiley-Interscience. Wiley, New Jersey (2006)

Paik, H.: The effect of prior probability on skill in two-group discriminant analysis. Qual Quant. **32**, 201–211 (1998)

Santafe, G., Inza, I., Lozano, J.A.: Dealing with the evaluation of supervised classification algorithms. Artif. Intell. Rev. **44**, 467–508 (2015)

Using EVT to Assess Risk on Energy Market

Alicja Ganczarek-Gamrot, Dominik Krężołek, and Grażyna Trzpiot

Abstract The aim of this paper is to describe and measure the risk of price changes in the energy market. The risk is estimated with Conditional Value-at-Risk (CVaR) and Median Shortfall (MS) based on some types of Value-at-Risk measures: VaR, stress VaR, Incremental Risk Charge (IRC) estimated using Extreme Value Theory (EVT). These measures are calculated for time series of daily and hourly rates of return of electric energy prices from the European Energy Exchange (EEX) spot market. Based on time series from 1st January 2002 to 31st December 2016, we attempt to answer the question: which measure is the most appropriate for risk estimation on the energy market.

Keywords Risk · Rates of return · VaR · Electric energy market

1 Introduction

The electric energy prices depend on demand and actual free volume of power, so during night hours and days off electric energy prices are low and even negative on some markets. On the other hand in peak hours of a day, they can be very high. Both phenomena are noted regularly. These changes increase the volatility of prices. High volatility of prices poses difficulties in exact prediction of price values and increases prediction errors.

A. Ganczarek-Gamrot (✉) · D. Krężołek · G. Trzpiot
University of Economics in Katowice, 1 Maja 50, Katowice, Poland
e-mail: alicja.ganczarek-gamrot@ue.katowice.pl

D. Krężołek
e-mail: dominik.krezolek@ue.katowice.pl

G. Trzpiot
e-mail: grazyna.trzpiot@ue.katowice.pl

T. Chadjipadelis et al. (eds.), *Data Analysis and Rationality in a Complex World*, Studies in Classification, Data Analysis, and Knowledge Organization,
https://doi.org/10.1007/978-3-030-60104-1_7

The aim of this paper is to verify whether extreme quantile measures which characterize the risk of improbable but catastrophic price changes Jajuga (1999) are suitable for the electric energy market.

This research is based on quotes from Day Ahead Market (DAM) on European Electric Energy Exchange (EEX) in the period from 1st January 2002 to 31st December 2016. Daily and hourly linear rates of return were considered in the analysis. DAM consists of 24 contracts on every hour in a day. Based on Principal Component Analysis (PCA), we chose four contracts to analyse daily risk of change prices. Hourly rates of return were analysed in three 5 years windows: (I) from 2008 to 2012; (II) from 2010 to 2014 and (III) from 2012 to 2016. Daily and hourly quotes of electric energy prices were smoothed by Seasonal and Trend decomposition using Loess STL (Cleveland et al. 1990).

2 Literature Review

Extreme Value Theory (EVT) is a branch of statistics, which studies extreme deviations from location parameters for given probability distributions. It is useful for analysing extreme observations of variables which appear with low probability. It may be applied for data with asymmetric distributions and fat tails and finds various uses in numerous disciplines such as finance, hydrology and meteorology.

From the pioneering publication (Gumbel 1958) where mathematical foundations of EVT were first described, the field has undergone a dynamic development. Our work was inspired by the work (Gilli and Këllezi 2006), whose authors use EVI to compute tail risk by quantile measures for major stock market indices. The authors of Embrechts et al. (1999), also estimated risk in the right tail for industrial fire insurance claims distribution and for simulated realizations of an idealized ARCH process. The EVT was also applied in the papers (McNeil and Frey 2000; Manel et al. 2015 to estimate time-dependent quantiles in the GARCH model fit to the financial market, and electric energy market data. In our approach to VaR estimation through EVT theory, we used Wakeby distribution considered earlier by Houghton (1978) for modelling water flows. In the quantile estimation we employed the R *extremeStat* package used earlier in Boessenkool et al. (2017) (see also Penalva et al. 2013).

3 Methodology

In this paper, we used Value-at-Risk (VaR) to estimate risk. This measure is often chosen in EVT to risk estimation (Manel et al. 2015). VaR is defined as such loss of value,which is not exceeded with the given probability α at the given time period h, and it is expressed by the formula (Jajuga 1999):

$$P(Y_{t+h} \leq Y_t - VaR_\alpha(Y)) = \alpha \tag{1}$$

where
Y_t is a present value,
Y_{t+h} is a random variable.

The filtering procedure: STL (Seasonal-Trend Decomposition based on Loess) Cleveland et al. (1990) is used to clear prices Y_t from outliers and impute missing data:

$$Y_t = \tau_t + S_t + u_t \tag{2}$$

where
Y_t—time series of prices,
τ_t—trend,
S_t—seasonality, u_t - residuals.

Due to negative values of electric energy prices, VaR was estimated using linear rates of return:

$$R_t = \frac{Y_t - Y_{t-1}}{Y_{t-1}} \tag{3}$$

where
R_t is a linear rate of return (daily or hourly),
Y_t is an electric energy price (EUR/MWh) in the period t (in day t or hour t),
Y_{t-1} is an electric energy price (EUR/MWh) in the period $t-1$ (in day $t-1$ or hour $t-1$).

We can express VaR_α at the given time period $h = 1$ according to Kou and Peng (2014) in the following form:

$$VaR_\alpha(R) = F_R^{-1}(\alpha) \tag{4}$$

where $F_R^{-1}(\alpha)$- is the α - quantile of rates of return.

To estimate VaR_α, we used Extreme Value Theory (EVT) in the right distribution tail (Gilli and Këllezi 2006). We were looking for a distribution that best matched to the right tail of empirical distribution. To measure the goodness of fit, Root Mean Square Error (RMSE) Chai and Draxler (2014) was used:

$$RMSE = \sqrt{\frac{1}{n}\sum_{i=1}^{n}(R_t - R_t^*)^2} \tag{5}$$

where R_t^*- theoretical rate of return.

We considered Wakeby distribution which is defined by the quantile function (inverse CDF) Houghton (1978), Busababodhin et al. (2016):

$$R^*(F) = \xi + \frac{\theta}{\beta}(1 - (1 - F)^{\beta}) - \frac{\gamma}{\delta}(1 - (1 - F)^{-\delta}) \tag{6}$$

where ξ is location parameter, θ is scale parameter and $\beta,\ \gamma,\ \delta$ define shape of the quantile function .

The Proportion of Failures Test (POF) proposed by Kupiec (1995) was used to assess VaR_{α} :

$$LR_{POF} = -2ln\frac{(1-\alpha)^{n-k}\alpha^k}{(1-w)^{n-k}w^k} \tag{7}$$

where
α—quantile order,
k—number of quantile excesses,
n—the number of observations,
w—percentage of quantile excesses in the whole data set (k/n).

When the probability of exceeding VaR_{α} equals α the statistic LR_{POF} follows a χ_1^2 distribution.

Afterwards, two coherent risk measures known as expected shortfall (ES) and median shortfall (MS) were calculated for the right tail of the distribution (Artzner et al. 1999) in order to measure expected losses for a situation when VaR is exceeded. They, respectively, take the following form:

$$ES_{\alpha} = E(R|R > VaR_{\alpha}) \tag{8}$$

where ES_{α} is the Expected Shortfall (ES) in the right tail of distribution describing average losses higher than VaR and

$$MS_{\alpha} = Median(R|R > VaR_{\alpha}) \tag{9}$$

where MS_{α} is the Median Shortfall (MS) in the right tail of distribution describing the median of losses higher than VaR.

4 Empirical Analysis

Daily and hourly quotes of electric energy prices were smoothed by the STL. Table 1 presents estimated parameters of chosen distributions of daily time series energy prices in hour 1, 7, 10 (AM) and 22 (10 PM) before (H1t, H7t, H10t, H22t) and after (H1tc, H7tc, H10tc, H22tc) smoothing and for hourly time series energy prices (DIt, DIIt, DIIIt) before and (DItc, DIItc, DIIItc) after smoothing by STL filter in each time window (I–III).

The results of smoothing by STL affected mainly the tails of the distribution, especially the left tail and consequently influenced skewness and kurtosis. The influence

Table 1 Parameters of price distributions

Price	Min	1stQu	Median	Mean	3rdQu	Max	Sd	Skewness	Kurtosis
H1t	−149.90	22.52	29.32	30.32	37.37	76.02	12.66	−0.6408	13.2043
H1tc	−13.49	22.66	29.41	30.46	37.38	70.26	11.79	0.4430	0.3737
H7t	−199.99	22.43	32.06	32.13	42.53	104.93	17.20	−1.3975	18.6147
H7tc	−12.31	22.94	32.24	32.66	42.64	95.04	15.39	0.2081	0.2305
H10t	−36.10	33.09	43.98	47.86	58.48	499.68	24.75	3.2400	32.3370
H10tc	6.56	33.13	43.99	47.68	58.44	200.78	22.36	1.6296	5.5154
H22t	−3.35	30.39	38.36	40.10	48.06	118.93	14.23	0.7882	1.2982
H22tc	8.51	30.43	38.36	40.14	48.05	110.02	14.09	0.7978	1.1574
DIt	−500.02	36.63	46.66	48.57	58.55	494.26	21.80	0.2505	24.2689
DItc	−0.01	36.76	46.72	48.85	58.56	175.61	19.75	1.0551	2.81289
DIIt	−221.99	31.63	41.54	41.75	52.15	210.00	16.47	3.2400	14.7981
DIItc	−7.71	31.64	41.54	41.86	52.15	111.01	15.26	0.0533	0.1259
DIIIt	−221.99	25.94	33.36	34.75	43.06	210.00	15.60	0.7882	18.0590
DIIItc	−21.30	25.94	33.36	34.85	43.06	110.02	14.36	0.3979	1.1574

Own calculations

in central measures like mean and standard deviation is marginal. After applying the STL algorithm the goodness of fit for the Wakeby distribution has improved.

For 24 rates of return of daily contracts (contracts for each of 24 h during a day), the Principal Component Analysis (PCA) was carried out. Based on the first two principal components we divide all variables into two groups:

- Rates of return weakly correlated with the first two principal components, which were rates in night hour from 24 to 7 and 14 ($RH1tc - RH7tc$, $RH14tc$)
- Rates of return strongly correlated with the first two principal components, which were rates in hour from 8 to 23 without hour 14 ($RH8tc - RH13tc$,$RH15tc - Rh23tc$)

Four variables were chosen based on the results of PCA to represent various distributions of return rates:
$RH1tc$—slightly negatively correlated with first principal component and positively correlated with second principal component
$RH7tc$— slightly negatively correlated with first and second principal component
$RH10tc$—strongly positively correlated with first principal component and moderation positively correlated with second principal component
$RH22tc$—strongly positively correlated with first principal component and negatively correlated with second principal component.

In Table 2, parameters of linear rates of return for daily and hourly distributions were presented (for STL—corrected prices). Distributions of daily rates of return in off-peak hours RH1tc and RH7tc are leptokurtic, very highly volatile and extremely asymmetric (RH1tc—right asymmetry, RH7t—left-hand asymmetry). Distributions of daily rates of return on 10 and 22 h (RH10tc and RH22tc) are characterized by high

Table 2 Parameters of rates of return distributions

Rates	n	Min	1stQu	Median	Mean	3rdQu	Max	Sd	Skewness	Kurtosis
RH1tc	5478	−477.25	−0.1147	0.0020	0.4400	0.1257	1001.0	23.96	23.1083	959.79
RH7tc	5478	−2795.0	−0.2330	−0.0285	−0.7224	0.1708	1576.7	69.39	−22.6401	957.57
RH10tc	5478	−0.8677	−0.2062	−0.0276	47.68	0.0993	0.1790	0.56	2.7728	12.79
RH22tc	5478	−0.8677	−0.1040	−0.0015	40.14	0.0240	0.1113	0.24	0.8507	9.55
RDItc	43847	−879.69	−0.0730	−0.0103	0.0207	0.0636	702.67	5.43	−47.2307	22080.5
RDIItc	43823	−555.00	−0.0684	−0.0088	0.0135	52.15	0.0596	538.75	−2.9601	5846.46
RDIIItc	43847	−1070.0	−0.0737	−0.0086	−0.0297	0.0652	765.50	12.59	−30.0566	3237.03

Own calculations

Table 3 Results of Wakeby distribution estimation

Rates	ξ	θ	β	γ	δ	*RMSE*
RH1tc	−55.17	7046.53	127.89	0.0563	0.9401	0.0967
RH7tc	−1391.96	830722.50	596.90	0.1787	0.9041	0.0751
RH10tc	−0.74	6.8178	15.1034	0.3091	0.2508	0.0243
RH22tc	−0.44	2.7422	8.3458	0.1422	0.1487	0.0075
RDItc	−1.33	34.12	26.96	0.0575	0.5670	0.0439
RDIItc	−5.83	348.84	60.30	0.0405	0.7511	0.0720
RDIIItc	−40.78	6264.81	153.79	0.0353	0.8770	0.0967

Own calculations

leptokurtosis, volatility and positive skew. However, these effects are weaker than in off-peak distributions. Distributions of hourly rates of return independent of the research period (RDItc, RDIItc, RDIIItc) are characterized by extreme leptokurtosis, very high volatility and extreme negative skew.

Median negative value (except RH1tc) suggests that in the observed period prices were falling.

In terms of RMSE, the Wakeby distribution fits the empirical data more reliably than other well-known distributions including exponential, gamma, generalized extreme value, generalized logistic, generalized Pareto, generalized normal, Gumbel (extreme value type I), kappa, lognormal, normal, Pearson type III, and Weibull. In Table 3, Wakeby distribution parameters and RMSE values were presented.

In Table 4, results of VaR ($VaR_{0.95}$), stress VaR ($sVaR(VaR_{0.99})$) and Incremental Risk Charge ($IRC(VaR_{0.999})$) were presented and calculated by formula (4) based on Wakeby distributions. Next to VaR empirical quantiles Q_α, results of LR_{POF} and additionally conditional *VaR* measures *ES* and *MS* were presented. For daily rates of return (daily rates for chosen contracts in hour 1, 7, 10 and 22), the percentage of *VaR* excesses usually does not differ significantly from the expected proportion of excesses. Except for VaR, SVaR for the contract in the night hour 1 (where calculated measures are overestimated), IRC in the morning hour 7 and VaR in hour 10 (where measures were underestimated). Analysing hourly data we can say that for a higher frequency of observations *VaR* were estimated incorrectly (independent in three periods of time). All incorrect VaR measures are overestimated. Only *IRC*

Table 4 Results of risk estimation

Rates	α	Q_α	VaR_α	k	w	LR_{pof}	p	ES	MS
RH1tc	0.95	0.5927	0.8667	159	0.0290	59.37	<0.001	28.05	1.42
	0.99	1.7588	4.4042	12	0.0022	49.45	<0.001	352.55	297.58
	0.999	363.21	39.403	10	0.0018	2.997	0.083	421.15	367.08
RH7tc	0.95	2.61	2.51	283	0.0517	0.3150	0.5746	29.689	4.8542
	0.99	15.59	12.25	61	0.0111	0.6881	0.4068	120.37	30.09
	0.999	308.33	101.44	13	0.0024	7.4358	0.0063	454.49	275.71
RH10tc	0.95	1.2413	1.0950	338	0.0617	14.744	<0.001	1.7685	1.5234
	0.99	2.2882	2.3945	45	0.0082	0.0082	0.1706	3.2037	3.0200
	0.999	4.0434	5.4525	2	0.0004	2.9278	0.0871	5.9475	5.9475
RH22tc	0.95	0.4241	0.4289	269	0.0491	0.0928	0.7606	0.7033	0.618
	0.99	0.8558	0.8326	58	0.0106	0.1876	0.6649	1.1555	1.0307
	0.999	1.7232	1.6071	6	0.0011	0.0483	0.8261	1.9428	1.8603
RDItc	0.95	0.3221	0.3860	1558	0.0355	213.98	<0.001	1.3995	0.5418
	0.99	0.7140	1.2121	139	0.0032	281.63	<0.001	9.7432	1.7546
	0.999	2.2509	4.9250	16	0.0004	23.45	<0.001	69.786	14.833
RDIItc	0.95	0.3076	0.4028	1329	0.0303	413.03	<0.001	2.9404	0.5906
	0.99	0.7394	1.6044	133	0.0030	295.43	<0.001	23.546	3.1092
	0.999	7.0388	9.5480	36	0.0008	1.4893	0.2223	78.13	52.963
RDIIItc	0.95	0.3208	0.4629	1054	0.0240	763.67	<0.001	8.7368	0.6911
	0.99	0.7717	2.1890	150	0.0034	257.06	<0.001	56.733	20.139
	0.999	58.73	17.1	81	0.0018	25.15	<0.001	100.55	62

Own calculations

from the second period was estimated properly. In Table 4, we have included Conditional *VaR*. Both conditional measures *ES* and *MS* show the average loss when *VaR* is exceeded. For extreme asymmetry distributions like *RH1tc*, *RH7tc*, *RDItc*, *RDIItc*, *RDIIItc* we can see clearly the difference between *ES* and *MS* even after cleaning the data from outliers.

5 Conclusions

To sum up the results of *VaR* estimation for various rates of return of electric energy prices in different periods of time, we can say that Wakeby distribution applied to the electricity returns exhibits the best fit in terms of RMSE criterion for both daily and hourly data. For daily frequency data, *VaR* was properly estimated, while for higher frequency data—hourly rates of return—almost all values of *VaR* were overestimated, despite the STL smoothing. For high-frequency data, this approach is not robust enough. The results for daily contracts can be used to estimate risk on Day Ahead Marked, where decisions are taken for one-day investment horizon independently for every hour. In the future, we plan to adapt EVT with quantile dynamic estimation based on GARCH models for higher frequency data and extremely skewed distributions following (McNeil and Frey 2000; Manel et al. 2015).

References

Artzner, P., Delbaen, F., Eber, J.M., Heath, D.: Coherent measures of risk. Math. Financ. **9**, 203–228 (1999)

Bardou, O., Frikha, N., Pagès, G.: Computation of VaR and CVaR using stochastic approximations and unconstrained importance sampling. Monte Carlo Methods Appl. **15**(3), 173–210 (2009)

Boessenkool, B., Brüger, G., Heistermann, M.: Effects of sample size on estimation of rainfall extremes at high temperatures. Nat. Hazard Earth Sys. **17**, 1623–1629 (2017)

Busababodhin, P., Seo, Y.A., Park, J.S., Kumphon, B.: LH-moment estimation of Wakeby distribution with hydrological applications. Stoch. Env. Res. Risk A. **30**, 1757–1767 (2016)

Chai, T., Draxler, R.R.: Root mean square error (RMSE) or mean absolute error (MAE)? - Arguments against avoiding RMSE in the literature. Geosci. Model Dev. **7**, 1247–1250 (2014)

Cleveland, R.B., Cleveland, W.S., McRae, J.E., Terpenning, I.J.: STL: a seasonal-trend decomposition procedure based on loess. J. Off. Stat. **6**, 3–73 (1990)

Embrechts, P., Resnick, S.I., Samorodnitsky, G.: Extreme value theory as a risk management tool. N. Am. Actuar. J. **3**, 30–41 (1999)

Gilli, M., Këllezi, E.: An application of extreme value theory for measuring financial risk. Comput. Econ. **27**, 207–228 (2006)

Gumbel, E.J.: Statistics of Extremes. Columbia University, New York (1958)

Houghton, J.C.: Birth of a parent: the Wakeby distribution for modeling flood flows. Water Resour. Res. **14**, 1105–1110 (1978)

Jajuga, K.: Zarzadzanie ryzykiem. PWN, Warszawa (2008)

Kou, S., Peng, X.: Expected shortfall or median shortfall. J. Fin. Eng. **1**, 1–6 (2014)

Kupiec, P.: Techniques for verifying the accuracy of risk management model. J. Deriv. **2**, 173–184 (1995)

Manel, Y., Lotif, B., Khaled, M.: Value-at-risk estimation of energy commodities: a long-memory GARCH-EVT approach. Energy Econ. **51**, 99–110 (2015)

McNeil, A, J., Frey, R.: Estimation of tail-related risk measures for heteroscedastic financial time series: an extreme value approach. J. Empir. Financ. **7**, 271–300 (2000)

Penalva, H., Neves, M., Nunes, S.: Topics in data analysis using R in extreme value theory. Metodolozski zvezki **10**, 17–29 (2013)

Measuring and Testing Mutual Dependence for Functional Data

Tomasz Górecki, Mirosław Krzyśko, and Waldemar Wołyński

Abstract In this paper, measures of mutual independence of many-vector random processes were defined. Based on these measures, permutation tests of mutual independence of these random processes were also given. The properties of the described methods were presented using simulation studies for univariate and multivariate processes.

Keywords Functional data · Mutual correlation · Measures of multiple independence

1 Introduction

Many processes currently used in different fields of science and research lead to random observations that can be analyzed as curves. We can also find a large amount of data for which it would be more appropriate to use some interpolation techniques and consider them as functional data. This approach turns out to be essential when data have been observed at different time intervals.

Earlier, Górecki et al. (2017, 2020) showed how to use commonly known measures of correlation for two sets of variables: ρV coefficient (Escoufier 1973), distance

T. Górecki (✉)
Faculty of Mathematics and Computer Science, Adam Mickiewicz University, Uniwersytetu Poznańskiego 4, Poznań 61-614, Poland
e-mail: tomasz.gorecki@amu.edu.pl

M. Krzyśko
Interfaculty Institute of Mathematics and Statistics, Calisia University, Nowy Świat 4, 62-800 Kalisz, Poland
e-mail: mkrzysko@amu.edu.pl

W. Wołyński
Faculty of Mathematics and Computer Science, Adam Mickiewicz University, Uniwersytetu Poznańskiego 4, Poznań 61-614, Poland
e-mail: wolynski@amu.edu.pl

T. Chadjipadelis et al. (eds.), *Data Analysis and Rationality in a Complex World*, Studies in Classification, Data Analysis, and Knowledge Organization,
https://doi.org/10.1007/978-3-030-60104-1_8

correlation coefficient (dCorr) (Székely et al. 2007), and HSIC coefficient (Gretton et al. 2005) for multivariate functional data.

In this paper, using ρV and dCorr coefficients, we define measures of mutual independence of vector random processes whose realizations are multidimensional functional data. Based on these measures, permutation tests of mutual independence of vector random processes $\boldsymbol{X}_1, \ldots, \boldsymbol{X}_K$, $K \geq 2$, $\boldsymbol{X}_i \in L_2^{p_i}(I)$, where $L_2(I)$ is a Hilbert space of square-integrable functions on the interval I, $i = 1, \ldots, K$ are also considered.

The rest of this paper is organized as follows. We first review the concept of transformation of discrete data to multivariate functional data (Sect. 2). Section 3 contains the functional version of the ρV and dCorr coefficients. Section 4 is devoted to measures of mutual independence of vector random processes and permutation tests of mutual independence associated with these measures. Section 5 contains the results of our simulation experiments.

2 Functional Data

Let us assume that $\boldsymbol{X} = (X_1, X_2, ..., X_p)^\top \in L_2^p(I)$ is p-dimensional random process, where $L_2(I)$ is the Hilbert space of square-integrable functions on the interval I. Moreover, assume that the kth component of the vector $\boldsymbol{X}$ can be represented by a finite number of orthonormal basis functions $\{\varphi_b\}$ of space $L_2(I)$:

$$X_k(t) = \sum_{b=0}^{B_k} \alpha_{kb}\varphi_b(t), \ t \in I, \ k = 1, \ldots, p.$$

Let $\boldsymbol{\alpha} = (\alpha_{10}, \ldots, \alpha_{1B_1}, \ldots, \alpha_{p0}, \ldots, \alpha_{pB_p})^\top$ and

$$\boldsymbol{\Phi}(t) = \begin{bmatrix} \boldsymbol{\varphi}_1^\top(t) & \boldsymbol{0} & \ldots & \boldsymbol{0} \\ \boldsymbol{0} & \boldsymbol{\varphi}_2^\top(t) & \ldots & \boldsymbol{0} \\ \ldots & \ldots & \ldots & \ldots \\ \boldsymbol{0} & \boldsymbol{0} & \ldots & \boldsymbol{\varphi}_p^\top(t) \end{bmatrix}, \tag{1}$$

where $\boldsymbol{\varphi}_k(t) = (\varphi_0(t), ..., \varphi_{B_k}(t))^\top$, $k = 1, ..., p$.

Using the above matrix notation, process $\boldsymbol{X}$ can be represented as

$$\boldsymbol{X}(t) = \boldsymbol{\Phi}(t)\boldsymbol{\alpha}.$$

This means that the realizations of a process $\boldsymbol{X}$ are in finite-dimensional subspace of $L_2^p(I)$.

We can estimate the vector $\boldsymbol{\alpha}$ on the basis of n independent realizations $\boldsymbol{x}_1, \boldsymbol{x}_2, \ldots, \boldsymbol{x}_n$ of the random process $\boldsymbol{X}$ (functional data).

Typically data are recorded at discrete moments in time. Let x_{kj} denote an observed value of the feature $X_k, k = 1, 2, \ldots, p$ at the jth time point t_j, where $j = 1, 2, ..., J$. Then our data consist of the pJ pairs (t_j, x_{kj}). These discrete data can be smoothed by continuous functions x_k and I is a compact set such that $t_j \in I$, for $j = 1, ..., J$.

Details of the process of transformation of discrete data to functional data can be found in Ramsay and Silverman (2005), Horváth and Kokoszka (2012), or in Górecki et al. (2014).

3 $K = 2$ Case

For two random vectors $\boldsymbol{X} \in R^p$ and $\boldsymbol{Y} \in R^q$, Escoufier (1973) introduced correlation coefficient ρV as a nonnegative number given by

$$\rho V_{\boldsymbol{X},\boldsymbol{Y}} = \frac{\|\boldsymbol{\Sigma}_{XY}\|_F}{\sqrt{\|\boldsymbol{\Sigma}_{XX}\|_F \|\boldsymbol{\Sigma}_{YY}\|_F}},$$

where $\|\cdot\|_F$ denoted the Frobenius norm and

$$\boldsymbol{\Sigma} = \begin{bmatrix} \boldsymbol{\Sigma}_{XX} & \boldsymbol{\Sigma}_{XY} \\ \boldsymbol{\Sigma}_{YX} & \boldsymbol{\Sigma}_{YY} \end{bmatrix}$$

is a covariance matrix of vectors $\boldsymbol{X}$ and $\boldsymbol{Y}$.

Correlation coefficient ρV has the following properties: $\rho V_{\boldsymbol{X},\boldsymbol{Y}} = 0$ if and only if random vectors $\boldsymbol{X}$ and $\boldsymbol{Y}$ are uncorrelated. Moreover, if the joint distribution of $\boldsymbol{X}$ and $\boldsymbol{Y}$ is $p + q$ dimensional normal distribution, random vectors $\boldsymbol{X}$ and $\boldsymbol{Y}$ are independent.

We may extend this coefficient to two random processes $\boldsymbol{X} \in L_2^p(I)$ and $\boldsymbol{Y} \in L_2^q(I)$ assuming that

$$\|\boldsymbol{\Sigma}_{XY}\|_F = \sqrt{\int_I \int_I \text{tr}(\boldsymbol{\Sigma}_{XY}^\top(s,t)\boldsymbol{\Sigma}_{XY}(s,t))dsdt}.$$

Moreover, if processes $\boldsymbol{X}$ and $\boldsymbol{Y}$ have the form

$$\mathbf{X}(t) = \boldsymbol{\Phi}_1(t)\boldsymbol{\alpha}, \quad \mathbf{Y}(s) = \boldsymbol{\Phi}_2(s)\boldsymbol{\beta}, \quad t, s \in I, \tag{2}$$

then Górecki et al. (2017)

$$\rho V_{\boldsymbol{X},\boldsymbol{Y}} = \rho V_{\boldsymbol{\alpha},\boldsymbol{\beta}}.$$

In this case, the problem of testing the correlation of processes $\boldsymbol{X}$ and $\boldsymbol{Y}$ is equivalent to the problem of zeroing the coefficient $\rho V_{\boldsymbol{\alpha},\boldsymbol{\beta}}$.

Note, that the coefficient $\rho V_{\boldsymbol{X},\boldsymbol{Y}}$ is appropriate only for linear dependence. It is useless for more complicated situations. It "cannot see" nonlinear dependencies. In such a situation, we ought to use some other measures of dependence.

One such measure is proposed by Székely et al. (2007) distance correlation. Let us denote by $\phi_{X,Y}$ and ϕ_X, ϕ_Y the joint and the marginals characteristic functions of random vectors $\boldsymbol{X} \in R^p$ and $\boldsymbol{Y} \in R^q$, respectively. Distance correlation of random vectors $\boldsymbol{X} \in R^p$ and $\boldsymbol{Y} \in R^q$ is a nonnegative number given by

$$\mathrm{dCorr}(\boldsymbol{X}, \boldsymbol{Y}) = \frac{\mathrm{dCov}(\boldsymbol{X}, \boldsymbol{Y})}{\sqrt{\mathrm{dCov}(\boldsymbol{X}, \boldsymbol{X})\,\mathrm{dCov}(\boldsymbol{Y}, \boldsymbol{Y})}},$$

where

$$\mathrm{dCov}(\boldsymbol{X}, \boldsymbol{Y}) = \|\phi_{X,Y}(\boldsymbol{l}, \boldsymbol{m}) - \phi_X(\boldsymbol{l})\phi_Y(\boldsymbol{m})\|_w,$$

and

$$\|f\|_w = \sqrt{\iint |f(\boldsymbol{l}, \boldsymbol{m})|^2 w(\boldsymbol{l}, \boldsymbol{m}) d\boldsymbol{l} d\boldsymbol{m}}.$$

The weight function $\boldsymbol{w}$ is chosen to produce scale free and rotation invariant measure that does not go to zero for dependent random vectors.

Defining the joint characteristic function of processes $\boldsymbol{X} \in L_2^p(I)$ and $\boldsymbol{Y} \in L_2^q(I)$ as

$$\phi_{\boldsymbol{X},\boldsymbol{Y}}(\boldsymbol{l}, \boldsymbol{m}) = \mathrm{E}\{\exp[i < \boldsymbol{l}, \boldsymbol{X} >_p + i < \boldsymbol{m}, \boldsymbol{Y} >_q]\},$$

where

$$< \boldsymbol{l}, \boldsymbol{X} >_p = \int_{I_1} \boldsymbol{l}'(s)\boldsymbol{X}(s)ds, \quad < \boldsymbol{m}, \boldsymbol{Y} >_q = \int_{I_2} \boldsymbol{m}'(t)\boldsymbol{Y}(t)dt$$

and assuming that processes $\boldsymbol{X}$ and $\boldsymbol{Y}$ have the form (2) we have

$$\mathrm{dCorr}(\boldsymbol{X}, \boldsymbol{Y}) = \mathrm{dCorr}(\boldsymbol{\alpha}, \boldsymbol{\beta})$$

Górecki et al. (2017).

Thus, we can reduce the problem of testing the independence of random processes $\boldsymbol{X}$ and $\boldsymbol{Y}$ to the problem of testing the significance of their distance correlation $\mathrm{dCorr}(\boldsymbol{X}, \boldsymbol{Y})$.

4 $K > 2$ Case

Let us now discuss the problem of testing mutual independence for more than two vector processes.

Let $\boldsymbol{X}_1 \in L_2^{p_1}(I), \boldsymbol{X}_2 \in L_2^{p_2}(I), \ldots, \boldsymbol{X}_K \in L_2^{p_K}(I)$ be random processes with the following representation:

$$\boldsymbol{X}_1(t) = \boldsymbol{\Phi}_1(t)\boldsymbol{\alpha}_1, \boldsymbol{X}_2(t) = \boldsymbol{\Phi}_2(t)\boldsymbol{\alpha}_2, \ldots, \boldsymbol{X}_K(t) = \boldsymbol{\Phi}_K(t)\boldsymbol{\alpha}_K,\ t \in I. \tag{3}$$

Additionally, let the covariance matrix for vectors $\boldsymbol{\alpha}_1, \boldsymbol{\alpha}_2, \ldots, \boldsymbol{\alpha}_K$ have the form:

$$\boldsymbol{\Sigma} = \begin{bmatrix} \boldsymbol{\Sigma}_{11} & \boldsymbol{\Sigma}_{12} & \cdots & \boldsymbol{\Sigma}_{1K} \\ \boldsymbol{\Sigma}_{21} & \boldsymbol{\Sigma}_{22} & \cdots & \boldsymbol{\Sigma}_{2K} \\ \vdots & \vdots & & \vdots \\ \boldsymbol{\Sigma}_{K1} & \boldsymbol{\Sigma}_{K2} & \cdots & \boldsymbol{\Sigma}_{KK} \end{bmatrix}.$$

Assuming joint $p_1 + p_2 + \cdots + p_k$ dimensional normal distribution of vectors $\boldsymbol{\alpha}_1, \boldsymbol{\alpha}_2, \ldots, \boldsymbol{\alpha}_K$, the problem of testing the null hypothesis

$$H_0 \colon \boldsymbol{X}_1, \boldsymbol{X}_2, \ldots, \boldsymbol{X}_K \text{ are independent}$$

is equivalent to the problem of testing the null hypothesis

$$H_0 \colon \sum_{i<j} \|\boldsymbol{\Sigma}_{ij}\|_F = 0.$$

Let us define coefficient of mutual correlation ρMV as a positive number given by

$$\rho^2 MV = \frac{2}{K(K-1)} \sum_{i<j} \rho^2 V(\boldsymbol{X}_i, \boldsymbol{X}_j).$$

Assuming that the processes meet the assumptions of model (3) and that the joint distribution of vectors $\boldsymbol{\alpha}_1, \boldsymbol{\alpha}_2, \ldots, \boldsymbol{\alpha}_K$ is normal, the problem of testing the mutual independence is equivalent to the problem of testing the significance of coefficient ρMV.

Another way to test the mutual independence is to reduce this problem to a problem using two processes.

Let $\text{Corr}(\boldsymbol{X}_i, \boldsymbol{X}_j)$ be some measure of dependence for vector processes $\boldsymbol{X}_i$ and $\boldsymbol{X}_j$ with property: $\text{Corr}(\boldsymbol{X}_i, \boldsymbol{X}_j) = 0$ if and only if vector processes $\boldsymbol{X}_i$ and $\boldsymbol{X}_j$ are independent, $i, j = 1, 2, \ldots, K, i \neq j$.

Note that in the place of Corr we may put, e.g., dCorr.

Let

$$\boldsymbol{X}_{c+} = (\boldsymbol{X}_{c+1}^\top, \ldots, \boldsymbol{X}_K^\top)^\top,\ c = 1, \ldots, K-1,$$

$$\boldsymbol{X}_{c-} = (\boldsymbol{X}_1^\top, \ldots, \boldsymbol{X}_{c-1}^\top, \boldsymbol{X}_{c+1}^\top, \ldots, \boldsymbol{X}_K^\top)^\top,\ c = 1, \ldots, K.$$

Following the idea from Jin and Matteson (2018), we may define the coefficients of multiple independence as

$$\mathcal{R}(\boldsymbol{X}) = \frac{1}{K-1} \sum_{c=1}^{K-1} \text{Corr}^2(\boldsymbol{X}_c, \boldsymbol{X}_{c+}),$$

and

$$\mathscr{S}(\mathbf{X}) = \frac{1}{K}\sum_{c=1}^{K} \mathrm{Corr}^2(\mathbf{X}_c, \mathbf{X}_{c-}).$$

Thus, the following theorem is true:

Theorem 1 *$\mathbf{X}_1, \mathbf{X}_2, \ldots, \mathbf{X}_K$ are independent if and only if $\mathscr{R}(\mathbf{X}) = 0$ or $\mathscr{S}(\mathbf{X}) = 0$.*

Hence, the problem of testing the null hypothesis

$$H_0\colon \mathbf{X}_1, \mathbf{X}_2, \ldots, \mathbf{X}_K \text{ are independent}$$

is equivalent to the problem of testing the null hypothesis

$$H_0\colon \mathscr{R}(\mathbf{X}) = 0\ (\mathscr{S}(\mathbf{X}) = 0).$$

To verify these hypotheses, we propose to use a permutation test.

5 Example

5.1 *Univariate Case*

Let

$$\begin{aligned} X_t &= \varepsilon_{1t}, \\ Y_t &= 3X_t + \varepsilon_{2t}, \\ Z_t &= X_t^2 + \varepsilon_{3t}, \end{aligned}$$

where ε_{1t}, ε_{2t} and ε_{3t} are independent random variables with $N(0, 0.25)$ distribution. We generated 1000 random realizations for each process with length 100 (Fig. 1). To smooth the data we used Fourier series with 15 elements. Clearly, processes X_t and Y_t are linearly dependent and processes X_t, Z_t and Y_t, Z_t are non-linearly dependent.

From Table 1 (third column), we see that all measures of correlation for functional data detect dependence (at significance level 5%) when at least one pair of linearly dependent processes exist. However, when we have nonlinear dependence only measures based on dCorr detect it.

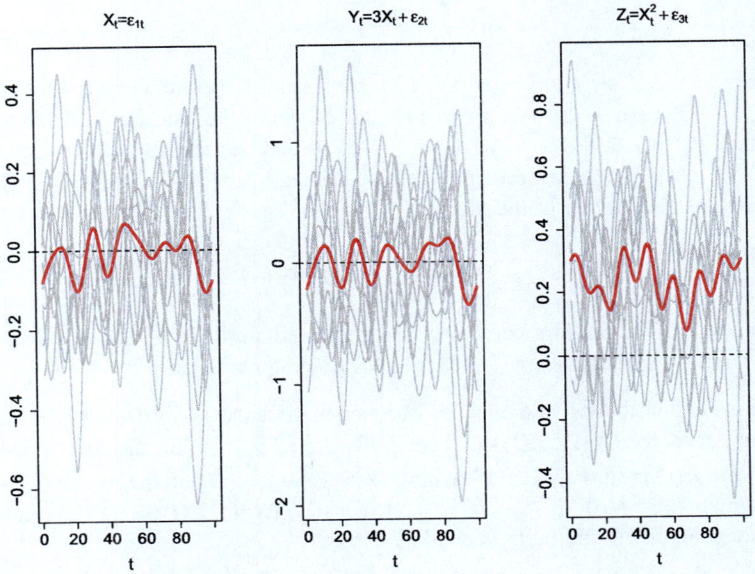

Fig. 1 10 realizations of univariate processes X_t, Y_t, and Z_t (functional means in red)

Table 1 Results of simulations (significant (5%) results are in bold)

Coefficient's name	Processes	p-value (Univ.)	p-value (Multi. – S1)	p-value (Multi. – S2)
ρMV	X_t, Y_t, Z_t	**0.014**	0.757	**0.036**
$\mathscr{R}-\rho V$	X_t, Y_t, Z_t	**0.006**	0.767	**0.015**
$\mathscr{S}-\rho V$	X_t, Y_t, Z_t	**0.027**	0.787	**0.024**
$\mathscr{R}-$dCorr	X_t, Y_t, Z_t	**0.007**	0.581	**0.017**
$\mathscr{S}-$dCorr	X_t, Y_t, Z_t	**0.016**	0.592	**0.006**
ρV	X_t vs Y_t	**0.001**	0.783	0.632
	X_t vs Z_t	0.367	0.568	0.203
	Y_t vs Z_t	0.481	0.566	0.526
dCorr	X_t vs Y_t	**0.001**	0.827	0.773
	X_t vs Z_t	**0.003**	0.486	0.094
	Y_t vs Z_t	**0.025**	0.457	0.470

5.2 *Multivariate Case*

Following Krzyśko and Smaga (2019) we consider the functional sample $\boldsymbol{x}_1(t), \ldots, \boldsymbol{x}_n(t)$ of size $n = 1000$ containing realizations of the random process $\boldsymbol{X}(t) = (X(t), Y(t), Z(t)), t \in [0, 1]$ of dimension $p = 3$. These observations are generated in the following discretized way:

$$\boldsymbol{x}_i(t_j) = \boldsymbol{\Phi}(t_j)\boldsymbol{\alpha}_i + \boldsymbol{\varepsilon}_{ij},$$

where $i = 1, \ldots, n$, t_j, $j = 1, \ldots, 100$ are equally spaced design time points in [0, 1], the matrix $\boldsymbol{\Phi}(t)$ is as in (1) and contains the Fourier basis functions only and $B_k = 5$, $k = 1, \ldots, p$, $\boldsymbol{\alpha}_i$ are $5p$-dimensional random vectors, and $\boldsymbol{\varepsilon}_{ij} = (\varepsilon_{ij1}, \ldots, \varepsilon_{ijp})^\top$ are measurement errors such that $\varepsilon_{ijk} \sim N(0, 0.025r_{ik})$ and r_{ik} is the range of the kth row of the matrix

$$\boldsymbol{\Phi}(t_1)\boldsymbol{\alpha}_i \ldots \boldsymbol{\Phi}(t_{100})\boldsymbol{\alpha}_i,$$

$k = 1, \ldots, p$. The random vectors $\boldsymbol{\alpha}_i$ are generated similarly to Todorov and Pires (2007) and Jin and Matteson (2018) in the following two setups:

S1 Normal distribution and equal covariance matrices: $\boldsymbol{\alpha}_i \sim N(\boldsymbol{0}_{5p}, \boldsymbol{I}_{5p})$.

S2 Part of $\boldsymbol{\alpha}_i$ for $(X(t), Y(t))$ is from $N(\boldsymbol{0}_{5(p-1)}, \boldsymbol{I}_{5(p-1)})$ and the first element of $\boldsymbol{\alpha}_i$ for $Z(t)$ is $\text{sgn}(\alpha_1\alpha_{5+1})W$, where $W \sim \text{Exp}(1/\sqrt{2})$ and the remaining $p-1$ elements are $N(\boldsymbol{0}_{5(p-1)-1}, \boldsymbol{I}_{5(p-1)-1})$. Clearly, $(X(t), Y(t), Z(t))$ is a pairwise independent but mutually dependent triplet.

Setup S1 is simple no dependence example. All tests correctly deal with this problem (Table 1, fourth column). Setup S2 is much harder to deal with. For whole triplet of data, all methods indicate dependence (Table 1, fifth column). For a pair of variables, all methods correctly detect independence for all pairs of processes.

6 Conclusions

We have considered the measuring and testing mutual dependence for multivariate functional data based on the basis functions representation of the data. We propose few measures of mutual dependence for multivariate functional data based on the equivalence to mutual independence through characteristic functions (Székely et al. 2007) and on ρV coefficient (Escoufier 1973). The performance of the proposed methods was studied in simulations. Their results have indicated that the proposed methods perform quite well. Finally, we can propose to use measures and tests based dCorr coefficient. Such methods correctly detect linear and nonlinear dependence structure both for univariate and multivariate processes.

References

Escoufier, Y.: Le traitement des variables vectorielles. Biometrics **29**(4), 751–760 (1973)

Górecki, T., Krzyśko, M., Waszak, Ł., Wołyński, W.: Methods of reducing dimension for functional data. Stat. Transit. New Ser. **15**(2), 231–242 (2014)

Górecki, T, Krzyśko, M., Wołyński, W.: Correlation analysis for multivariate functional data. In: Palumbo, F., Montanari, A., Vichi, M. (eds.) Data Science, Studies in Classification, Data Analysis, and Knowledge Organization, pp. 243–258. Springer International Publishing (2017)
Górecki, T., Krzyśko, M., Wołyński, W.: Independence test and canonical correlation analysis based on the alignment between kernel matrices for multivariate functional data. Artif. Intell. Rev. **53**, 475–499 (2020)
Gretton A., Bousquet O., Smola A., and Schölkopf B.: Measuring statistical dependence with Hilbert-Schmidt norms. In: Jain, S., Simon, H.U., Tomita, E. (eds.) Algorithmic Learning Theory. ALT 2005. Lecture Notes in Computer Science, vol. 3734, pp. 63–77. Springer, Berlin, Heidelberg (2005)
Horváth, L., Kokoszka, P.: Inference for Functional Data with Applications. Springer (2012)
Jin, Z., Matteson, D.S.: Generalizing distance covariance to measure and test multivariate mutual dependence via complete and incomplete V-statistics. J. Multivariate Anal. **168**, 304–322 (2018)
Krzyśko, M., Smaga, Ł.: A multivariate coefficient of variation for functional data. Stat. Interface **12**, 647–658 (2019)
Ramsay, J.O., Silverman, B.W.: Functional Data Analysis, 2nd edn. Springer (2005)
Székely, G.J., Rizzo, M.L., Bakirov, N.K.: Measuring and testing dependence by correlation of distances. Ann. Stat. **35**, 2769–2794 (2007)
Todorov, V., Pires, A.M.: Comparative performance of several robust linear discriminant analysis methods. Revstat. Stat. J. **5**, 63–83 (2007)

Single Imputation Via Chunk-Wise PCA

Alfonso Iodice D'Enza, Francesco Palumbo, and Angelos Markos

Abstract The straightforward application of Principal Component Analysis (PCA) to incomplete data sets is not possible and practitioners often remove or ignore observations that contain at least one missing value. Three different strategies can be mainly distinguished to apply PCA on a data set with missing entries: (i) imputation of the missings prior to the application of PCA; (ii) obtain the PCA solution and ignore the missings; and (iii) obtain the PCA solution and deal explicitly with missings. Methods implementing the latter strategy have been reviewed and, among them, the iterative PCA (iPCA) approach has been shown to be preferable. This paper proposes a chunk-wise implementation of iPCA, suitable for tall data sets, that is, with many observations. In the proposed approach, each data chunk is imputed according to the insofar analyzed data. The proposed procedure is compared to the batch iPCA and to a naive implementation, which imputes each data chunk independently. In a series of experiments, we consider different data sets and missing data mechanisms.

Keywords Principal component analysis · Missing data · Imputation

1 Introduction

Principal Component Analysis (Jolliffe 2002) is a well-known unsupervised learning method that can be used to describe the correlation structure characterizing a continuous data set, or as a preprocessing step for supervised learning methods. However,

---

A. Iodice D'Enza (✉) · F. Palumbo
Universitá degli Studi di Napoli Federico II, Napoli, Italy
e-mail: iodicede@unina.it

F. Palumbo
e-mail: fpalumbo@unina.it

A. Markos
Democritus University of Thrace, Xanthi , Greece
e-mail: amarkos@eled.duth.gr

T. Chadjipadelis et al. (eds.), *Data Analysis and Rationality in a Complex World*, Studies in Classification, Data Analysis, and Knowledge Organization,
https://doi.org/10.1007/978-3-030-60104-1_9

the standard implementation of PCA is not suitable for data sets with missing values. Real data sets, however, often present missing entries, and practitioners willing to apply standard PCA have two different options: to remove observations that contain at least one missing value, or to impute the missing entries before applying PCA. It is not advisable to delete each observation that presents at least one missing value or, alternatively, discard one or more attributes with many missings, because of the loss of information. Even prior imputation, for example, mean imputation, has drawbacks: it not only reduces the variance of an attribute, but also modifies its correlation with the other attributes (Little and Rubin 2019). A joint modeling approach to prior imputation, assuming data to be generated by a multivariate normal distribution, does not alter the correlation structure, albeit it is independent of the PCA solution (for details, see, e.g., Schafer (1997)). Another approach to obtain a PCA solution of a data set with missings was proposed by Gower (1971). It can be applied either on the observations pair-wise distance matrix, or on the correlation matrix: in both cases, the missing entries are skipped while computing distances or correlations. The former case leads to PCA scores, the latter to PCA loadings. The drawback of this approach lies in the possibility of negative eigenvalues when it comes to the decomposition of the data matrices.

Due to the limitations of the prior imputation or skipping missing values strategies, PCA algorithms have been proposed that deal explicitly with missing data. Recent contributions review and compare methods/approaches for PCA with missings: Dray and Josse (2015), Folch-Fortuny et al. (2015), Van Ginkel et al. (2014), Geraci and Farcomeni (2018), and Loisel and Takane (2019). Most of the best performing methods for PCA with missings are based on iterative procedures, but this strategy is not desirable when dealing with a large number of observations. Such tall data sets, however, can be suitably analyzed chunk-wise by *splitting* the observations into chunks, *analyzing* each chunk separately, and *combining* the chunk-based solution to obtain the full solution. The so-called *chunk averaging* method (Matloff 2014) eases parallelization. In this paper, a chunk-wise iterative PCA-based single imputation method is proposed for the analysis of tall data sets with missings. The proposed procedure is compared with the batch (full) counterpart and to a naive implementation that imputes each data chunk independently: the experiments will refer to different data sets and missing data mechanisms.

The rest of the paper is structured as follows: Sect. 2 recalls the PCA definition, and it describes a suitable procedure to handle missing values in PCA; we introduce a chunk-wise PCA implementation in Sect. 3, and a generalization to the case of the missing values of the chunk-wise PCA is defined in Sect. 4. A comparative application on a process data set reported in Sect. 5 concludes the paper.

2 Dealing with Missing Data in PCA

Let $\mathbf{X}$ be an $n \times Q$ data matrix, where n is the number of observations, and Q is the number of quantitative attributes that we assume to be scaled to a unit variance.

When $\mathbf{X}$ has no missing entries, the PCA solution consists of the singular value decomposition (SVD) of

$$\mathbf{S}_{pca} = n^{-1/2} (\mathbf{X} - \mathbf{M})\, Q^{-1/2} = \mathbf{V}\boldsymbol{\Sigma}\mathbf{U}^{\mathsf{T}} \tag{1}$$

where $\mathbf{M} = n^{-1}\mathbf{1}\mathbf{1}^{\mathsf{T}}\mathbf{X}$ is the centering operator and $\mathbf{1}$ is an n-dimensional vector of ones; $\mathbf{V}$ is a $n \times Q$ orthonormal matrix with left singular vectors on columns, $\boldsymbol{\Sigma}$ is a diagonal matrix containing the Q singular values $\sqrt{\lambda_j}$, $j = 1, \ldots, Q$, and $\mathbf{U}$ is a $Q \times Q$ matrix of right singular vectors. The jth singular value corresponds to the standard deviation along the direction of the jth singular vector, $j = 1, \ldots, Q$.

Let $\hat{\mathbf{V}}$, $\hat{\mathbf{U}}$, and $\hat{\boldsymbol{\Sigma}}$ be the first d singular vectors and values; by the Eckart and Young theorem, the object scores $\hat{\mathbf{F}} = n^{1/2}\hat{\mathbf{V}}\hat{\boldsymbol{\Sigma}}$ and the loadings $\hat{\mathbf{G}} = Q^{1/2}\hat{\mathbf{U}}$ are such that $\hat{\mathbf{F}}\hat{\mathbf{G}}^{\mathsf{T}}$ represents the best rank d approximation of $\mathbf{X}$ (centered) in the least squares sense, Greenacre (2010). The PCA loss function

$$\|\mathbf{X} - \hat{\mathbf{X}}\|^2 = \|\mathbf{X} - \mathbf{M} - \hat{\mathbf{F}}\hat{\mathbf{G}}^{\mathsf{T}}\|^2 \tag{2}$$

is referred to as the low-rank approximation criterion.

In order to account for the presence of missing values in PCA, a so-called $n \times Q$ *shadow matrix* $\mathbf{W}$ is defined that has general element $w_{ij} = 0$ if the value for the ith observation of the jth attribute is missing, and $w_{ij} = 1$ otherwise. Then the criterion is

$$\|\mathbf{W} * \left(\mathbf{X} - \mathbf{M} - \hat{\mathbf{F}}\hat{\mathbf{G}}^{\mathsf{T}}\right)\|^2 \tag{3}$$

where the operator '$*$' indicates the Hadamard product. Equivalently, by defining

$$\tilde{\mathbf{X}} = \mathbf{W} * \mathbf{X} + (1 - \mathbf{W}) * \hat{\mathbf{F}}\hat{\mathbf{G}}^{\mathsf{T}},$$

the loss function can be further re-stated as

$$\|\tilde{\mathbf{X}} - \mathbf{M} - \hat{\mathbf{F}}\hat{\mathbf{G}}^{\mathsf{T}}\|^2, \tag{4}$$

as pointed out by Loisel and Takane (2019).

The optimization of the criterion in Formula 3 is not possible via a direct solution and it takes the iterative procedure proposed by Kiers (1997) and is further described by Josse and Hussin (2012). Here, follows a brief description of the steps of the algorithm.

- Step 1. initialize the iteration counter $\ell = 0$. Replace each missing entry in $\mathbf{X}$ with some initialization values, e.g., the mean of the complete values of the jth attribute, obtaining the starting $\tilde{\mathbf{X}}^{\ell}$;

- Step 2. perform a PCA on $\tilde{\mathbf{X}}^{\ell}$ to obtain $\hat{\mathbf{F}}^{\ell}$ and $\hat{\mathbf{G}}^{\ell}$. Use the reconstruction formula

$$\hat{x}_{ij}^{\ell} = \sum_{d=1}^{\mathcal{D}} \sqrt{\hat{\lambda}_d^{\ell}} \hat{v}_{id}^{\ell} \hat{u}_{jd}^{\ell} = \sum_{d=1}^{\mathcal{D}} \hat{f}_{id} \hat{g}_{jd} \quad \forall i, j \tag{5}$$

  to obtain $\hat{\mathbf{X}}^{\ell}$;
- Step 3. impute the missing entries in $\tilde{\mathbf{X}}^{\ell}$ with the corresponding values of $\hat{\mathbf{X}}^{\ell}$, formally $\tilde{\mathbf{X}}^{\ell} = \mathbf{W} * \mathbf{X} + (1 - \mathbf{W}) * \hat{\mathbf{X}}^{\ell}$, and update the counter $\ell = \ell + 1$;
- Step 4. Repeat steps 2 and 3 until convergence, that is, when the value of Formula 3 does not decrease from an iteration to the next one.

We refer to this procedure as iterative PCA (iPCA) as in Dray and Josse (2015), but in the literature it is also referred to as weighted low-rank approximation Kiers (1997), and expectation-maximization PCA (Josse and Hussin 2012; Geraci and Farcomeni 2018). The iPCA procedure estimates the PCA parameters and it also provides missing value imputation. Therefore, it is a single imputation method that takes into account both the similarities among individuals and the correlation structure characterizing the attributes. The iPCA method may suffer from overfitting, as both the number of missing entries and the dimensionality of the underlying structure increase. A regularized version of the iterative PCA algorithm (RPCA) has been proposed by Josse and Hussin (2012) to tackle the overfitting problem. Since regularization affects the singular values, the RPCA algorithm differs from iPCA in the reconstruction formula used to impute the missing entries at each iteration.

A review by Loisel and Takane (2019) compared the performance of iPCA (and RPCA) with other methods for PCA with missings that proved to perform well in previous comparative reviews. The use of either iPCA or RPCA can be problematic for tall data sets. In fact, when the number of observations is large, it can be profitable to analyze the dataset chunk-wise and update the solution as new chunks are analyzed, more so if each chunk is analyzed iteratively.

3 Chunk-Wise PCA

The general idea of the chunk-wise PCA (CW-PCA) approach is to apply the PCA to each chunk separately and then merge the obtained solutions to get the PCA of the whole matrix. Consider the case where the quantitative $n \times Q$ data matrix $\mathbf{X}$ is split into chunks

$$\mathbf{X} = \begin{bmatrix} \mathbf{X}_1 \\ \mathbf{X}_2 \\ \dots \\ \mathbf{X}_k \end{bmatrix}. \tag{6}$$

A CW-PCA is implemented by embedding the *eigenspace arithmetics*, Hall et al. (2002), to update the current PCA solution as new data chunks are analyzed, Iodice D'Enza et al. (2018).

Let $\Omega_j = \{\mathbf{V}_j, \Sigma_j, \mathbf{U}_j, \mu_j, n_j\}$ be the eigenspace of the jth data chunk $\mathbf{X}_j$, with mean μ_j, size n_j , and singular vectors and values $\mathbf{V}_j, \mathbf{U}_j, \Sigma_j$, respectively. If $\mathbf{X}_3 = [\mathbf{X}_1; \mathbf{X}_2]$, then

$$\Omega_3 = \Omega_1 \oplus \Omega_2,$$

where '$\oplus$' is the 'merge' operator. In particular, the SVD of $\mathbf{X}_3$, that is $\mathbf{X}_3 = \mathbf{V}_3\Sigma_3\mathbf{U}_3^{\mathsf{T}}$, can be obtained using the quantities in Ω_j, $j = 1, 2$. Consider the decomposition of the following block matrix:

$$\begin{bmatrix} \Sigma_1\mathbf{U}_1^{\mathsf{T}} & \mathbf{V}_1^{\mathsf{T}}\mathbf{V}_2\Sigma_2\mathbf{U}_2^{\mathsf{T}} \\ 0 & \mathbf{v}^{\mathsf{T}}\mathbf{V}_2\Sigma_2\mathbf{U}_2^{\mathsf{T}} \end{bmatrix} + \begin{bmatrix} \mathbf{V}_1^{\mathsf{T}}(\mu_1 - \mu_3)\mathbf{1}_{n_1} & \mathbf{V}_1^{\mathsf{T}}(\mu_2 - \mu_3)\mathbf{1}_{n_2} \\ \mathbf{v}^{\mathsf{T}}(\mu_1 - \mu_3)\mathbf{1}_{n_1} & \mathbf{v}^{\mathsf{T}}(\mu_2 - \mu_3)\mathbf{1}_{n_2} \end{bmatrix} = \mathbf{R}\Sigma\mathbf{U}^{\mathsf{T}},$$

where $\mathbf{1}_n$ is an n-dimensional vector of ones. The singular vectors and values are then given by

$$\mathbf{V}_3 = [\mathbf{V}_1, \mathbf{v}]\,\mathbf{R},\ \Sigma_3 = \Sigma \text{ and } \mathbf{U}_3 = \mathbf{U}$$

where $\mathbf{v} = orth\,(\psi\,[\mathbf{H}, \mathbf{h}])$, the $orth$ operator stands for a Gramm–Schmidt orhogonalization procedure, and ψ discards very small column vectors from the matrix; furthermore, $\mathbf{H} = \mathbf{V}_2 - \mathbf{V}_1\mathbf{V}_1^{\mathsf{T}}\mathbf{V}_2$ and $\mathbf{h} = (\mu_1 - \mu_2) - \mathbf{V}_1\mathbf{V}_1^{\mathsf{T}}(\mu_1 - \mu_2)$.

Therefore, the eigenvectors in $\mathbf{V}_3$ are linear combinations of the already available $\mathbf{V}_1$. In order to deal with a change in dimension, a basis sufficient span $\mathbf{V}_3$ is constructed, that is $\mathbf{V}_1$ augmented by $\mathbf{v}$.

The PCA observation scores and loadings are given by $\mathbf{F} = n_3^{1/2}\mathbf{V}_3\Sigma_3$ and $\mathbf{G} = Q^{1/2}\mathbf{U}_3$, respectively. It is worth pointing out that the CW-PCA solution using the method described above is *exact*, in the sense that it collapses into the ordinary PCA solution on the covariance matrix and then it can also be defined as chunk averaging PCA.

4 Chunk-Wise Single Imputation Via PCA

When the data matrix containing missings is tall, it can be suitably processed chunk-wise, but the issues due to the presence of missings do hold. In fact, if the general chunk $\mathbf{X}_i$ contains missings, it has to be imputed before the current PCA solution can be updated. A straightforward strategy is to *(i)* impute the current chunk with iPCA, *(ii)* update the current PCA solution as described in Sect. 3, and process a further chunk: we refer to this approach as *naive* CW-iPCA since the iPCA-based imputation of a chunk is independent of the other chunks. As opposed to the naive implementation, a non-naive CW-iPCA is introduced to impute each chunk analyzed using the underlying data structure, gathered by previously analyzed chunks.

In order to ease the description of the procedure, we will assume the data to be centered and equally scaled.

The CW-iPCA procedure, for the general block $\mathbf{X}_i$ and $i > 1$, can be summarized, according to a split-apply-combine paradigm, as follows:

split the rows of $\mathbf{X}_i$ in $\mathbf{X}_i^o$, containing complete rows (with no missing entries), and $\mathbf{X}_i^m$, with the remaining rows; compute Ω_i^o, the eigenspace of $\mathbf{X}_i^o$, and merge it with Ω, the eigenspace of all the blocks insofar analyzed; formally, $\Omega = \Omega \oplus \Omega_i^o$;
apply a properly modified version of the iPCA algorithm on $\mathbf{X}_i^m$ to obtain $\tilde{\mathbf{X}}_i^m$ and the corresponding Ω_i^m;
combine the obtained $\tilde{\mathbf{X}}_i$-based PCA solution with the current one, to update $\Omega = \Omega \oplus \Omega_i^m$.

For $\mathbf{X}_1$, the same procedure applies; obviously, in step 1 it will be $\Omega = \Omega_1^o$.

This CW-iPCA is equivalent to append each new chunk to the previously imputed version of the already processed chunks.

5 Application

The Tennessee Eastman Problem (TEP) data is a sensor data benchmark simulating an industrial chemical process (see, e.g., Severson et al. 2017). PCA is successfully applied to process data as a tool for multivariate process control. Missing values in process control data are fairly common, and they might be due to sensor failures, for example. The benchmark data we refer to is available in `.RData` format, Rieth et al. (2017), and it consists of observations of both normal operation of the process and different failures. The attributes mostly (52 out of 55) refer to sensors that monitor the process in question. While the faulty process detection is beyond the scope of this paper, we used the fault-free training data set that contains a total of 250 thousand observations.

In a preprocessing phase, we selected a subset of 24 attributes: in particular, only one attribute was selected from each collinear pair of attributes; furthermore, the attributes with limited to no correlation were discarded. The preprocessing phase led to a mildly defined correlation structure. Then a proportion of missing values per attribute ranging from 0 to above 50% is introduced that alters the existing correlation structure.

We considered a chunk size of 500 observations, and an increasing number of analyzed chunks. Specifically, the number of chunks is such that $n_{ch} \in \{5, 10, \ldots, 25\}$. The total number of observations in each scenario is, therefore, $500 \times n_{ch}$. An appraisal of CW-iPCA performance compared to naive CW-iPCA and iPCA was carried out concerning the imputation error, given by the mean absolute difference between the *true* and *imputed* values. The imputation error results refer to 10 replicates of the experiments and are reported in Fig. 1: the error increases with the number

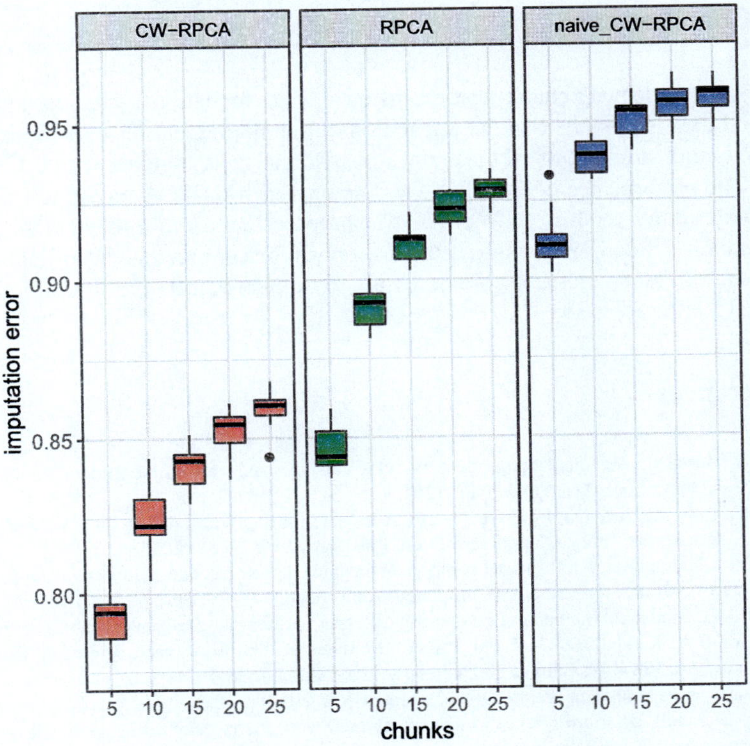

Fig. 1 Imputation error distributions over ten replicates: results on 5–25 analyzed chunks. Left-to-right panels show the performances of CW-iPCA, iPCA, and naive CW-iPCA, respectively

of analyzed chunks. Such a result is not surprising as the proportion of observations with defined correlation structure decreases as more chunks with missings are analyzed. Furthermore, CW-iPCA is characterized by a lower imputation error compared to iPCA and, more so, to the naive CW-iPCA.

The CW-iPCA *learns* the correlation structure by analyzing the first chunk and takes it into account when processing the forthcoming chunks. Instead, iPCA processes the observations as a whole, and so the correlation structure information carried by complete chunks is diluted in the full set of observations. Finally, the naive CW-iPCA that processes each chunk independently has the highest imputation error. We also compared the PCA solutions of the methods, measuring their congruency to the PCA solution on data without missings. The RV results are similar to the imputation error results (not reported due to space limitations).

6 Conclusion

This work presented a chunk-wise extension of iPCA for data sets with missings. The general idea is grounded on the imputation of the missing entries of a chunk using the low-rank structure of all the chunks insofar analyzed, together with the current one. The performance of CW-iPCA was compared with iPCA on the full data set and with a *naive* version of CW-iPCA. The proposed approach resulted in the lowest imputation error on a real-world data set. Applications on simulated data, taking into account different parameters, are necessary to further evaluate CW-iPCA.

References

Dray, S., Josse, J.: Principal component analysis with missing values: a comparative survey of methods. Plant Ecol. **216**(5), 657–667 (2015)

Folch-Fortuny, A., Arteaga, F., Ferrer, A.: PCA model building with missing data: new proposals and a comparative study. Chemometr. Intell. Lab. Syst. **146**, 77–88 (2015)

Geraci, M., Farcomeni, A.: Principal component analysis in the presence of missing data. In: Naik, G.R. (ed.) Advances in Principal Component Analysis, pp. 47–70. Springer (2018)

Gower, J.C.: Statistical methods of comparing different multivariate analyses of the same data. In: Hodson F.R., Kendall, D. G., Tautu, P. (eds.) Mathematics in the Archaeological and Historical Sciences, pp. 138–149. Edinburgh University Press, Edinburgh (1971)

Greenacre, M.J.: Biplots in practice, Fundacion BBVA (2010)

Hall, P., Marshall, D., Martin, R.: Adding and subtracting eigenspaces with eigenvalue decomposition and singular value decomposition. Image Vision Comput. **20**(13–14), 1009–1016 (2002)

Iodice D'Enza, A., Markos, A., Buttarazzi, D.: The idm package: incremental decomposition methods in R. J. Stat. Softw. **86**(1), 1–24 (2018)

Jolliffe, I.T.: Principal Component Analysis. Springer, New York, NY (2002)

Josse, J., Hussin, F.: Handling missing values in exploratory multivariate data analysis methods. J. Société Française Statistique **153**(2), 79–99 (2012)

Kiers, H.: Weighted least squares fitting using ordinary least squares algorithms. Psychometrika **62**(2), 251–266 (1997)

Little, R., Rubin. D.: Statistical Analysis with Missing Data. Wiley (2019)

Loisel, S., Takane, Y.: Comparisons among several methods for handling missing data in principal component analysis (PCA). Adv. Data Anal. Classi. **13**(2), 495–518 (2019)

Matloff, N.: Software alchemy: turning complex statistical computations into embarrassingly-parallel ones. arXiv preprint arXiv:1409.5827 (2014)

Rieth, C.A., Amsel, B.D., Tran, R., Cook, M.B.: Additional Tennessee Eastman Process Simulation Data for Anomaly Detection Evaluation. Harvard Dataverse (2017)

Schafer, J.L.: Analysis of Incomplete Multivariate Data. CRC Press (1997)

Severson, K.A., Molaro, M.C., Braatz, R.D.: Principal component analysis of process datasets with missing values. Processes **5**(3), 38 (2017)

Van Ginkel, J.R., Kroonenberg, P.M., Kiers, H.: Missing data in principal component analysis of questionnaire data: a comparison of methods. J. Stat. Comput. Sim. **84**(11), 2298–2315 (2014)

Clustering Mixed-Type Data: A Benchmark Study on KAMILA and K-Prototypes

Jarrett Jimeno, Madhumita Roy, and Cristina Tortora

Abstract Benchmarking in cluster analysis is the process of analyzing which clustering techniques give the best result for different types of data structures as well as setting a standard for evaluation of newer clustering methods. There are many instances of benchmarking in cluster analysis for continuous data, but only a few for mixed-type data, i.e. data sets with nominal and continuous variables. Therefore, we explore the process for benchmarking various clustering methods on simulated mixed-type data sets with varying proportions of continuous and nominal variables. For this purpose, we test a newer clustering algorithm, KAMILA, against K-prototypes and tandem analysis where data are preprocessed using multiple correspondence analysis and then clustered using K-means, fuzzy K-means, probabilistic distance clustering (PD), and Student-*t* mixture models.

Keywords Mixed-type data clustering · Multiple correspondence analysis · KAMILA · K-prototypes

1 Introduction

Clustering, an unsupervised learning technique, is used to find homogeneous groups of units in a data set. A cluster can be defined in many different ways, for example, clusters can be defined as high-density areas in data space, or clusters could be fitted by certain homogeneous probability models, Hennig (2015). Each definition led to the development of several clustering techniques, which were suited for different

J. Jimeno (✉) · M. Roy · C. Tortora
Department of Mathematics and Statistics, San José State University, One Washington Square, San José, CA 95192, USA
e-mail: jarrett.jimeno@sjsu.edu

M. Roy
e-mail: madhumita.roy@sjsu.edu

C. Tortora
e-mail: cristina.tortora@sjsu.edu

T. Chadjipadelis et al. (eds.), *Data Analysis and Rationality in a Complex World*, Studies in Classification, Data Analysis, and Knowledge Organization,
https://doi.org/10.1007/978-3-030-60104-1_10

problems and data sets. This calls for exploring which clustering techniques give the best result according to the problem and the type of data.

A type of data which is currently explored in the statistical literature is mixed-type data. As its name suggests, mixed-type data is any set of data that contains multiple kinds of variables. These kinds of data sets are of great interest because many real-world data are mixed-type, many of which are high dimensional, spanning hundreds of variables. Clustering mixed-type data is relatively new within cluster analysis; for reviews of mixed-type data clustering technique, see, for example, Hunt and Jorgensen (2011) and Ahmad and Khan (2019). A simple strategy would be to convert all the variables into categorical, but this would determine a loss of information. A common, more efficient, strategy called tandem analysis is based on two steps: a quantification of the categorical variables into continuous ones using dimension reduction techniques, followed by a clustering technique (Hubert and Arabie 1985). This approach has some drawbacks, namely, the results are impacted by the choice of the number of factors and the two steps optimize different criteria (van de Velden et al. 2019). An alternative approach is to measure the distance among units in the data set using a distance measure for mixed-type data. A common distance uses Euclidean distance for continuous variables and Hamming distance for categorical variables. K-prototypes (Huang 1998) is one of the reference algorithms in this group. After selecting k prototypes from a data set, each unit is allocated in the cluster whose prototype is the nearest, the prototypes are updated based on the new clusters, and the algorithm is reiterated until convergence. The weight of the categorical versus continuous variables in the distance measure can affect the results. In 2016, Foss et al. (2018) developed and published a new semi-parametric method for clustering mixed-type data. Short for KAy-means for MIxed LArge data, KAMILA does not require any pre-preparation of mixed-type data into purely numeric data to find cluster membership. When using KAMILA, continuous data clustering behaves like the K-means algorithm not making strong parametric assumptions about the data, while categorical variable clustering is based on Gaussian-multinomial mixture models. And unlike K-prototypes, KAMILA does not require the choice of weights. Other clustering techniques have been proposed and this emphasizes the need for standardized methods for benchmarking (Boulesteix et al. 2018). The goal of this paper is to focus on KAMILA and its performance when compared to the state-of-the-art techniques, like K-prototypes, or simpler techniques like tandem analysis.

2 Experimental Design and Methodology

All analyses done for our benchmarking study were performed in R (Version 3.5.0) R Core Team (2018). The function `kamila`, package (Foss and Markatou 2018), performs cluster analysis using KAMILA. We set the number of random initializations to 20 to ensure stability of the solution. To perform K-prototypes, we used the `kproto` function found in Szepannek (2018). Much like KAMILA, we increased the number of initializations in the algorithm, setting them to 10. The data quantification for the tandem analysis is obtained using multiple correspondence analysis (MCA) Greenacre and Hastie (1987) R package `FactoMineR` Lê et al. (2008). MCA was chosen since it is one of the most common quantification techniques in categorical data clustering (van de Velden et al. 2017). One of the advantages of MCA is that it represents the data as points in a new low-dimensional space where distances can be interpreted as similarities among points, the more similar are the points, the closer are represented in the space (Greenacre and Hastie 1987), and Euclidean distance can be used in the new space. For simplicity, we set the number of factors to 5 in each variable keeping each categorical variable with the same number of factors so that our simulation results remained consistent. Though, studying the effects of factors on clustering performance is something that certainly can be explored in further studies. Many different techniques can be used for the clustering step of the tandem analysis; we used four different ones. The first clustering method we consider is K-means (MacQueen et al. 1967). It is the most commonly used technique to cluster purely numerical sets. In R, this algorithm is the `kmeans` function in the `stats` package. To ensure the stability of results, we increased the number of random starts to 20 and increased the maximum iterations to 25. In small test trials on similar simulations, the algorithm typically took 15 iterations to complete. The second method we considered is the fuzzy K-means algorithm (also known as fuzzy C-means) (Bezdek 2013); this technique has the advantage of being more robust to outliers. The C-means algorithm assigns to each point a vector of probabilities to belong to each cluster. In R, this algorithm is the function `FKM` in the package `fclust` (Ferraro and Giordani 2015). We adjust the number of random starts for C-means to 20 and keep the preset maximum iterations of one million. Another common soft clustering method we considered is probabilistic distance (PD) clustering (Iyigun and Ben-Israel 2008). PD clustering has the advantage of working with non-spherical clusters, outliers, or noisy data (Tortora et al. 2016). To limit the number of parameters in our benchmarking design, we assumed the clusters to be the same size. Without this assumption, we would rather employ PD clustering adjusted for cluster size (PDQ) (Iyigun and Ben-Israel 2008). In R, the function for PD clustering is `PDclust` from the package `FPDclustering` (Tortora et al. 2019). No other changes to the input were made as the preset maximum number of iterations performed was one thousand. Finally, we considered model-based clustering. To guarantee flexibility and robustness, we chose the Student-t mixture models (Andrews et al. 2011) instead of the classic Gaussian mixture models. The algorithm that performs cluster analysis using a mixture of Student-t distributions is `teigen` in the homonym package (Andrews et al. 2018).

2.1 *Evaluating Clustering Performance*

Since each clustering method optimizes different criteria, the accuracy of each clustering method is measured using techniques that are invariant of the clustering method. Testing against the known clustering partition of our simulations, the accuracy of a clustering technique can be measured with the adjusted Rand index (ARI)

(Hubert and Arabie 1985) (available in R package MixGHD Tortora et al. 2019). The ARI is based on the pair-wise agreement between partitions, and it corrects the Rand index for its expectation; it is equal to 1 when there is perfect agreement between two partitions and the expected value is 0 for random classification. For stability of results, we simulate data sets using every combination of parameters 10 times each and take the average ARI and the standard deviation.

Since the speed of any clustering technique used is affected by the programming language or computer used, measuring computation performance by completion time becomes a complex task to interpret. Instead, we choose to calculate the average number of iterations a clustering technique takes on our data sets for any given set of parameters as a proxy of the convergence speed. Since the algorithms used optimize different criteria, we did not change the default convergence setup. As for the ARI, we report the average number of iterations and standard deviation over 10 simulations.

2.2 *Simulation Design*

We create 27 simulated scenarios based on the following parameters: number of clusters, amount of overlap in each cluster, and proportion of continuous variables. We chose 2, 5, and 7 clusters, levels of overlap of each cluster as 30, 60, and 80%, and the ratio of continuous variables to nominal variables as 1:3, 1:1, and 3:1. In each case, we fix each data set with a total of 128 variables and 1920 observations. The level of overlap is obtained moving the cluster means, where 0% means the clusters are completely separated, 100% completely overlap, i.d. same mean, Fig. 1 in the Appendix[1] shows an example of a two-dimensional data set with different levels of overlapping.

Following McParland and Gormley (2016), the continuous and nominal variables were simulated from multivariate Gaussian mixture models (Genz et al. 2019) fixing the covariance matrix as the identity matrix and varying the mean to determine the overlap level. For non-continuous data, we considered only nominal variables as binary ones are a special case with two factors. For each of our nominal variables, we set the number of factors to five. In the appendix, Fig. 2 shows an example of a smaller data set generation and transformation with a 30% overlap.

In addition to the 27 scenarios from Gaussian mixture models, we also considered the performance of our previous clustering methods on three extra scenarios. In these cases, we restrict the parameters of the simulated data set to 5 clusters, a ratio of numeric to nominal variables set to 3:1, and the overlap of clusters set to 60%. We kept the number of variables to 128, and the nominal level at 5. In the first scenario, we add a correlation structure among the continuous variables where each of the 5 clusters is simulated using a different covariance matrix. The first cluster has a covariance matrix with diagonal elements equal to 2 in each variable, and no correlation. The second cluster also has no correlation, but the diagonal elements are

[1] https://cristinatortora.github.io/Benchmark-on-Clustering-Mixed-Type-Data/.

equal to 0.5. The last three clusters have covariance matrices with unitary diagonal and either a strong positive correlation of 0.9, a weak negative correlation of -0.3, or a weak positive correlation of 0.5. In the second scenario, we generated the continuous data from a skew-t distribution. The function that was used in R was `rdmst` from the package `EMMIXskew` (Wang et al. 2018). While we can adjust the strength and direction of skewness in every variable, we skewed only the numeric variables with a skewness set to 3 in every variable to maintain consistent results in our data. The nominal variables were normally generated and concatenated with our new simulated data set. In the third scenario, we reduce the size of each cluster to 30 observations each. For each scenario, we simulated 10 data sets and performed cluster analysis with all the clustering methods, and we report the average and standard deviation of ARI of each technique against true labels and of the number of iterations. For the sake of space, all the results are available in Appendix,[2] only the most significant results are reported in Tables 1, 2. The code is available in Roy et al. (2019).

3 Results and Discussion

Results for 80% overlap are shown in Table 1. For 2 clusters, all the clustering techniques perfectly cluster the data when the overlap is 30 and 60%, the corresponding ARI values are equal 1. For 80% overlap, K-means poorly cluster the data with an ARI of about 0.22, followed by C-means with an ARI of about 0.66. All the other techniques' ARI values are about 1.

The performances have more variability for 5 clusters; surprisingly C-means and PD clustering perform better when the overlap is 60% compared to 30%, while the other techniques have a close to perfect ARI. When the overlap increases to 80%, C-means has the worst performances followed by K-means and PD clustering; those techniques have also the highest standard deviation.

For 7 clusters, C-means and PD clustering have lower ARI compared to the other techniques when the overlap is 30% and 60%, and the ARI values are no longer all equal to 1 for KAMILA, K-prototypes, and Student-t. When the overlap increases to 80%, C-means has the worst performances followed by K-means and PD clustering, and those techniques have also the highest standard deviation.

The results for the three extra cases are shown in Table 2. On the data set with correlated clusters, we see that all the techniques, with the exception of PD clustering, have similar performance, Student-t has a smaller standard deviation while C-means requires more iterations. For the skewed case, we see a significant drop in performance in most of the clustering methods when compared to the results of the Gaussian counterpart. Although Student-t assumes symmetric clusters, it performed well on skewed clusters. All the other techniques have lower ARI values; C-means requires more iterations to converge, followed by Student-t and PD clustering. When

[2] https://cristinatortora.github.io/Benchmark-on-Clustering-Mixed-Type-Data/.

Table 1 Average and standard deviation of ARI of each clustering method against the true labels for 2, 5, and 7 clusters

Number of clusters: 2	Cluster overlap	80%					
	Variable Proportion continuous: nominal	1:3		1:1		3:1	
Clustering method		Mean	Std. Dev.	Mean	Std. Dev.	Mean	Std. Dev.
KAMILA		1.000	0.000	1.000	0.000	1.000	0.000
K-prototypes		0.994	0.003	1.000	0.000	1.000	0.000
K-means		0.221	0.438	0.222	0.441	0.222	0.441
C-means		0.661	0.495	0.666	0.500	0.667	0.500
PD		0.995	0.003	1.000	0.000	1.000	0.000
Student-*t*		0.998	0.002	1.000	0.000	1.000	0.000
Number of clusters: 5	Cluster overlap	80%					
	Variable Proportion continuous: nominal	1:3		1:1		3:1	
Clustering method		Mean	Std. Dev.	Mean	Std. Dev.	Mean	Std. Dev.
KAMILA		1.000	0.000	1.000	0.000	1.000	0.000
K-prototypes		0.994	0.003	0.994	0.003	0.994	0.003
K-means		0.837	0.184	0.837	0.184	0.837	0.184
C-means		0.675	0.228	0.675	0.228	0.675	0.228
PD		0.809	0.228	0.809	0.228	0.809	0.228
Student-*t*		1.000	0.000	1.000	0.000	1.000	0.000
Number of clusters: 7	Cluster overlap	80%					
	Variable Proportion continuous: nominal	1:3		1:1		3:1	
Clustering method		Mean	Std. Dev.	Mean	Std. Dev.	Mean	Std. Dev.
KAMILA		0.938	0.092	0.915	0.100	0.938	0.093
K-prototypes		0.993	0.004	1.000	0.000	1.000	0.003
K-means		0.837	0.184	0.773	0.147	0.758	0.148
C-means		0.687	0.054	0.708	0.073	0.676	0.021
PD		0.848	0.180	0.857	0.170	0.857	0.170
Student-*t*		1.000	0.000	0.957	0.084	0.980	0.060

Table 2 Average ARI with corresponding standard deviation on simulated data sets with correlated clusters, skewed clusters, and clusters with fewer observations

Fixed parameters	60% Overlap, 3:1 Variable ratio, 5 Clusters					
Variable parameters	Correlated continuous variables		Skew t Continuous variables		Small sample size	
Clustering method	Mean	Std. Dev.	Mean	Std. Dev.	Mean	Std. Dev.
KAMILA	0.828	0.012	0.283	0.011	1.000	0.000
K-prototypes	0.827	0.013	0.309	0.011	1.000	0.000
K-means	0.828	0.014	0.249	0.010	1.000	0.000
C-means	0.836	0.011	0.252	0.009	1.000	0.000
PD	0.460	0.021	0.266	0.005	0.940	0.133
Student's t	0.819	0.009	0.722	0.003	1.000	0.000

the number of elements per cluster is smaller, each of the techniques perform similarly well compared to the data set simulation with more observations, and only PD clustering experiences a small drop in clustering accuracy.

Overall, KAMILA and K-prototypes have similar performances and they perform better than tandem analysis when the clustering step is based on K-means, C-means, or PD clustering. A tandem approach that uses a mixture of Student-*t* distributions in the clustering step performs as well as KAMILA and K-prototypes in most scenarios and performs better when the clusters are skewed. This is probably explained by the increase in tail flexibility compared to a technique based on Gaussian distributions or distance. However, Student-*t* performs better than the other two robust techniques used in the comparison: C-means and PD clustering. Considering the average number of iterations (see Appendix), KAMILA and K-prototypes have similar performance while the tandem analysis using the mixture of Student-*t* distributions in a few cases requires more iterations.

4 Conclusion

We simulated several data sets, varying the number of clusters, overlaps, and proportions of continuous to nominal variables to study the effect of these parameters on clustering performance. The data were simulated from Gaussian distributions which implies that simulated clusters are spherical in structure. Additionally, we simulated data with a correlation structure, with skewed clusters, and with fewer observations per cluster. The results show that K-prototypes and KAMILA performed consistently well both in terms of ARI and number of iterations for spherical clusters. As the number of clusters increased, tandem analysis performance worsened when using C-Means or PD clustering in the clustering step. The mixture of Student-*t* distributions, instead, performed comparably well for both spherical and skewed clusters. Overall, K-prototypes and KAMILA are comparatively efficient in terms of both cluster quality and the iterations needed for completion, when clusters are not skewed. A tandem approach that uses multiple correspondence analysis followed by a mixture of Student-t distributions performed well in all the analyzed scenarios.

References

Ahmad, A., Khan, S.S.: Survey of state-of-the-art mixed data clustering algorithms. IEEE Access **7**, 31883–31902 (2019)

Andrews, J.L., McNicholas, P.D., Subedi. S.: Model-based classification via mixtures of multivariate t-distributions. Comput. Stat. Data An. **55**(1), 520–529 (2011)

Andrews, J.L., Wickins, J.R., Boers, N.M., McNicholas, P.D.: teigen: an R package for model-based clustering and classification via the multivariate t distribution. J. Stat. Softw. **83**(7), 1–32 (2018)

Bezdek, J.C.: Pattern Recognition with Fuzzy Objective Function Algorithms. Springer Science & Business Media (2013)

Boulesteix, A.L., Dangl, R., Dean, N., Guyon, I., Hennig, C., Leisch, F., Steinley, D., Van Mechelen, I.: Benchmarking in cluster analysis: a white paper. arXiv preprint arXiv:180910496 (2018)

Ferraro, M.B., Giordani, P.: A toolbox for fuzzy clustering using the R programming language. Fuzzy Set Syst. **279**, 1–16 (2015)

Foss, A.H., Markatou, M.: kamila: Clustering mixed-type data in R and Hadoop. J. Stat. Softw. **83**(13), 1–45 (2018)

Genz. A., Bretz, F., Miwa, T., Mi, X., Leisch, F., Scheipl, F., Hothorn, T.: mvtnorm: Multivariate Normal and t Distr. R package version 1.0-10 (2019)

Greenacre, M., Hastie, T.: The geometric interpretation of correspondence analysis. J. Am. Stat. Ass. **82**(398), 437–447 (1987)

Hennig, C.: What are the true clusters? Pattern Recognit. Lett. **64**, 53–62 (2015)

Huang, Z.: Extensions to the k-means algorithm for clustering large data sets with categorical values. Data Min. Knowl. Discov. **2**(3), 283–304 (1998)

Hubert, L., Arabie, P.: Comparing partitions. J. Classif. **2**(1), 193–218 (1985)

Hunt, L., Jorgensen, M.: Clustering mixed data. Wiley Int. Rev. Data Min. Knowl. Disc. **1**(4), 352–361 (2011)

Iyigun, C., Ben-Israel, A.: Probabilistic distance clustering adjusted for cluster size. Probab. Eng. Inform. Sci. **22**(4), 603–621 (2008)

Lê, S., Josse, J., Husson, F.: FactoMineR: a package for multivariate analysis. J. Stat. Softw. **25**(1), 1–18 (2008)

MacQueen, J., et al.: Some methods for classification and analysis of multivariate observations. In: Proceedings of the 5th Berkeley Symposium on Mathematical Statistics and Probability, CA, USA, vol. 1, pp. 281–297 (1967)

McParland, D., Gormley, I.C.: Model based clustering for mixed data: clustmd. Adv. Data Anal. Classi. **10**(2), 155–169 (2016)

R Core Team: R: A Language and Environment for Statistical Computing. R Foundation for Statistical Computing, Vienna, Austria (2018). https://www.R-project.org/

Roy, M., Jimeno, J., Tortora, C.: (2019). https://github.com/cristinatortora/Benchmark-on-Clustering-Mixed-Type-Data

Szepannek, G.: clustmixtype: user-friendly clustering of mixed-type data in R. R J. **10**(2), 200–208 (2018)

Tortora, C., Summa, M.G., Marino, M., Palumbo, F.: Factor probabilistic distance clustering (FPDC): a new clustering method. Adv. Data Anal. Classi. **10**(4), 441–464 (2016)

Tortora, C., ElSherbiny, A., Browne, R.P., Franczak, B.C., McNicholas, P.D.: MixGHD: Model Based Clustering, Classification and Discriminant Analysis Using the Mixture of Generalized Hyperbolic Distributions. R package version 2.3.1 (2019)

Tortora, C., Vidales, N., McNicholas, P.D.: FPDclustering: PD-Clustering and Factor PD-Clustering. R package version 1.3 (2019)
van de Velden, M., Iodice D'Enza, A., Palumbo, F.: Cluster correspondence analysis. Psychometrika **82**(1), 158–185 (2017)
van de Velden, M., Iodice D'Enza, A., Markos, A.: Distance-based clustering of mixed data. Wiley Interdiscip. Rev. Comput. Stat. **11**(3), e1456 (2019)
Wang, K., Ng, A., McLachlan, G.: EMMIXskew: The EM Algorithm and Skew Mixture Distribution. R package version 1.0.3 (2018)

Exploring Social Attitudes Toward the Green Infrastructure Plan of the Drama City in Greece

Vassiliki Kazana, Angelos Kazaklis, Dimitrios Raptis, Efthimia Chrisanthidou, Stella Kazakli, and Nefeli Zagourgini

Abstract A complex Green Infrastructure (GI) plan has been recently put into operation in the city of Drama located in the northeastern part of Greece, aiming at the upgrading of the environmental, bioclimatic, and economic conditions of the city downtown area. Within the project, a governance network has been established to promote active social participation and increase the project's social acceptability. This work presents the preliminary results of the first of a series of social surveys carried out within the governance network's function to explore the attitudes of the entrepreneurs of the area, who are expected to be heavily affected by the GI plan. A total of 117 responses were collected using a questionnaire and joint dimension reduction, and clustering of the data was conducted to identify the main factors comprising the entrepreneurs' attitudes patterns toward the GI plan. These factors involve the perceived negative impacts during the project implementation phase, the potential usefulness of the GI infrastructure, and the perceived benefits after the project

V. Kazana (✉)
Department of Forestry and Natural Environment 1st km Drama-Mikrohori, International Hellenic University, Drama 66100, Greece
e-mail: vkazana@for.ihu.gr

A. Kazaklis
OLYMPOS Non Profit Integrated Centre for Environmental Management, Drama, Greece
e-mail: akaz98@otenet.gr

D. Raptis · E. Chrisanthidou · S. Kazakli · N. Zagourgini
International Hellenic University, Drama, Greece
e-mail: d_rapt@for.ihu.gr

E. Chrisanthidou
e-mail: efchrisanthidoy@gmail.com

S. Kazakli
e-mail: stkaz98@gmail.com

N. Zagourgini
e-mail: nefeliza@gmail.com

T. Chadjipadelis et al. (eds.), *Data Analysis and Rationality in a Complex World*, Studies in Classification, Data Analysis, and Knowledge Organization,
https://doi.org/10.1007/978-3-030-60104-1_11

completion. Three groups of entrepreneurs were identified in terms of their attitudes toward the GI plan: (a) negative to change, (b) utilitarians, and (c) positive to change. Each group was profiled according to its sociodemographic characteristics.

1 Introduction

Since 2013, the European Commission has adopted an EU-wide strategy promoting investments in GI (EU/COM/2013/0249). According to the Landscape Institute (2009), the crucial feature of GI projects is that they promote the interaction between the quality of natural elements and environmental, social, and economic performance. In recent years, there has been a growing interest in special GI projects, which in general aim to deliver an extensive range of eco-facilities including water purification, climate mitigation, air superiority, special zones for recreation, improved energy use efficiency, noise reduction, and aesthetic improvement, Chen and Jim (2008), Dunnett and Kingsbury (2008). These projects aim also at improving the environmental conditions, and people's health and quality of life, as well as enhancing the green economy and creating job opportunities, Mell et al. (2016). However, several barriers to GI projects have been noted in the international literature. These are mainly connected to (i) lack of awareness and knowledge of the content of GI and the benefits that it provides, (ii) lack of data demonstrating benefits, costs and performance, and (iii) inadequate technical knowledge and experience, Matthews et al. (2015), Naumann et al. (2011), and Tzoulas et al. (2007). In all cases, community engagement and social consensus are considered vital for GI project implementation (Baptiste et al. (2015), Barnhill and Smardon (2012)).

This study concerns a complex GI plan that has been recently put into operation in the city of Drama located in the northeastern part of Greece, which aims at upgrading the environmental and bioclimatic conditions, as well as the economy of the city downtown area. The GI plan includes a number of different projects, such as the reconstruction of the area water supply and sewerage network, the construction of underground power supply cables, street lighting, road and pavement reconstruction, tree planting, special bioclimatic constructions for energy saving and reduction of air temperature during the hot summer months, and traffic signing. One major concern of the GI planning authorities is related to the impact of the GI plan on the different types of stakeholders and in particular, the entrepreneurs of the GI intervention area, as these are considered to be heavily affected due to the long-expected GI plan implementation period. Moreover, no social consensus or any other form of active community participation had preceded the GI plan implementation. In this context, the authors initiated a project to set up and operate a governance network to contribute to the successful implementation of the GI plan by increasing its social acceptability and community involvement. The network was established in 2018 and its members include the entrepreneurs of the intervention area, representatives of public and private agencies with concern to the GI plan, the residents of the intervention area, and a moderating team.

This article presents the preliminary results of the first of a series of social surveys carried out to explore the attitudes of the entrepreneurs of the GI plan area, who are expected to be heavily affected by the plan. Specifically, the research aims of this study were to (i) identify the factors comprising the attitudes of the entrepreneurs of the GI intervention area, (ii) classify the entrepreneurs according to their attitudes toward acceptance of the GI plan, and (iii) profile each group of entrepreneurs according to sociodemographic characteristics, such as age, educational status, family status, position in the business, and membership in a trade union.

2 Methods

Data were collected via face-to-face interviews on-site, using a structured questionnaire. Purposive sampling including the entire target population, that is all the entrepreneurs of the GI plan intervention area, was applied. The final sample consisted of the entrepreneurs that accepted to participate in the survey.

The questionnaire contained items organized in five sections: (i) sociodemographic questions, (ii) questions related to whether the entrepreneurs had participated in any prior consultation about the GI plan and how they would prefer to be informed about the GI plan, (iii) questions about the usefulness of the different GI plan items, (iv) questions about the potential benefits after the completion of the GI plan, and (v) questions about the potential impacts during the implementation stage of the GI plan. Questions of parts (iii), (iv), and (v) were Likert-type formatted with four scale ratings from very important to not important. The questionnaire was piloted in July 2018 to 15 entrepreneurs, Fowler and Floyd (2013), Naumann et al. (2011). The results of the pilot survey indicated that the instrument required no modification and therefore, the main survey was conducted in August and September 2018.

Joint dimension reduction and clustering were employed for data analysis. The reason for adopting this approach instead of the most often used tandem approach, which includes the application of Principal Component Analysis (PCA) for dimension reduction, followed by cluster analysis of the PCA scores, was to avoid potential problems reported in the literature, Markos et al.–Barnhill and Smardon (2019), regarding the achievement of an optimal cluster allocation. This has been mainly attributed to the fact that the two methods, PCA and cluster analysis, optimize different criteria. The data was analyzed with the R package `clustrd`, Markos et al.–Barnhill and Smardon (2019). We used the function cluspca, which corresponds to the Reduced K-means (RKM) algorithm. The aim of RKM is to achieve simultaneous dimension reduction and cluster allocation of entrepreneurs by maximizing the between-cluster variance in the reduced space Markos et al.–Barnhill and Smardon (2019).

The data set includes 15 variables: the reconstruction of the area water supply and sewerage network, the construction of underground power supply cables, the road and pavement reconstruction, tree planting, special bioclimatic constructions for energy savings and reduction of air temperature during the hot summer months, the street lighting, the traffic signing, an increase in customers' number at the end of the GI plan, turnover increase at the end of the GI plan, quality of life improvement at the end of the GI plan, aesthetic improvement at the end of the GI plan, a decrease in customers' number during the project implementation, turnover decrease, noise (Noise), and dust (Dust) also during project implementation.

Both the number of clusters and the number of dimensions were set to 3, based on ease of interpretation. The number of random starts was set to 100. The variables were centered and standardized prior to analysis and a varimax rotation of the factors was performed.

3 Results

In the Drama city, GI intervention area of a total of 138 mainly small businesses is in operation. A sample of 117, that is 84.78%, accepted to participate in the survey. Most of the participating entrepreneurs (86%) are self-employed business owners. Moreover, 22% of the corresponding businesses employ one person only, 7% employ 2 persons, and only 12% employ 3 or more persons. Most businesses (58%) have been in operation in the GI intervention area for over 20 years, while 73% of these for over 10 years. The shortest period of business operation recorded in the area was 1 year, while the longest period was 65 years. About 73% of the entrepreneurs stated that they were aware of the Drama city GI plan at the time of the survey. However, none of these knew what the plan actually involved, as they never participated in any consultation meeting during the setup and funding attraction phase of the GI plan. Almost 62% of the participants stated that they were willing to receive information about the on-going progress of the plan and about 68% of them stated that they would be willing to become members of the governance network. The type and number of businesses in the GI intervention area are presented in detail in Table 1.

The solution of RKM included 3 clusters in three dimensions. The entrepreneurs' attitude patterns toward acceptance of the Drama city GI plan reflected through the three clusters involved: (i) a negative attitude toward the changes induced by the GI plan, (ii) a utilitarian attitude related to the utility value involved in the GI constructions, and (iii) a positive attitude toward the changes induced by the GI plan. The variable scores or loadings on the three dimensions are displayed in Table 2. Table 3 shows the mean values for the three clusters of entrepreneurs in the reduced space.

Factor reliability was assessed by calculating Cronbach's alpha values for the key variables of each group. These were 0.873, 0.894, and 0.905, respectively. The results show that groups 1 and 3 are clearly separated from each other. Dimension 1 is mainly characterized by the variables related to the perceived impacts of the GI plan after its completion, namely the increase in customers' number, the turnover increase, the aesthetic improvement of the area, and the improvement of quality of life. Dimension 2 is influenced by the variables related to the perceived utility of the GI plan and particularly the reconstruction of the water supply and sewerage network and

Table 1 Number and type of businesses in the intervention area

Business sector	Number of enterprises
Wholesale and retail[a]	52
Catering services[b]	27
Activities related to human health and social care[c]	6
Agricultural and livestock products[d]	9
Financial and insurance services[e]	5
Other services[f]	17

[a]Pet stores, sanitary stores, church supplies, beverages, plumbing, carpets trade, clothing trade, shoe stores, furniture stores, butchers, electricians and electrical products shops, grocery stores, nuts store, kiosk, bakeries and pastry raw materials trade, pyrotechnics shop, warehouse, awning shop, and glassware/lumber trade
[b]Bakeries, pastry shops, cafes, hotels, restaurants, and grill shops
[c]Dental clinic, pediatric clinic, and pharmacies
[d]Agricultural and veterinary stores
[e]Insurance offices and accounting offices
[f]Barbers, real estate agency, foreign language schools, pawnshops, tattoo shop, OPAP store, tourist agency and photography

Table 2 Comprising the entrepreneurs' attitude patterns toward acceptance of the Drama city GI plan on the three dimensions

Variable scores	Dimension 1	Dimension 2	Dimension 3
Water supply and sewerage	0.0468	−0.7103	0.1349
Underground power supply	−0.0012	−0.3764	−0.0894
Road and pavement reconstruction	−0.0669	−0.0264	−0.4663
Tree planting	−0.1370	−0.1869	−0.1684
Special bioclimatic constructions	−0.2261	−0.2950	0.0825
Street lighting	−0.0247	−0.0024	−0.7328
Traffic signs	−0.0110	−0.2914	−0.2554
Aesthetic improvement	−0.3043	−0.2462	−0.0017
Increase of number of customers	−0.5727	−0.0517	0.2865
Turnover increase	−0.5953	0.2492	−0.0692
Quality of life improvement	−0.3839	0.0480	−0.1577
Decrease of number of customers	0.0016	−0.0174	−0.0444
Turnover decrease	0.0073	−0.0040	−0.0639
Noise	0.0118	−0.0953	−0.0245
Dust	0.0118	−0.0953	−0.0245

Within cluster sum of squares by cluster: 114.1421 71.9853 1.0292 (between_SS / total_SS = 76.05%)

Table 3 Clusters of entrepreneurs according to their attitudes toward acceptance of the Drama city GI plan. Values represent the cluster means

	Perceived impacts after the end of the GI plan	Perceived utility of the GI plan infrastructure	Perceived impacts during the GI plan implementation
Negative to change	1.0173	0.3036	0.2221
Utilitarians	−2.4879	0.1862	0.1849
Positive to change	−4.8822	−6.5653	−5.0702
No. of entrepreneurs ($n = 99$)	73	22	4

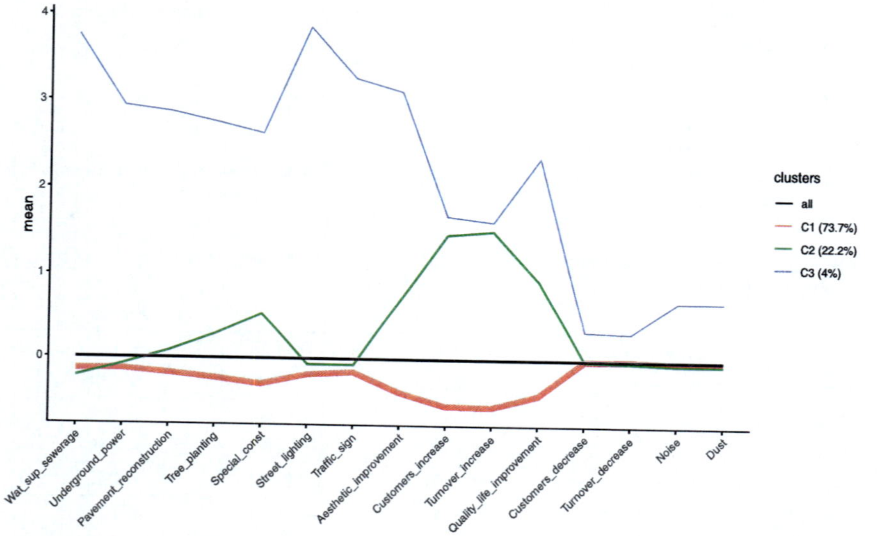

Fig. 1 Parallel coordinate plot of the three cluster means (C1: negative to change, C2: utilitarians, C3: positive to change)

construction of power supply cables, whereas dimension 3 is characterized mostly by the variables related to the road and pavement reconstruction and the street lighting, which are considered to generate the most negative economic impacts, such as the decrease in customers' number and turnover decrease and the most negative health and environmental impacts due to increased levels of noise and dust. In Fig. 1, the variable means in each cluster provide a better insight into the differences between the identified attitude patterns of the entrepreneurs toward acceptance of the GI plan.

The cluster named "negative to change" entrepreneurs (cluster 1) contains 73.7% of the respondents. They are pessimists and oppose the implementation of the GI plan because they think that they will be affected greatly both from the economic and environmental points of view. In particular, they perceive that their business turnover and the number of their customers will decrease during the GI implementation phase,

Table 4 Entrepreneurs' profile according to sociodemographic characteristics

		Negative to change	Utilitarians	Positive to change	Chi-Square
	Percent entrepreneurs				
Position in the business	Owner	64%	31%	4%	$\chi^2 = 52.654$, df = 1
	Employee	73%	20%	7%	$p < 0.001$
Membership in a trade union	Yes	43%	57%	0%	$\chi^2 = 75.922$, df = 1
	No	67%	27%	5%	$p < 0.001$
Age	18–29	80%	20%	0%	$\chi^2 = 65.154$, df = 3
	30–44	77%	20%	2%	$p < 0.001$
	45–64	53%	41%	6%	
	>65	67%	17%	17%	
Educational status	Primary school	33%	33%	33%	$\chi^2 = 56.846$, df = 2
	High school	56%	44%	0%	$p < 0.001$
	Lyceum	69%	24%	8%	
	Bachelors	67%	33%	0%	
	Post-graduate	50%	50%	0%	
Marital status	Married	61%	34%	4%	$\chi^2 = 91.320$, df = 4
	Single	75%	17%	8%	$p < 0.001$
	Divorced	70%	30%	0%	

while at the same time suffering from dust and noise. The entrepreneurs, who were named "utilitarians" (cluster 2), comprise 22.2% of the respondents. Their attitudes toward the Drama city GI plan is considered mostly positive in particular to the bioclimatic GI as they value their usefulness for the intervention area. They also think that there will be positive impacts on the intervention area after the GI project completion, although they share the same concerns about the negative economic and environmental impacts during the implementation phase of the GI plan. The named "positive to change" entrepreneurs comprise the smallest part of the respondents, only 4%. They are purely optimists, they like changes, and they value highly the positive impacts that they believe will accrue for the intervention area after the GI plan will be completed. These involve mainly the improvement of their economic status and quality of life and aesthetic improvement.

Finally, each group of the entrepreneurs was profiled in relation to sociodemographic characteristics (Table 4). The most negative ones to change were business employees, aged between 18 and 44, singles, and bachelor degree holders. On the other hand, utilitarians were aged between 45 and 64, high school graduates, married, and members in trade unions. Finally, the positive ones to change appeared to be older entrepreneurs (over 65 years of age), singles, and primary school graduates.

4 Conclusions

The application of Reduced K-means indicated three groups with different attitude patterns of entrepreneurs operating in the Drama city GI intervention area toward acceptance of the GI plan: the negative to change, the utilitarians, and the positive to change. The majority of entrepreneurs in the area appeared negative due to the perceived negative economic, health, and environmental damages during the GI plan implementation taking also into account the expected long duration of the project. About one-fourth of the respondents were mostly positive toward the GI plan, placing particular value on the perceived usefulness of GI utilities, although they shared the same skepticism with the negative to change group regarding the perceived negative impacts during the GI plan implementation. A very small part of the respondents appeared very optimistic and positive to change due to the perceived potential positive economic and quality of life benefits after the end of the GI plan. The study was conducted within the governance network that was set up to promote social participation and acceptance of the Drama city GI plan through an adaptive process. The results of the study will be used to implement appropriate awareness and communication actions in order to increase the entrepreneurs' engagement in the network and the social acceptability of the GI plan. The study results will also be used in combination with follow-up surveys to monitor potential changes in the entrepreneurs' social acceptability of the Drama city GI plan implementation and could be useful for planning agencies in other places, where GI plans are to be implemented and stakeholders' involvement is sought.

Acknowledgements The research reported in the current article received funding from the European Regional Fund and Eastern Macedonia and Thrace NSRF (ESPA) 2014–2020.

References

Baptiste, A.K., Foley, C., Smardon, R.: Understanding urban neighborhood differences in willingness to implement green infrastructure measures: a case study of Syracuse. NY. Landsc Urban Plan. **136**, 1–12 (2015)

Barnhill, K., Smardon, R.: Gaining ground: green infrastructure attitudes and perceptions from stakeholders in Syracuse. New York. Environ Pract. **14**(1), 6–16 (2012)

Chen, W.Y., Jim, C.Y.: Assessment and valuation of the ecosystem services provided by urban forests. For. Ecol. Manag. **53–83** (2008)

Dunnett, N., Kingsbury, N.: Planting Green Roofs and Living Walls. Timber Press Portland, OR (2008)

Fowler Jr., Floyd J.: Survey Research Methods. Sage Publications (2013)

Hair, J.F., Black, W.C., Babin, B.J., Anderson, R.E., Tatham, R.L.: Multivariate Data Analysis. Pearson Education Limited, Essex (2014)

Institute, Landscape: Green Infrastructure: Connected and Multifunctional Landscapes. Landscape Institute, London (2009)

Markos, A., Iodice D'Enza, A., van de Velden, M.: Beyond tandem analysis: joint dimension reduction and clustering in R. J. Stat. Softw. **91**, 1–24 (2019)

Matthews, T., Lo, A.Y., Byrne, J.A.: Reconceptualizing green infrastructure for climate change adaptation: barriers to adoption and drivers for uptake by spatial planners. Landsc Urban Plan. **138**, 155–163 (2015)

Mell, I.C., Henneberry, J., Hehl-Lange, S., Keskin, B.: To green or not to green: establishing the economic value of green infrastructure investments in The Wicker. Sheffield. Urban For Urban Green. **18**, 257–267 (2016)

Naumann, S., Davis, M., Kaphengst, T., Pieterse, M., Rayment, M.: Design, implementation and cost elements of Green Infrastructure projects. Final report, European Commission, Brussels, 138 (2011)

Tzoulas, K., Korpela, K., Venn, S., Yli-Pelkonen, V., Kaźmierczak, A., Niemela, J., James, P.: Promoting ecosystem and human health in urban areas using Green Infrastructure: a literature review. Landsc Urban Plan. **81**(3), 167–178 (2007)

Spatial Perception for Structured and Unstructured Data In topological Data Analysis

Yoshitake Kitanishi, Fumio Ishioka, Masaya Iizuka, and Koji Kurihara

Abstract Recent years have witnessed the accumulation of vast amounts of data and information. It is difficult to capture the characteristics of these data spatially or visualize them robustly and stably with respect to data updates and increases using conventional methods. The purpose of this study is to systematically visualize the relationships among drugs using diverse information. While studies have conducted visualization research using structured data, such as chemical descriptors, research has not yet been performed from comprehensive viewpoints using unstructured data on efficacy, adverse events, and other phenomena. Therefore, we use a topological data analysis mapper and a spatial perception method to obtain and visualize data based on the integrated principal component score of quantitative and qualitative data. Consequently, a network composed of characteristic clusters according to drug class was shown. Findings show that heterogeneous compounds in the cluster may indicate the potential for drug repositioning. Our proposed method is an effective means of obtaining new knowledge of pharmaceuticals.

Keywords Topological data analysis · Topological data analysis mapper · Spatial perception · Quantitative and qualitative data · Structured and unstructured data

Y. Kitanishi (✉) · F. Ishioka · M. Iizuka · K. Kurihara
Okayama University, Okayama, Japan
e-mail: ykitanishi@s.okayama-u.ac.jp

F. Ishioka
e-mail: fishioka@okayama-u.ac.jp

M. Iizuka
e-mail: iizuka@okayama-u.ac.jp

K. Kurihara
e-mail: kurihara@okayama-u.ac.jp

T. Chadjipadelis et al. (eds.), *Data Analysis and Rationality in a Complex World*, Studies in Classification, Data Analysis, and Knowledge Organization,
https://doi.org/10.1007/978-3-030-60104-1_12

1 Introduction

Determining how to create satisfactory hypotheses is the key to scientific progress. The vast data and information accumulated in recent years should be used efficiently for hypothesis creation. However, it has become difficult to capture features of the data spatially and visualize the data robustly against data updates and increase using conventional methods (e.g., hierarchical cluster analysis). Therefore, the topological data analysis (TDA) mapper, first introduced by Singh (2007), is attracting attention as a new visualization method. The TDA mapper creates an easy-to-understand graph with the same topological structure as the original data and displays data features. Nevertheless, there is the issue of target data. Creating innovative hypotheses using the traditional approach of gaining knowledge only from quantitative data stored in structured databases is becoming challenging. Therefore, to increase the amount of information, qualitative data such as texts and images are also included in the analysis. The problem-solving approach to quantify qualitative data for feature extraction and feature value calculation has a variety of applications and has several calculation methods. Luhn (1957) was the first to use term weighting for quantifying qualitative data. The quantification of natural language was studied by Bates (1995). Torres-Tramon (2015) applied the TDA mapper to text data as qualitative data. In the healthcare industry, the TDA mapper has been applied actively, especially for big data such as genome data (Rizvi et al. 2017) and magnetic resonance imaging (Nielson et al. 2015; Saggar et al. 2018). The pharmaceutical industry has similar data issues; drugs have been classified and target variables have been predicted using chemical descriptors and physical property data, which are quantitative. However, to ensure discontinuous changes such as new mechanisms and new indications in the pharmaceutical industry, it is important to promote the use of more information, including various qualitative data. In other words, it is useful to integrate and classify text data as phenotypic data, such as drug indications and adverse event information, and chemical descriptor data as structural data of the drug itself. Simultaneously evaluating the origin of the drug and the phenotype in humans will help in accumulating more knowledge and hypotheses. Nevertheless, previous studies have analyzed qualitative (unstructured) and quantitative (structured) data separately. Therefore, in this study, we integrate qualitative and quantitative data and apply the TDA mapper. In other words, the purpose of this study is to classify drugs continuously based on the combined information from two types of data, and show the possibility of developing new hypotheses. We report the results of the spatial perception using the TDA mapper on quantitative data (mainly chemical descriptors and physical data on drugs) and qualitative data (mainly text data, including drug information).

2 TDA Mapper

The TDA mapper algorithm is a visualization method for analyzing huge data. In addition, a classification method is used as a part of the data analysis process to distinguish between several groups. Thus, using the mapper algorithm techniques, huge data sampled from shapes of known topology can be identified and visualized robustly, even in the presence of white noise. This section introduces a technique that is commonly researched and developed for TDA (Singh et al. 2007; Zomorodian and Carlsson 2005). The TDA mapper technology makes it easy to visually understand the characteristic features of data that capture the details of data shapes as numerical values. For example, the TDA mapper is suitable for scenarios wherein features are extracted from huge data. The TDA mapper process comprises four steps: inputting the target data, adjusting the distance function, using the filter function, clustering, and visualizing each parameter. First, input the data as in Step 1. The input is N data with M-dimensional features. Next, in Step 2, the distance between the two points in the input data is obtained, and an N-by-N distance matrix, X, is created. Moreover, using a filter function for this distance matrix, mapping can be performed in a low dimension. The value obtained by mapping in a low dimension is called the filter value. Then, adjust the parameters of the filter function used at this time. In Step 3, the data set is divided into each interval based on this filter value. In the TDA mapper, clustering is performed at each interval. Set an overlap when dividing the intervals. By setting the overlap, during visualization, it is possible to draw a line between clusters with overlapping nodes and show the connection between them. In Step 4, clustering is performed for each interval, and a line is drawn between clusters containing the same nodes to represent the connection. Note that this method allows overlap, therefore, if one object is in the overlap section, it belongs to multiple nodes. Figure 1 shows the analysis process (Chazal and Michel 2017; Singh et al. 2007) with specific parameters as an example.

We discuss the comparison of the hierarchical clustering and stability of the results of the TDA Mapper. As an example, the result of the iris data clustering of 150 records and 4 variables are shown in Fig. 2. The figure shows how these results

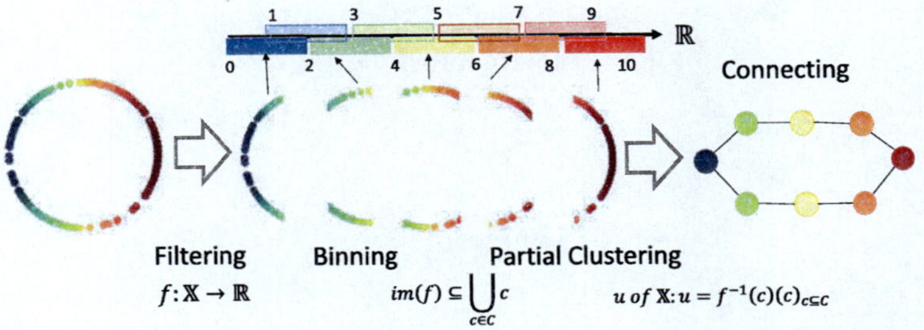

Fig. 1 Topological data analysis mapper process: The filtering parameters are as follows. Range: 0–10, Interval length: 2, Intervals: 5, Overlap: 50pct

change when records are deleted at random. The appearance of the TDA mapper does not change much when the records are deleted. On the contrary, the results of hierarchical clustering vary greatly. In other words, the TDA mapper classification method allows for overlap; thus, the results are stable.

3 Application to Drug Data

This section describes the process and data applied by the TDA mapper. The English version of Wikipedia was adopted as the source of qualitative data, and drug data stored in DrugBank was converted into quantitative chemical descriptor data using Mold2. The details of the data are as follows:

- Wikipedia (Ver. 01Oct2017): www.dumps.wikimedia.org: only articles on drug information,
- National Drug Code Directory: www.fda.gov/drugs/drug-approvals-and-databases/national-drug-code-directory (NDC Database File, Ver. 01Jul2019): the list of drugs to recognize articles on drugs published on Wikipedia, and
- DrugBank (Ver. 5.1.4): www.drugbank.ca/: structured data for calculating chemical descriptors.

Details of the software used for data processing and analysis are as follows:

- Python Mapper: www.danifold.net/mapper (Müllner and Babu 2013): for implementing the TDA mapper algorithm in Python,
- Tidytext: www.cran.r-project.org/web/packages/tidytext/index.html: R package for text mining, a

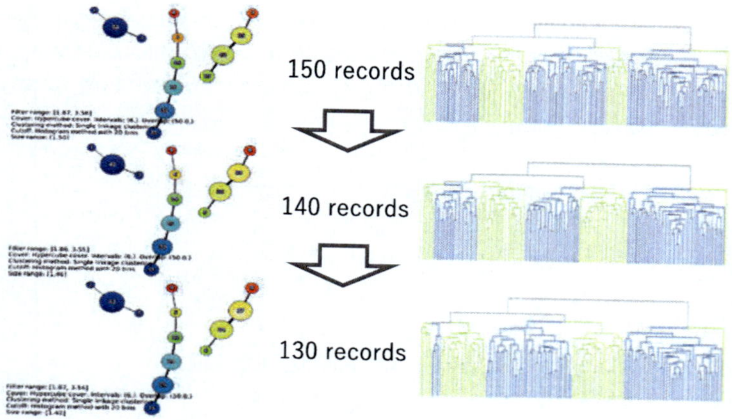

Fig. 2 Topological data analysis mapper versus hierarchical clustering [*Data* Iris data Fisher 1936]

- Mold2: www.drugbank.ca: for calculating chemical descriptors from structure data, and
- Spotfire: www.drugbank.ca: for summarizing the result.

3.1 *Qualitative Data on Drugs*

Information gathered from Wikipedia is used as text data that includes the history of each drug and other important peripheral information. Specifically, about 1,000 pharmaceutical articles are extracted from an XML file with 5 million articles and a table (https://dumps.wikimedia.org/enwiki/) with information on the category to which the articles belong. The extracted text data were quantified by excluding stop words. When quantifying, the local weight is multiplied by the term frequency (*TF*) value, which represents the frequency of each word in the document, and the global weight inverse document frequency (*IDF*) value, and it is normalized by the length of the sentence. The $TF - IDF$ values are used to capture characteristic words. The index was proposed by Jones Sparck Jones (1972) and discussed in detail by Salton Salton and McGill (1983). The *IDF* value and the normalized $TF - IDF$ value are expressed by Eqs. (1) and (2), respectively.

$$IDF = \log_2 \frac{N}{n_i} + 1 \tag{1}$$

(N:Number of all documents, n_i : Number of documents containing the words)

$$TF - IDF = \frac{TF_{ij} \times IDF_i}{\sqrt{\sum_{i=1}^{m} \left(TF_{ij} \times IDF_i\right)^2}} \tag{2}$$

(Normalized $TF - IDF$ for word j in document i)

3.2 *Quantitative Data on Drugs*

The chemical structure of compounds is quantified to capture the chemical structure characteristics of pharmaceutical data. DrugBank (collecting more than 13,000 compounds, including unapproved drugs) is used as a database of drug compounds approved by the US Food and Drug Administration (FDA). Among them, pharmaceutical drug compounds with low molecular weight that have been approved are extracted (2,596 compounds), and 777 descriptors are calculated using the chemical descriptor conversion software, Mold2, which is published by the FDA. By using these, an information matrix of compounds (rows) and chemical descriptors (columns) was created.

3.3 Integration Between Qualitative and Quantitative Data

Wikipedia data as qualitative data and DrugBank data as quantitative data were integrated limitedly to those with matching drug names to classify compounds by fusing and grasping the heterogeneous data. Figure 3 shows the data integration process. To integrate the two types of data equally, they were normalized using the score calculated by principal component analysis and integrated using the scores up to the tenth principal component.

3.4 Result of the Classification

This study conducted principal component analysis after standardization of each data, as well as mapping using the 10 principal components (see Fig. 4a and b). Figure 4c shows the result of applying the TDA mapper to the data obtained by integrating the two types of data via principal component analysis. The data shape now included mixed elements of Wikipedia and DrugBank data. Figure 4d illustrates the composition ratio as a pie chart according to the indications of the component drugs. Moreover, Fig. 5 shows the composition of each part as a pie chart to confirm whether each part of the shape contains many indication drugs and has the characteristic composition. It can be confirmed that the infection treatment drugs are gathered in the cluster near the edge (i.e., their proportion is high within the cluster).

In Fig. 6, the principal component scores were plotted for this cluster to confirm the degree of influence of Wikipedia and DrugBank data. While certain words are thought to be affected, the influence of structural information is also confirmed. For example, the drug indicated by the red color is for treating diarrhea and is said to have antibacterial effects.

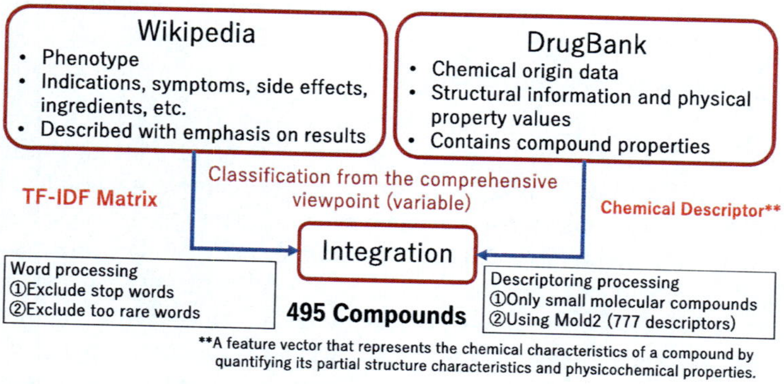

Fig. 3 Process of integration between the data from Wikipedia and DrugBank

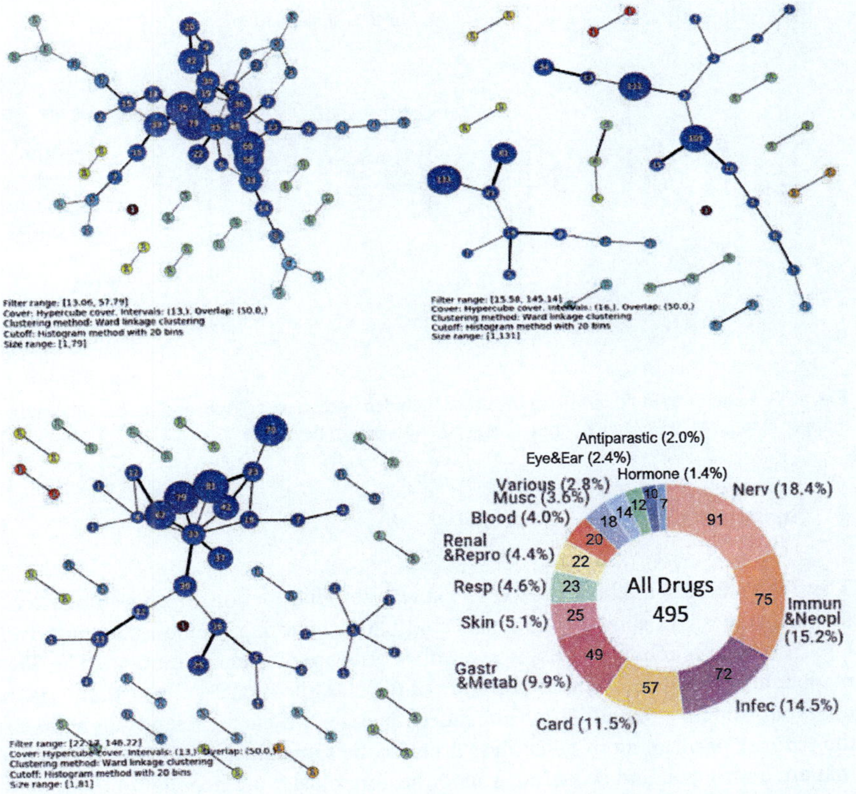

Fig. 4 **a** Application to Wikipedia data. **b** Application to DrugBank data. **c** Application to integrated data. **d** Pie chart showing the composition ratio

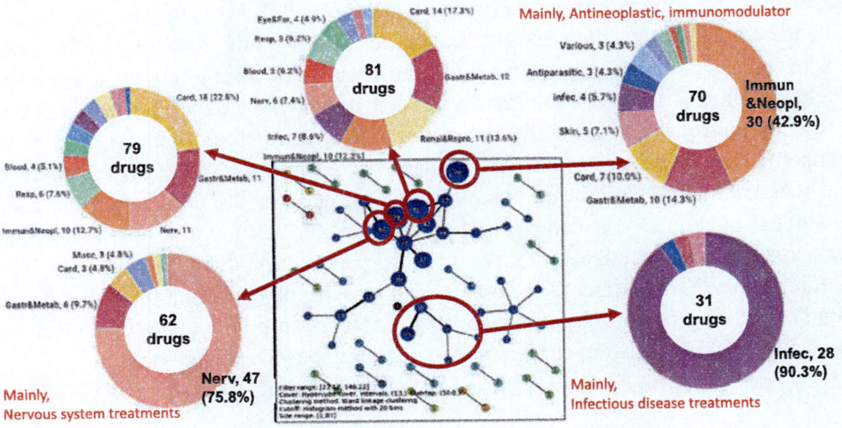

Fig. 5 Percentage of indication-specific drugs included in each part of the topological data analysis mapper results

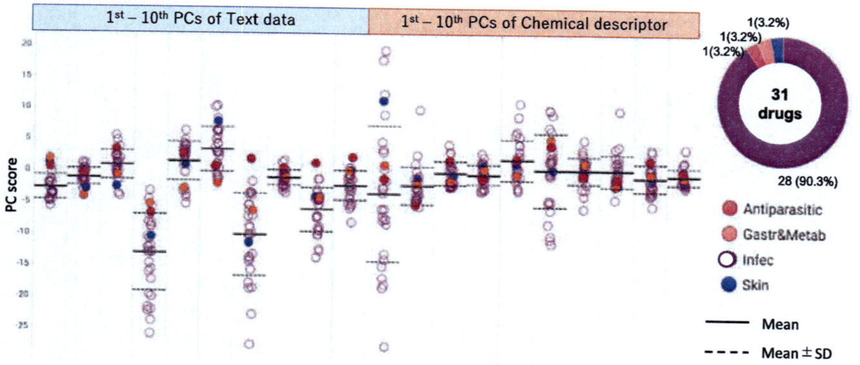

Fig. 6 Percentage of indication-specific drugs included in each part of the topological data analysis mapper results: PC = principal component, SD = standard deviation

4 Summary

The TDA mapper method is used to express the distribution of high-dimensional data as shapes. As shown in the case study, the clustering method allowing overlap, the shape is relatively robust even if records are omitted at random. The TDA mapper also facilitates the interpretation of the classification results visually. As the drugs are roughly classified by indication disease, we can efficiently hypothesize the purpose or meaning of other class drugs in the same cluster. Alternatively, drugs that are ungrouped and scattered around the core cluster are peculiar outliers in the article and/or structure, and it is possible to consider why they are outliers, and then develop hypotheses. Therefore, the TDA mapper results based on Wikipedia data (on drug articles) and the chemical structure (based on DrugBank and Mold2) are reasonable. In addition, as with hierarchical clustering, it is necessary to consider the distance calculation method and parameters according to the data and application. It is necessary to devise parameters to set the shape to be as continuous as possible. Then, the "shape" and "position" are interpreted as the whole data. The results of this study can clarify the positional relationship among drugs (compounds). The TDA mapper is characterized by its ability to classify continuously rather than discretely, and it is useful for hypothesis planning that emphasizes the connection between meanings. Specifically, it can be applied to elements, such as the positional relationships of pharmaceuticals (compounds) and patents, and drug repositioning. The feature of the TDA Mapper is to connect and classify objects continuously without clearly grouping objects. In other words, the relationship between clusters can be considered, which is useful for hypothesis planning based on huge data.

References

Bates, M.: Models of natural language understanding. Proc. Natl. Acad. Sci. USA **92**(22), 9977–9982 (1995)

Chazal, F., Michel, B.: An introduction to Topological Data Analysis: fundamental and practical aspects for data scientists, https://arxiv.org/pdf/1710.04019.pdf (2017)

Fisher, R. A.: The use of multiple measurements in taxonomic problems. Ann. Eugen. **7**, Part II, 179–188 (1936)

Luhn, H.P.: A statistical approach to mechanized encoding and searching of literary information (PDF). IBM J. Res. Dev. **1**(4), 309–317 (1957)

Müllner, D., Babu, A.: Python Mapper: An open-source toolchain for data exploration, analysis and visualization (2013). http://danifold.net/mapper/introduction.html

Nielson, J., Paquette, J., Liu, A., et al.: Topological data analysis for discovery in preclinical spinal cord injury and traumatic brain injury. Nat. Commun. **6**, 8581 (2015)

Rizvi, A., Camara, P., Kandror, E.K., Roberts, T.J., Schieren, I., Maniatis, T., Rabadan, R.: Single-cell topological RNA-seq analysis reveals insights into cellular differentiation and development. Nat. Biotechnol. **35**, 551–560 (2017)

Saggar, M., Sporns, O., Gonzalez-Castillo, J., et al.: Towards a new approach to reveal dynamical organization of the brain using topological data analysis. Nat. Commun. **9**, 1399 (2018)

Salton, G., McGill, M. J. Introduction to Modern Information Retrieval. McGraw-Hill Computer Science Series, vol. XV, 448 p. McGraw-Hill, New York (1983)

Singh, G., Mémoli, F., Carlsson, G. E.: Topological methods for the analysis of high dimensional data sets and 3D object recognition. In: Eurographics Symposium on Point-Based Graphics, pp. 91–100 (2007)

Sparck Jones, K.: A statistical interpretation of term specificity and its application in retrieval. J. Doc. **28**(1), 11–21 (1972)

Torres-Tramon P., Hromic H., Heravi, B.R.: Topic detection in Twitter using topology data analysis. In: Daniel F., Diaz O. (eds.) Current Trends in Web Engineering. ICWE 2015. Lecture Notes in Computer Science, pp. 186–197 (2015)

Zomorodian, A., Carlsson, G.: Computing persistent homology. Discrete Comput. Geom. **33**(2), 249–274 (2005)

Text, Content and Data Analysis of Journal Articles: The Field of International Relations

Nikos Koutsoupias and Kyriakos Mikelis

Abstract Term frequencies is the basic, if not the main, measure that springs out, during the process of mapping latent information in a text corpus. This paper addresses the issue of exploring a set of textual documents based on their metadata and term frequencies, by introducing the mixed use of text mining and data analysis methods for analyzing social science journal articles. In particular, this survey links the quantitative research of scientific discourse—through specific tools of data analysis—with research on the development of a scientific field, namely International Relations. Preliminary results on field-related published journal articles demonstrate the effectiveness of the proposed methodology.

Keywords Text mining · Multiple correspondence analysis · Term frequencies · Content analysis · International relations

1 Introduction

This study explores the issue of proper clustering of a collection of documents based on their metadata and word frequencies, by employing a combined use of text mining and data analysis methods to review social science journal articles. Specifically, this paper connects scientific discourse quantitative analysis to work on the growth of the specific field International Relations (IR), which widely explores the relationships between states and organizations or groups that impact or are affected by states. The chosen case-study corpus originates from 'Perceptions: Journal of International Affairs', a Turkish journal of International Studies published in English.

In light of huge amounts of journal articles, academic papers, etc., new taxonomies are needed toward a better understanding of the status or trajectory of themes, content

N. Koutsoupias (✉) · K. Mikelis
University of Macedonia, Thessaloniki, Greece
e-mail: nk@uom.gr

K. Mikelis
e-mail: kmikelis@uom.edu.gr

T. Chadjipadelis et al. (eds.), *Data Analysis and Rationality in a Complex World*, Studies in Classification, Data Analysis, and Knowledge Organization,
https://doi.org/10.1007/978-3-030-60104-1_13

and communication patterns. Such a process is presented here through a combination of text mining and multivariate data analysis methods. Eventually, there is growing capacity in the interpretation and discussion of the results on article clusters as well as in the presentation of quantitative and qualitative findings.

2 Corpus Context

Similarly to other disciplines, IR is subject to a multifaceted discussion regarding the conceptual assessment of the observed phenomena and its sociological or (meta)theoretical characteristics. This discussion includes the emergence and evaluation of the role of discourse and the margin for the (re)interpretation of concepts, through various methods referring to genealogy, deconstruction or contextual framing. The complexity of the matter is also illustrated from the vantage point of Linguistics, taking into consideration the emphasis of the field to specialized translation and to how international (e.g., legal) language semantics emerge and evolve through the translating process (Saridakis 2013) as well as to the utilization of text corpora for the development of a multilingual terminological resource in IR subfields, like Geopolitics (Saridakis 2016).

Indeed, the rising IR discipline's self-reflection over its identity led to the emergence of the subfield of historiography and sociology of IR (Gofas et al. 2018; Schmidt and Guilhot 2019). Regarding the development of the field, the relevant empirical research includes methods within the qualitative and/or quantitative analysis of various aspects, such as indicatively (a) the characteristics and opinions of scholars, (b) the validity or popularity of hypotheses or arguments, (c) the themes, the (meta)theoretical content and the reference patterns in journals or textbooks (Kristensen 2012; Maliniak et al. 2018), and (d) teaching programs. These are complemented employing qualitative methods, addressing myths, toward the direction of the reconstruction or deconstruction of reified concepts (e.g., 'Westphalia', i.e., the anarchic international system of sovereign states, or 'security' or 'diplomacy'). While these efforts are worthy, they currently suffer limitations, as the input data is mostly qualitative by nature, aggregated and subjective.

According to a historical account of the use of content analysis in IR, the period from the 1940s to the 1960s signifies the initiation of the first wave which predominantly if not exclusively included quantitative and manual terms. The second one is traced in the twenty-first century, characterized by the emphasis to computer-aided content analysis. Subsequently, the possibility for a fully integrated content analysis is recognized, in the sense of utilizing all shorts: quantitative, qualitative, manual and computer-aided (Pashakhanlou 2017: esp. 459). This work builds on past research presented by Koutsoupias and Mikelis (2019) focusing solely on journal article term and metadata analysis.

3 Data and Methodology

In this context and with confidence for the use of multivariable analysis, the adoption of the respective methods for the exploration of the IR literature is here advanced, in light of the explosion of openly available texts and online documents and of the need to 'let the data speak'.

In particular, this work investigates terms and metadata related to corpora from the scientific journal Perceptions: Journal of International Affairs by employing text mining and methods of multivariate data analysis (Principal Component Analysis, Hierarchical Clustering, and simple and Multiple Correspondence Analysis). The explored journal is linked to a research center of the Turkish Ministry of Foreign Affairs (Center for Strategic Research/Stratejik Araştırmalar Merkezi, SAM). The choice of a Turkish journal is grounded on the particular interest drawn in the development of IR in Turkey. Overall, the relevant scientific work was Turkish-centric, with emphasis on security, but it also emerged in a differentiated manner (indicatively Bilgin and Tanrisever 2009; Yazgan 2012; Aydin and Yazgan 2013). Moreover, the extensive use of quantitative methods in the relevant research of both domestic IR and Turkish foreign policy was vividly proposed. The central assumption here is that, despite the noticeably increased international presence, the country's IR community continues to be characterized as 'fragmented', without sufficient active participation in scholarly debates (Aydinli and Biltekin 2017).

At first, a decision had to be taken, whether to include parts of the contents or the entire corpus in each issue. We concluded that it would be appropriate to focus only on complete articles and omit editorials and book reviews since the object in question, i.e., a record in the analyzed data table, had to be a contributor's comprehensive view, extended thesis or argument.

Since modern multivariate exploratory methods are generally acceptable in a wide variety of similar applications, this study proposes the use of MCA (Greenacre and Blasius 2006; Le Roux and Rouanet 2010) as the main step for visualizing the inherent structure in our data, which is not directly observable but exists in dormant form. MCA analyzes the data in factorial axes that may be considered new, composite variables/parameters summarizing our multivariate data table into a 2- or 3-dimensional space, with the least possible loss of information. This method allows for the simultaneous depiction of both variables/parameters and objects on a factorial map satisfying certain optimization criteria. On the basis of these criteria, the produced coordinates of each object and variable/parameter category may form an intermediate table ready to be used for constructing taxonomies on objects, i.e., the examined articles. This, combined with text mining, may aid in revealing latent structures among variables/parameters and articles, without making any a priori assumptions about the procedures (the article types in our case) by which the input table is constructed. The selected implementation of the method, via the R package FactoMineR by Kristensen (2012), generates a dendrogram suggesting clusters along with tables for the taxonomy's interpretation, in order to better understand latent structures.

As mentioned previously, the examined data set was formed by employing both metadata and content analysis of journal article related information in the field of IR. In order to form the article matrix, content analysis was initially carried out so as to extract field-related metadata. The emerged categorical data vector was of the form:

$$V_j = \{\text{Au, Ar, Th, Re, Em}\} \text{ (Rel. 1)}$$

where

Au (origin of author): Au.T/Au.nT, Turkish or non-Turkish, reflected in author's address information (following the common by now practice of emphasizing workplace rather than national origin). In a few cases, the phenomenon 'author with a presumably Turkish name and an affiliation with a non-Turkish institution' is observed, as well as vice versa ('author with a presumably foreign name and an affiliation with a Turkish institution'). The two dimensions of the phenomenon are almost offset, with some primacy of the first dimension.

Ar (scholarly or political orientation of author): Ar.Sc/Ar.P, based on declared affiliation. An academic is considered as a politician if (s)he contributes with the latter capacity. For the period considered, the participation of politicians is extremely small, compared to earlier phases of the journal.

Th (Turkish- or non-Turkish-centered agenda): ThT/Th.nT, primarily based on the inclusion of the term Turk in the title of the article.

Re (adoption of realist IR perspective or not): Re/nRe, based on the explicit or implicit acceptance (or, in reverse, critique) of IR realism or more broadly state-centrism. Realism considers IR as predominantly interstate relations, primarily driven by the power, survival and self-help considerations of states.

Em (orientation of article): EmTh/EmPo, reflected in the degree of theoretical embeddedness or of the emphasis on policy-relevant issues and on empirical/historical analysis. Articles are classified as 'theory', as long as they deal with an issue theoretically or as they record empirical material with an initial and extensive theoretical reference. They are classified as 'policy' if they focus on policy issues or empirical/historical analysis, without particular discussion of their theoretical framework.

Following the above schema, two trained coders, a field expert, as well as an experienced reader recognized and composed a V_j tuple for each examined article with a high reported agreement (Krippendorff's alpha, $K\alpha = 0.92$). In order to include term-related content in the analysis, in the next phase of input data construction, an additional categorical variable was concatenated pertaining to terms cluster (Tc). The aim was to find latent formations in article corpora, considering both manuscript content and metadata. In an attempt to integrate a parameter related to text content and thematic orientation, the corpus was preprocessed (Vijayarani et al. 2015), with functions such as conversion to plain lower-case text, formation of verbal roots and deletion of non-alphanumeric characters, punctuation, numerals, intermediate terms—stopwords and gaps.

Text mining techniques were also utilized to extract frequently related terms in the corpus of IR articles. For this, the well-known term frequency times inverse document frequency or tf-idf (Joachims 1996) algorithm was applied (Salton 1989), as in all text mining procedures described in this study, utilizing R package tm (Feinerer et al. 2008). Thus, out of more than 13 thousand terms in the corpus, this study focused on the top 287.

On the resulting 125 × 287 article-term matrix (tf-idf matrix), a Principal Component Analysis (PCA) was next performed, using function PCA in R package FactoMineR (Kristensen 2012). The results of PCA were validated and found to be similar to those of simple CA when applied to the same dataset. Subsequently, Hierarchical Clustering on PCA coordinates as suggested in Zhao et al. (2005) and MCA were employed from the same package and their output was investigated with R package FactoInvestigate (Thuleau and Husson 2017).

Next, the inertia of the first dimension was examined in order to determine whether there are strong relationships among variables (terms). PCA showed that the first two dimensions express 8.29% of the total matrix inertia; that means that 8.29% of the individual (or variable) cloud total variability is explained by the 1st PCA plane. This is a very small percentage and the first plane represents a small part of the data variability, although significant, as it is greater than the reference value that equals 4.09% (the reference value is the 0.95-quantile of the inertia percentages distribution obtained by simulating $n > 500$ data tables of equivalent size on the basis of a normal distribution).

PCA output was next passed to function HCPC for Hierarchical Clustering (on the Euclidean distance dissimilarity matrix using Ward's method, followed by k-means consolidation) so as to obtain the terms cluster for each object (IR journal article). Hence, six clusters were suggested by the method, focusing on IR terms as follows (sorted from the most popular):

Tc1 ($n = 24$): characterized by high occurrence of terms like market, invest, economy, industry, supply, trade, energy, exports, technology and finance. The cluster combines issues of economic, energy and regional cooperation, with frequent non-Turkish reference (Th.nT).

Tc2 ($n = 38$): with most common terms such as alliance, tension, stability, Iraq, PKK, Bush, Obama, NATO, security and neighbor. It covers foreign policy and international political issues, sometimes with reference to bilateral relations and national security, with a focus on Turkey (Th.T).

Tc3 ($n = 16$): showing above-average frequency of the terms migration, labor, German, legal, abroad, action, immigration, law, visa and citizen. The cluster is overwhelmingly Turkish-centric (Th.T, in Turkey or Turkish immigrants in the EU and especially in Germany), with the issues of migration, social integration, mobility and freedom of movement (as well as the Cyprus issue, with respect to two articles) being dominant.

Tc4 ($n = 31$): that emphasizes more on subjects that mostly include terms like history, violence, civil, social, democracy, minority, ethnography, discourse, world and constitution. It relates to topics such as identity, democracy, intercultural relations and criticism of Eurocentrism (Thry, nRe), with relatively little Turkish-centerd focus (Th.nT).

Tc5 ($n = 12$): expressing interest in IR areas that put emphasis on the terms individual, analysis, scholar, method, attitude, theory, inform, psychology, behavior and Islam. It involves thematics related to foreign policy analysis, political communication, political psychology (e.g., leadership, images, discourse), with a slight lead in emphasis in Turkey.

Tc6 ($n = 4$): focusing mostly on the IR terms Nagorno, arm, Azerbaijan, normalize, protocol, statement, principle, territory, conflict and Arab. It basically includes relations, conflict and disputes between Armenia and Azerbaijan (3/4 articles, while the other refer generally to conflicts).

Thus, the resulting articles vector V_j' becomes

$$V_{j'} = \{\text{Au, Ar, Th, Re, Em, Tc}\} \text{ (Rel. 2)}$$

(see V_j in Rel. 1) where

Tc (terms cluster): one of Tc1...Tc6 frequent term clusters. Subsequently, in order to extract latent knowledge from the total corpus of IR articles under consideration, MCA was performed using the corresponding R function (FactoMineR package) on the matrix of $V_{j'}$ vectors (see Rel. 2).

In the results, the first MCA axis expressed 35.82% of the total matrix inertia, which means that 35.82% of the individual (or variable) cloud total variability is explained by the axis. The first axis represents part of the total inertia, however, this value is greater than the reference value that equals 28.6%; the variability explained by this axis is thus significant (the reference value is the 0.95-quantile of the inertia percentages distribution obtained by simulating $n > 1000$ data tables of equivalent size on the basis of a uniform distribution).

Indicatively, MCA's first axis carrying the most latent knowledge load opposes its 'positive' values' high frequency for the factors CLUST = Tc4, Tc1, Thry.Plcy = Thry, Real.nReal = nRe and Authr.T.nT = Au.T (factors are sorted from the most common) and low frequency for the factors CLUST = Tc2, Thry.Plcy = Plcy, Authr.T.nT = Au.nT and Real.nReal = Re (factors are sorted from the rarest). On the 'negative' side, axis 1 shows high frequency for the factors Them.T.nT = Th.T, Authr.T.nT = Au.T, CLUST = Tc2, Real.nReal = Re and Thry.Plcy = Plcy (factors are also sorted from the most common) and low frequency for the factors Them.T.nT = Th.nT, Authr.T.nT = Au.nT, Real.nReal = nRe, Thry.Plcy = Thry and CLUST = Tc4 (factors here are sorted from the rarest).

The key juxtaposition of journal articles on the first MCA axis in the analyzed data set refers to the one between the second group (Tc2) and the fourth and fifth (Tc4, Tc5), which in turn are characterized by a certain affinity. Also, tension appears between the articles which adopt state-centrism with a rather small critique of criticism (RE, near cl. Tc2) and articles that have extensive and explicit theoretical impregnation (Thry, near cl. Tc4 and Tc5). On the same axis, the proximity of groups Tc6 and Tc2 is observed, which seems reasonable in the sense that they refer to inter-state relations and the relevant national security issues. Figure 1 below pictures the flow of methods and processes of the described methodology.

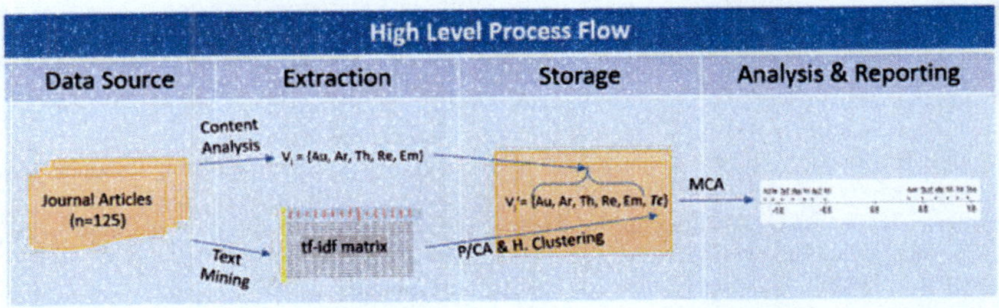

Fig. 1 Workflow of methods and processes

4 Conclusions

As the number of documents produced in social sciences has grown vast, our approach to exploring content and metadata in the respective texts produced readable results with the use of openly accessible tools that implement text mining and multivariate data analysis methods. In this study, the main research contribution is the utilization of text mining of IR articles, combined with the application of MCA on document terms and metadata.

The methodology produced six latent formations. A major emerging juxtaposition is that of authors adopting state-centrism and mainly refers to the interstate relations and issues of national security usually concerning Turkey, with mainly non-Turkish writers characterized by extensive theoretical impregnation and tends to highlight the ideational aspects of international politics. This finding is compatible with the general concern on the theoretical differentiation of IR in the country and at the same time on the existence of still a strong margin for the active involvement of the local IR community with the global one (Yazgan 2012; Aydinli and Biltekin 2017).

In general, it is found that topics of scientific articles are better reflected when mining scientific terms within their corpus and the analysis is combined with the employment of PCA and MCA. Consequently, the incorporation of the latter into the research of not only the scientific work (as it is reflected in scientific journals) of IR, but also broadly of social sciences, opens up new margins in the understanding of the scientific content of each field and its development, in theoretical as well as meta-theoretical/epistemological terms. However, some limitations are worth pointing out. The initially examined sparse tf-idf matrix may pose methodological questions (pointing toward the use of simple CA or Sparse PCA in future work). In action, bootstrapping showed that albeit low variability, PCA results are significant. Further on, in the near-future agenda and within the present scope, the investigation of additional areas of text mining is planned, such as the use of bi-gram IR terms.

References

Aydin, M., Yazgan, K.: Türkiye'de Uluslararası İlişkiler Akademisyenleri. Eğitim, Arastırma ve Uluslararası Politika Anketi - 2011. Uluslararası İlişkiler, **9**, 3–44 (2013)

Aydinli, E., Biltekin, G.: Time to quantify Turkey's foreign affairs: setting quality standards for a maturing international relations discipline. Int. Stud. Perspect. **18**(3), 267–287 (2017)

Bilgin, P., Tanrisever. O.: A telling story of ir in the periphery: telling Turkey about the world, telling the world about Turkey. J. Int. Relat. Dev. **12**(2), 174–179 (2009)

Feinerer, I., Hornik, K., Meyer, D.: A text mining infrastructure in R. J. Stat Softw. **25**)(5), 1–54 (2008)

Gofas, A., Hamati-Ataya, I., Onuf, N. (eds): The SAGE Handbook of the History, Philosophy and Sociology of International Relations. SAGE (2018)

Greenacre, M., Blasius, J. (eds.): Multiple Correspondence Analysis and Related Methods. CRC Press (2006)

Joachims, T.: A Probabilistic Analysis of the Rocchio Algorithm with TFIDF for Text Categorization (No. CMU-CS-96-118). Carnegie-Mellon Univ Pittsburgh PA (1996)

Kristensen, P.: Dividing discipline: structures of communication in international relations. Int. Stud. Rev. **14**(2), 32–5 (2012)

Koutsoupias, N., Mikelis, K.: Exploring International Relations Journal Articles: A Multivariate Approach. SAGE Research Methods Cases (2019)

Kristensen, P.: Dividing discipline: structures of communication in international relations. Int. Stud. Rev. **14**(2), 32–5 (2012)

Le Roux, B., Rouanet, H.: Quantitative Applications in the Social Sciences: Multiple Correspondence Analysis. SAGE (2010)

Maliniak, D., Peterson, S., Powers, R., Tierney, M.J.: Is international relations a global discipline? Hegemony, insularity, and diversity in the field. Secur. Stud. **27**(3), 448–484 (2018)

Pashakhanlou, A.H.: Fully integrated content analysis in international relations. Int. Relat. **31**(4), 447–465 (2017)

Salton, G.: Automatic Text Processing. Addison-Wesley (1989)

Saridakis, I.: Text Corpora and Specialized Terminology: The Terminology of Geopolitics (in Greek). https://geoterm.turkmas.uoa.gr/wp-content/uploads/2016/07/Geopolitics_dictionary_Saridakis_2016.pdf. Accessed 21 June 2019 (2016)

Saridakis, I.: Cross-linguistic Semantics of International Law. A Corpus-informed Translation of A. Cassese's International Law into Greek. Linguistica Antverpiensia, New Series— hemes in Translation Studies, **12**, 197–215 (2013)

Schmidt, B., Guilhot, N. (eds.) Historiographical Investigations in International Relations. Palgrave Macmillan (2019)

Thuleau, S., Husson, F.: Factoinvestigate: automatic description of factorial analysis. R package version 1.1. https://CRAN.R-project.org/package=FactoInvestigate. Accessed 21 June 2019 (2017)

Vijayarani, S., Ilamathi, M.J., Nithya, M.: Preprocessing techniques for text mining-an overview. Int. J. Comput. Commun. **5**(1), 7–16 (2015)

Yazgan, K.: The Development of International Relations Studies in Turkey. Ph.D. Thesis. University of Exeter (2012)

Zhao, Y., Karypis, G., Fayyad, U.: Hierarchical clustering algorithms for document datasets. Data Min. Knowl. Disc. **10**(2), 141–168 (2005)

Quantile Measures of Extreme Risk on Metals Market

Dominik Krężołek and Grażyna Trzpiot

Abstract During the period of dynamic economic development, certain events that may cause disruptions at the level of various economic processes are observed. These disruptions usually bring significant consequences. There are many methods for identifying rare events. Some of them include the so-called extreme statistics. From a statistical point of view, rare events are associated with high order quantiles for probability distributions that allow for determining the level of risk for which the probability of occurrence of a risky event is extremely low. The paper focuses on the possibility of using the Hill estimator and its modifications to assess the risk of rare events. Results of the analysis for selected theoretical models are compared in the paper. The empirical analysis was conducted on the example of assets from the precious metals market, i.e. gold and silver.

Keywords Hill estimator · Extreme statistics · High order quantiles · VaR · Precious metals

1 Introduction

Economic processes observed in the modern world are remarkably diverse, and their volatility is often independent of their nature. In the context of uncertainty and unpredictability, there occur the so-called rare events that have a significant impact on many market processes, which may consequently result in increased risk. The analysis of the volatility of gold and silver return rates over the last few decades indicates their significant diversity, especially during crises and periods of market instability. Events that significantly affect the strength and direction of these changes

D. Krężołek (✉) · G. Trzpiot (✉)
Department of Demographics and Economic Statistics, University of Economics in Katowice, ul. 1-go Maja 50, 40-287 Katowice, Poland
e-mail: dominik.krezolek@ue.katowice.pl

G. Trzpiot
e-mail: grazyna.trzpiot@ue.katowice.pl

T. Chadjipadelis et al. (eds.), *Data Analysis and Rationality in a Complex World*, Studies in Classification, Data Analysis, and Knowledge Organization,
https://doi.org/10.1007/978-3-030-60104-1_14

are observed. This is reflected in the shape of the empirical distribution of return rates, which are leptokurtic, asymmetric, and heavy-tailed. The latter feature is important from the point of view of measuring risk. Considering the fact that investments in gold and silver are approached as the so-called "safety haven" primarily in the periods of instability in financial markets and increased uncertainty in the global economy, investors perceive the value of these ores as more stable on comparison with other assets. Therefore, considering the occurrence of rare events, it is necessary to accurately measure the risk.

2 Extreme Risk

Rare events generate a risk at a level far from the expected one. This type of risk is called extreme risk and is associated with events that occur with an extremely low probability, but once they occur, they cause significant consequences, usually losses (Jajuga 2008). Extreme risk theory and extreme statistics are used to analyse extreme risk. They allow for estimating various parameters that can be used to define such events. These parameters are, for example, high order quantiles for empirical distributions of the studied phenomena or parameters determining the periods in which the analysed processes take extreme values (Gumbel 2004). In economic terms, there are many examples of events that have caused various kinds of drastic changes on the market, for example, Black Thursday (October 24, 1929), Black Monday (October 19, 1987), World Trade Centre (September 11, 2001), the recent financial crisis (2008–09), and the oil crisis (2014).

In the theory of extreme events, two parameters that relate to rare events can be distinguished. The first is the high order quantile of the probability distribution of the analysed process, while the other one is the tail thickness index of this distribution, which indirectly also determines the probability of a rare event. The extreme value distribution (EVD) should be defined. It can be described with the use of the following density function (Gumbel 2004):

$$EVD_{\gamma}(x) = \begin{cases} \exp\left[-(1+\gamma x)^{-1/\gamma}\right], & 1+\gamma x \geq 0 \text{ for } \gamma \neq 0 \\ \exp\left[-\exp(-x)\right], & x \in \mathbb{R} \quad \text{ for } \gamma = 0 \end{cases} \tag{1}$$

where parameter γ determines extreme value index (EVI). This is the most important parameter in this distribution. It measures the thickness of the tail, and thus the probability of the occurrence of an extreme event. The tail index takes real values, and the heavier the tail, the higher the index value. There are many methods for estimating the tail index value. They include extreme value theory models, alpha-stable models, regression models, etc. (see Embrechts et al. 1997; Beierlant et al. 2006; Reiss and Thomas 2007; Clauset et al. 2009). In this analysis, Hill estimator and its modifications are used as a measure for determining the thickness of the tail of the distribution.

3 Hill Estimator and Its Modification

Consider the model with heavy tails. For the sample $\underline{X}_n = (X_1, X_2, \ldots, X_n)$ and the associated sample of order statistics arranged in ascending order $(X_{1:n} \leq X_{2:n} \leq \cdots \leq X_{n:n})$ the Hill estimator is the classic EVI estimator is Hill (1975):

$$H_{k,n} \equiv H_{k,n}\left(\underline{X}_n\right) := \frac{1}{k}\sum_{i=1}^{k}\left[\ln\left(\frac{X_{n-i+1:n}}{X_{n-k:n}}\right)\right] \tag{2}$$

where k represents the number of excesses over random level $X_{n-k:n}$ meeting the condition that $k = k_n \to \infty$ and $k/n \to 0$ as $n \to \infty$.

Hill's estimator has some interesting features. Nonetheless, it can be stated that it is a scale-invariant, but not a location-invariant parameter. In addition, the Hill estimator usually shows a high asymptotic bias, but despite this, its distribution is asymptotically normal with an appropriate variance value and a non-zero mean. Therefore, it is not robust to significant diversity as regards data. For this reason, its various modifications are used in practice. The first is the so-called PORT estimator (excesses over the random threshold), designated as PORT-Hill (Araújo Santos et al. 2006). The PORT-Hill estimators are functionals of the sample exceedances in relation to the random level $X_{[nq]+1:n}$, i.e. functionals of the form $\underline{X}^{(q)}_{k,n} := (X_{n:n} - X_{[nq]+1:n}, \ \ldots, X_{n-k:n} - X_{[nq]+1:n})$ with $1 \leq k < n - [nq] - 1$. Therefore, the PORT-Hill estimator has the following form:

$$H^{(q)}_{k,n} := H_k\left(\underline{X}^{(q)}_{k,n}\right) = \frac{1}{k}\sum_{i=1}^{k}\left[\ln\left(\frac{X_{n-i+1:n} - X_{[nq]+1:n}}{X_{n-k:n} - X_{[nq]+1:n}}\right)\right] \tag{3}$$

for a certain tuning parameter q meeting the condition $0 \leq q < 1$. The PORT-Hill tail index estimator is location and scale invariant, and the tuning parameter q makes it more flexible in the case of disturbed data.

For high order quantiles, Gomes et al. (2010) proposed another modification of the classic Hill estimator—the so-called quasi-PORT-Hill estimator. Assume that the regularly varying function $L_U(t) = t^{-\gamma}U(t)$ tends to a finite non-zero constant with $\gamma > 0, \ \rho < 0, \ \beta \neq 0$, where $U(t) = Ct^{\gamma}(1 + \frac{\gamma\beta t^{\rho}}{\rho} + o(t^{\rho}))$ as $t \to \infty$. Hence, the quasi-PORT-Hill estimator is expressed by the following formula:

$$\overline{H}^{(q)}_{k,n} \equiv \overline{H}^{(q)}_{k,n}(\widehat{\beta}, \widehat{\rho}) = H^{(q)}_{k,n}\left[1 - \frac{\widehat{\beta}\left(\frac{n}{k}\right)^{\widehat{\rho}}}{1 - \widehat{\rho}}\right] \tag{4}$$

where (β, ρ) are second-order parameters. These parameters have been comprehensively described by Caeiro et al. (2005). A certain extension of the estimator presented by formula (4) was proposed by Gomes et al. (2008). It is the so-called Hill estimator with minimum-variance reduced bias, MVRB-Hill, in which reducing the load without increasing the variance of the estimator is suggested. This estimator can be expressed by the following formula:

$$\overline{H}_{k,n} \equiv \overline{H}_{k,n}\left(\widehat{\beta}, \widehat{\rho}\right) = H_{k,n}\left[1 - \frac{\widehat{\beta}\left(\frac{n}{k}\right)^{\widehat{\rho}}}{1-\widehat{\rho}}\right] \tag{5}$$

The MVRB-Hill estimator, compared to the previous ones, is location invariant, but only approximately invariant, yet it has all the asymptotic properties as those provided in formulas (3)–(4).

The Hill estimator allows for estimating the thickness of the tail of distribution, which is important for proper assessment of the extreme risk measured by high order quantiles. In our case, a modification of the classic Hill estimator to estimate VaR is applied, while using an approach that represents the quantile of any probability distribution. For the estimators presented by the (2)–(5) formulas, the formulas for calculating high order quantiles with the use of Hill, PORT-Hill, quasi-PORT-Hill, and MVRB-Hill estimators are presented below:

$$Q_{p|H_{k,n}} := \widehat{\chi}_{p|H_{k,n}} := X_{n-k+1:n}\left(\frac{k}{np}\right)^{H_{k,n}} \tag{6}$$

$$Q_{p|H_{k,n}^{(q)}} := \widehat{\chi}_{p|H_{k,n}^{(q)}} := \left[X_{n-k:n} - X_{[nq]+1:n}\right]\left(\frac{k}{np}\right)^{H_{k,n}^{(q)}} + X_{[nq]+1:n} \tag{7}$$

$$Q_{p|\overline{H}_{k,n}^{(q)}} := \widehat{\chi}_{p|\overline{H}_{k,n}^{(q)}} := \left[X_{n-k:n} - X_{[nq]+1:n}\right]\left(\frac{k}{np}\right)^{\overline{H}_{k,n}^{(q)}} + X_{[nq]+1:n} \tag{8}$$

$$\ln Q_{p|\overline{H}_{k,n}} := \ln \widehat{\chi}_{p|\overline{H}_{k,n}} := \ln X_{n-k+1:n} + \overline{H}_{k,n}\left[\ln\left(\frac{k}{np}\right) + C_p\right] \tag{9}$$

where $H_{k,n}$, $H_{k,n}^{(q)}$, $\overline{H}_{k,n}^{(q)}$, $\overline{H}_{k,n}$, respectively, represent Hill, PORT-Hill, quasi-PORT-Hill, and MVRB-Hill estimators of the extreme value index, whereas $C_p = \widehat{\beta}\left(\frac{n}{k}\right)^{\widehat{\rho}} \frac{\left(\frac{k}{np}\right)^{\widehat{\rho}}-1}{\widehat{\rho}}$ are second-order parameters (β, ρ).

4 Empirical Study

The aim of the study is to verify the possibility of using the Hill estimator and its modification to assess the risk of investment in the market of precious metals, i.e. old and silver. Extreme risk measures determined as high order quantiles with the use of Hill estimators are used in the empirical part of the study. The advantage of extreme risk measurement with the use of the modified Hill estimators is assumed in comparison with the classic approach (such as the Risk Metrics methodology). Considering the high order quantiles, it is necessary to have a sufficiently large data

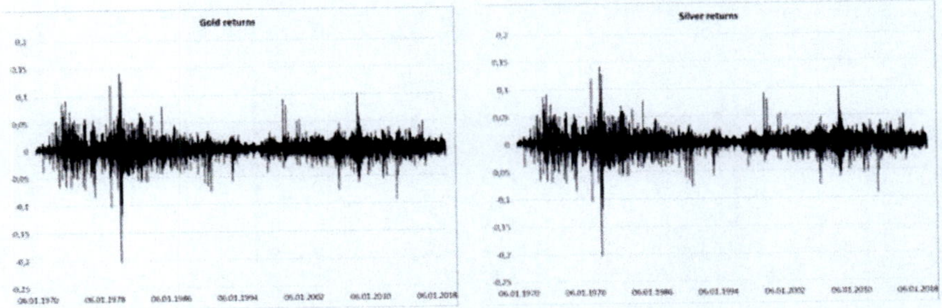

Fig. 1 Time series of gold and silver log-returns

set. In our case, a set of daily logarithmic rates of gold and silver return observed from January 1970 to March 2019 (about 50 years) was used. It gives a total of 12,521 observations. To measure the risk using value at risk (VaR), high order quantiles (0.001 and 0.0005, respectively) estimated with the use of Hill estimator and its modifications are applied. In addition, unconditional models for normal, t-Student, and asymmetric Laplace distribution (ALD), as well as GARCH and APARCH conditional volatility models with non-Gaussian residuals are used. The GARCH model class has been described in detail by Bollerslev (Bollerslev 1986), while APARCH models by Ding et al. (1993). Considering the error term, ε_t of the GARCH/APARCH model, a description of its conditional distribution using Student's t-distribution and ALD was proposed:

$$f_{t\text{-}Stud}\left(\varepsilon_t, \sigma_t^2; \theta\right) = \frac{\Gamma\left(\frac{v+1}{2}\right)}{\sigma_t \Gamma\left(\frac{v}{2}\right)\sqrt{\pi\left(v-2\right)}}\left(1+\frac{\varepsilon_t^2}{\left(v-2\right)\sigma_t^2}\right)^{\frac{v+1}{2}} \tag{10}$$

$$f_{ALD}\left(\varepsilon_t, \sigma_t^2; \theta\right) = \frac{\kappa}{\sigma_t^2\left(1+\kappa^2\right)}\begin{cases} \exp\left\{-\frac{\kappa}{\sigma_t^2}\varepsilon_t\right\}, & \text{for } \varepsilon_t \geq 0 \\ \exp\left\{\frac{1}{\kappa\sigma_t^2}\varepsilon_t\right\}, & \text{for } \varepsilon_t < 0 \end{cases} \tag{11}$$

where $\{\varepsilon_t\}$ is a sequence of random variables iid, σ_t^2 is a conditional variance, θ is a vector of estimated parameters, v is the number of degrees of freedom, and $\Gamma\left(k\right) = \int_0^{+\infty} x^{k-1}e^{-x}dx$ is the gamma function with k parameter. If t-Student distribution is applied, v parameter should be estimated. For ALD distribution, Kozubowski and Podgórski (2000) scale-invariant parameter is $\kappa = \frac{2\tau}{\left(\xi+\sqrt{4\tau^2+\xi^2}\right)}$ and additionally $\xi = \tau\left(\frac{1}{\kappa}-\kappa\right)$.

Empirical time series of logarithmic rates of gold and silver return for the whole period are presented in Fig. 1.

There are two subperiods with significant jumps in the case of both gold and silver rates—the first refers to the turn of the 1970s and the early 1980s, and the second for 2011–2012. In addition, the periods with greater variability in rates of return for both metals can easily be found. The phenomenon of clustering of variance is also

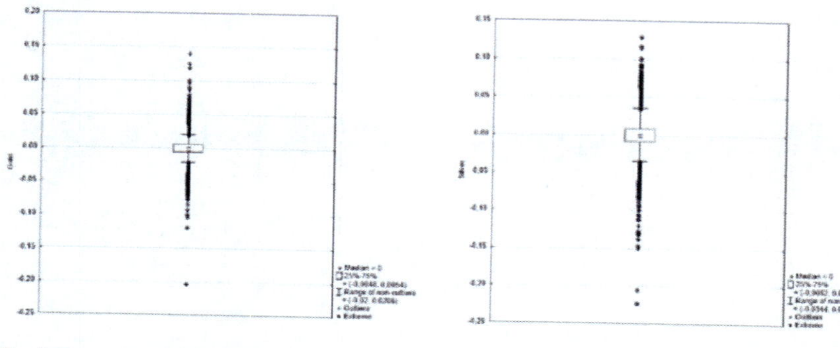

Fig. 2 Box-plots for gold and silver distribution

observed. The average rates of return for gold and silver over the studied period are positive. Empirical distributions are leptokurtic and left-skewed, and the outliers also occur (see graphs in Fig. 2).

In the final stage of the analysis, high order quantiles were calculated using Hill, PORT-Hill, quasi-PORT-Hill, and MVRB-Hill estimators. They were used to estimate VaR. Models of conditional variability of GARCH and APARCH were selected according to the AIC information criterion. In addition, theoretical quantile values were calculated for the unconditional distributions, i.e. normal, t-Student and ALD. The results of risk measurement for the rate of gold and silver returns are presented in Tables 1 and 2.

Table 1 Hill-based estimators for VaR between the levels 0.001 and 0.0005—gold

Model	Quantile 0.001				Quantile 0.0005			
	VaR	Kupiec LR	Kupiec p	RMSE	VaR	Kupiec LR	Kupiec p	RMSE
Empirical	−0.06606	0.01811	0.89294	–	−0.07733	0.08415	0.77176	–
Hill	−0.07805	4.21763	0.04001	0.01199	−0.09208	0.27291	0.60139	0.01476
PORT-Hill	−0.07632	1.09970	0.29433	0.01026	−0.08422	0.01100	0.91648	0.00689
Q-PORT-Hill[a]	−0.09744	7.91890	0.00489	0.03138	−0.08826	0.27291	0.60139	0.01094
MVRB-Hill	−0.07726	2.90348	0.08839	0.01120	−0.09038	0.27291	0.60139	0.01306
GARCH-stud	−0.08155	4.21763	0.04001	0.01549	−0.08923	0.27291	0.60139	0.01190
APARCH-ALD	−0.08023	4.21763	0.04001	0.01417	−0.09019	0.27291	0.60139	0.01287
normal	−0.03888	–	–	0.02718	−0.04142	296.9499	0.00000	0.03591
t-Student	−0.07889	2.90348	0.08839	0.01282	−0.09254	0.93768	0.33287	0.01522
ALD	−0.07722	1.87618	0.17077	0.01115	−0.09162	0.27291	0.60139	0.01429

[a]For $p = 0.5$

Table 2 Hill-based estimators for VaR between the levels 0.001 and 0.0005—silver

Model	Quantile 0.001				Quantile 0.0005			
	VaR	Kupiec LR	Kupiec p	RMSE	VaR	Kupiec LR	Kupiec p	RMSE
Empirical	−0.11028	0.02201	0.88207	–	−0.13139	0.08415	0.77176	–
Hill	−0.12456	0.54607	0.45993	0.01429	−0.15817	3.95800	0.04665	0.02678
PORT-Hill	−0.10382	4.76659	0.02902	0.00646	−0.14773	0.93768	0.33287	0.01634
Q-PORT-Hill[a]	−0.11213	2.11112	0.14623	0.00185	−0.15228	3.95800	0.04665	0.02089
MVRB-Hill	−0.12174	0.54607	0.45993	0.01146	−0.15784	3.95800	0.04665	0.02645
GARCH-stud	−0.10741	4.76659	0.02902	0.00287	−0.14285	0.27291	0.60139	0.01146
APARCH-ALD	−0.12084	0.54607	0.45993	0.01057	−0.14632	0.93768	0.33287	0.01493
normal	−0.05875	200.5554	0.00000	0.05153	−0.06257	0.00884	0.00000	0.06882
t-Student	−0.12517	0.54607	0.45993	0.01489	−0.15885	3.95800	0.04665	0.02746
ALD	−0.12396	0.19292	0.66050	0.01368	−0.15748	2.10796	0.14653	0.02609

[a]For $p = 0.5$

Tables 1 and 2 summarize the results, respectively, for gold and silver returns. The RMSE column estimates the root mean square error (RMSE). This allowed identifying the estimators that best approximate the empirical value of VaR. It can be noticed that the value at risk calculated in accordance with the assumption of the normal distribution is inadequately estimated within the meaning of the proportion of failure test by Kupiec, regardless of the level of the quantile. In other cases, the percentage of excesses is permissible. If we compare VaR estimates with all the presented methods. The results differ depending on the metal and the quantile level. For gold at 0.001, the best estimates were obtained for PORT-Hill and unconditional ALD estimators, while at 0.0005 for PORT-Hill and quasi-PORT-Hill estimators. For silver at the 0.001 quantile, the best estimates of the empirical VaR were obtained using the quasi-PORT-Hill estimator and the GARCH model with the error described by t-Student distribution, while for the quantile at 0.0005 for estimators based on the GARCH model with the error described by t-Student distribution and APARCH with the error described in the ALD distribution. If we look at the level of risk, some measures overestimate the empirical level of VaR, while others do not. Of course, the worst results were obtained for models based on the normal distribution.

5 Conclusions

The paper presents the approach to estimating extreme risk using the Hill estimator and its modifications. As examined, the empirical distributions of silver and gold return rates are leptokurtic, skewed, and thick-tailed in comparison with the normal distribution. Analysing the level of risk associated with investments in gold and silver, it was observed that silver is riskier, regardless of the order of the quantile. If we compare all models, it proves that the normal distribution is incorrect in terms of the test by Kupiec. To estimate VaR for appropriately low quantiles, models based on the modifications of the Hill estimator (PORT-Hill and quasi-PORT-HILL) should be used, but models of conditional variability with non-Gaussian error distributions can also be used, because they are more accurate in terms of RMSE error. It has been found that parameterization of the Hill estimator improves the accuracy of the VaR estimation. It has also been noted that some measures overestimate, while others underestimate the level of risk (primarily models based on the normal distribution).

It is recommended for investors investing in gold or silver that in times of uncertainty and occurrence of unpredictable random events such as the recent global financial crisis, they should estimate market risk using models built on the basis of modified Hill estimators or on the basis of the models of conditional volatility, which properly estimate the level of loss at very low probability values. The models based on classical unconditional probability distributions (such as the normal distribution) should be avoided.

References

Araújo Santos, P., Fraga Alves, M.I., Gomes, M.I.: Peaks over random threshold methodology for tail index and quantile estimation. Revstat. Stat. J. **4**(3), 227–247 (2006)

Beierlant, J., Goegebeur, Y., Segers, J., Teugels, J.: Statistics of Extremes: Theory and Applications. Wiley (2006)

Bollerslev, T.: Generalised autoregressive conditional heteroskedasticity. J. Econom. **31**, 307–327 (1986)

Caeiro, F., Gomes, M.I., Pestana, D.: Direct reduction of bias of the classical Hill estimator. Revstat. Stat. J. **3**(2), 111–136 (2005)

Clauset, A., Shalizi, C.R., Newman, M.E.: Power-law distributions in empirical data. SIAM Rev. **51**(4), 661–703 (2009)

de Haan, L., Ferreira, A.: Extreme Value Theory: An Introduction. Springer Series in Operations Research and Financial Engineering, Springer-Verlag, New York (2006)

Ding, Z., Granger, C.W.J., Engle, R.F.: A long memory property of stock market returns and a new model. J. Empir. Financ. **1**, 83–106 (1993)

Embrechts, P., Kluppelberg, C., Mikosch, T.: Modelling Extremal Events. Springer, New York (1997)

Gomes, M.I., de Haan, L., Henriques-Rodrigues, L.: Tail index estimation for heavy-tailed models: accommodation of bias in weighted log-excesses. J. R. Stat. Soc. Serie B **70**(1), 31–52 (2008)

Gomes, M.I., Figueiredo, F., Henriques-Rodrigues, L., Miranda, M.C.: A quasi-PORT methodology for VaR based on second-order reduced-bias estimation. Notas e Comunicações CEAUL (2010)

Gumbel, E.J.: Statistics of Extremes. Dover Publications Inc, Mineola, New York (2004)

Hill, B.M.: A simple general approach to inference about the tail of the distribution. Ann. Stat. **3**(5), 1163–1174 (1975)
Jajuga, K.: Zarządzanie ryzykiem. Polskie Wydawnictwo Naukowe PWN, Warszawa (2008)
Kozubowski, T.J., Podgórski, K.: Asymmetric Laplace distributions. Math. Scientist. **25**, 37–46 (2000)
Reiss, R., Thomas, M.: Statistical Analysis of Extreme Values: With Applications to Insurance, Finance. Birkhauser, Hydrology and Other Fields (2007)

Evaluation of Text Clustering Methods and Their Dataspace Embeddings: An Exploration

Alain Lelu and Martine Cadot

Abstract Fair evaluation of text clustering methods needs to clarify the relations between (1) preprocessing, resulting in raw term occurrence vectors, (2) data transformation, and (3) method in the strict sense. We have tried to empirically compare a dozen well-known methods and variants in a protocol crossing three contrasted open-access corpora in a few tens dataspaces with different metrics and/or matrix decompositions. We compared the resulting clusterings to their supposed "ground-truth" classes by means of four usual indices. The results show both a confirmation of well-established implicit combinations and good performances of unexpected ones, mostly in spectral or kernel dataspaces. The rich material resulting from these some 600 runs includes a wealth of intriguing facts, which needs further research on the specificities of text corpora in relation to methods and dataspaces.

Keywords Text clustering evaluation · Distance metrics · Spectral clustering · Graph partition · Kernel clustering

1 Introduction: Motivations and Goals

Evaluation of text clustering methods is one of the key issues in the problem of bibliometric delineation of scientific fields. As co-authors of Zitt et al. (2018), we have tried to test 17 clustering methods with a publicly available real-life test set, the Reuters' test bench (Lewis et al. 2004). This dataset adds up several difficulties of text clustering, i.e. mainly strongly unbalanced man-made classes (the targeted "ground truth"), and texts of unbalanced sizes. An unexpected result was that of antique agglomerative methods, especially Ward hierarchical clustering, which performed

A. Lelu
Université de Franche-Comté (rtd), Besançon, France
e-mail: alelu@orange.fr

M. Cadot (✉)
LORIA Nancy France, Vandoeuvre-lès-Nancy, France
e-mail: martine.cadot@loria.fr

T. Chadjipadelis et al. (eds.), *Data Analysis and Rationality in a Complex World*, Studies in Classification, Data Analysis, and Knowledge Organization,
https://doi.org/10.1007/978-3-030-60104-1_15

better than many more recent ones. Was it the case for all types of corpora? Above all, we realized that for the sake of fair comparisons, as well as conceptual clarity, we should clearly separate the transformations of the raw word-count data (for example, into Salton tf-idf vector representation, Laplacian spectral space, etc.) from the algorithms in the strict sense, instead of using longtime accepted implicit combinations. For example, no rationale forbids using Non-negative Matrix Factorization in a spectral space. This consideration is in line with the conceptual clarifications operated in Van Mechelen et al. (2018); last, but not least, unexpected recommendations may proceed from non-classic combinations. This clarification is one of our guiding threads, and led us to the study and report (Lelu and Cadot 2019) we submitted to the Neutral Cluster Benchmarking Challenge, organized by the Cluster Benchmarking Task Force of the IFCS, which won the Challenge.

Though restricting our scope to text clustering, it is clear that many types of texts now need to be processed: abstracts or plain texts of scientific papers, which are our primary scientific interest, or journals, literary or legal texts, or texts originating in the social nature of Internet communications, such as contributions to forum discussions, or social networks. We decided to base our present survey on three typical and contrasted test sets: a full-text scientific database, a wire of press agency, and an Internet discussion forum. It is clear that the complete text preprocessing chain is out of research goal, so we have to rest on one same linguistic—or weakly linguistic—term, lemma, or stem extraction scheme, and the same elimination of infrequent or too frequent words. This point must not keep us from exploring the influence of truncating the resulting vocabulary in chosen distribution quantiles, contrary to usual benchmark studies which merely mention an absolute occurrence threshold. All these specifications led us to the choices we expose in the methodology section.

Of course, the options on methods and types of dataspaces to be considered are inevitably somehow arbitrary: we tried to take account of the most usual algorithms, or method families, such as K-means, hierarchical agglomerative clustering, spectral clustering, graph clustering, and kernel clustering, and we added two more specific methods, i.e. Non-negative Matrix Factorization (NMF) and Latent Dirichlet Allocation (LDA), which amounts to a dozen methods and variants.

Concerning dataspaces, we chose to add to the plain term occurrence vector space the transformed spaces by Salton's and Okapi's tf-idf weighting schemes, by chi-square metrics, by Laplacian spectral decomposition, by Correspondence Analysis factors, and last by order-2 polynomial kernel expansion.

Given the combinatorics of the three main elements—text types, dataspaces, and algorithms—our research study could be nothing but an exploration, strongly constrained by available resources. However, some interesting conclusions will be drawn out of this exploration. In the final conclusion, we will deal with what may be continued and deepened in our perspective, given the results.

Let us close this introduction saying that we are indebted to the remarkable initiative of the Brazilian LABIC team (Rossi et al. 2013) who homogeneously preprocessed (LABIC stemmer 2020) some forty text collections and made the documents-by-terms matrices available online on their site (LABIC data 2020).

2 Methodology

To evaluate non-supervised classifications via a methodology devoted to supervised ones is an imposed solution, for want of anything better, it may at least result, in default of a universal ranking of methods, in fruitful reflections on the typology of texts, or the nature of the human categorization and abstraction process and its similarities and differences with methods mostly optimizing an intrinsic objective function. Another core imperative we have set is transparency and reproducibility, in addition to the direct link to the documents-by-terms matrices we have provided above, the complementary material in HAL site, Lelu and Cadot (2019), gathers the links to the public-access code we used. Though most algorithms are theoretically insensitive to the ordering of input vectors, in practice, we experienced that tied effects, among others, could affect the results. This is why we have randomly scrambled the data vectors.

2.1 *Choice of Test Corpora*

The three "prototype" test corpora mentioned in the introduction are, first, Reuters' "ModApté Split", Apté (1994), limited to the eight most important classes ("Re8" in the present study, 7674 documents, and 8901 terms), second, the ACM collection made of the proceedings of 40 conferences in different computer science areas (3493 papers and 60 768 terms), third, the "20 Newsgroups" collection ("Ng20") composed of 18 808 messages posted in 20 Usenet groups (45 434 terms). The size of the man-made reference classes is strongly unbalanced in the case of Re8 (two of them constitute 81% of the documents), roughly equal in the case of ACM and Ng20. It is to be noted that the sole Reuters' class labels are issued from direct manual indexing. The two others originate in the concatenation of sub-corpora of comparable size. They could therefore be considered "semi-real-world data", not really representative of real-life non-annotated corpora.

2.2 *Truncating the Vocabularies*

As the size of the vocabularies is unbalanced (Re8: about 8900 terms; ACM: 60 800 terms; Ng20: 45 000 terms) but extensive (the hapaxes, i.e. terms of total occurrence one, are included in this count), we decided a common scheme for a vocabulary-independent truncation by thresholds: in addition to the basic option of retaining the whole vocabulary, we built two sub-corpora per test corpus retaining the third quartile of the term distribution (25% of the total occurrences), and the seventh "octile" (12.5%).

2.3 Choice of Clustering Methods

For the bibliographic references to the methods, see Lelu and Cadot (2019). We have affected a lower priority to algorithms with two parameters (DbScan, Affinity Propagation, and Smart Local Moving Algorithm), or one parameter with deceptive results on Reuters' corpus (Density Peaks, Independent Component Analysis, Fuzzy C-means, K-Means++). We selected the following:

Plain K-means clustering ("KM"). We implemented 20 elementary runs, or "replicates", per run, selecting the best one in terms of the local optimum of the K-means intrinsic objective function.

Hierarchical agglomerative clustering with two linkage variants: average link ("HCa"), and Ward ("HCw"). Originally in O(#documents^3) time complexity, more recent contributions have lowered this constraint to O(#documents^2), Murtagh (1984).

Spectral clustering. We not only used the "standard" combination K-means/ Laplacian spectral dataspace, but also explored (with success, see below) many other combinations.

Graph clustering methods. We chose the two most broadly recognized ones, i.e. Louvain and Infomap. Note that these methods, in contrast to all the tested other ones, do not need fixing a required number of clusters, hence a major operational advantage when no idea of the "true number of clusters" is known beforehand, hierarchical clustering being in an intermediate position, as in one run it leaves the choice of the cluster number to the user.

Non-negative Matrix Factorization ("NMF"). This decomposition is akin to be used as a clustering method, when the label of a document is attributed as the axis number of its maximum projection. As this method converges to local optima of its objective function, we implemented the same "20 runs" strategy as for K-means.

Latent Dirichlet Allocation ("LDA") is well-known and much respected as deeply founded in theoretic grounds.

Kernel clustering. Thanks to the "kernel trick", a documents-by-documents similarity matrix ("Gram matrix") is built without explicit expansion of the raw dataspace by a kernel function. Here, we used an order-2 polynomial kernel, which amounts to take into account the wholeness of the 2-term itemsets in each document when comparing one to another. In this case, the raw dataspace is not made of numeric occurrence vectors, but of binary existence ones.

2.4 Choice of Dataspaces

For the mathematical formulations, see Lelu and Cadot (2019). In addition to the plain term-occurrence vector space, we have considered and built

- Salton's vector space, weighted by the classic tf-idf scheme;

- Okapi (also coined BM25) vector space, with a more cryptic, but statistically grounded, weighting scheme, Robertson (1994);
- Chi-square metrics, which amount to a Euclidean vector space with transformed vectors as specified in Legendre and Gallagher (2001);
- Laplacian spectral space, Von Luxburg (2007);
- Correspondence Analysis spectral space ("CA space"), Benzécri (1973);
- Kernel space : Given the much contrasted values in the Gram similarity matrix, the cosine distance is well-suited to this dataspace, Girolami (2002).

Note that Euclidean distances in the complete CA factor space equal chi-square distances (Benzécri 1973). Therefore, truncating this space by considering the sole most informative factors amounts to considering "partial chi-square" distances, a priori more relevant than chi-square distances. These six transformations of a documents-by-terms matrix are convenient for the KM, NMF, LDA, and Spectral Clustering methods. Other methods, such as Hierarchical Clustering, Graph methods, and Kernel methods, need a documents-by-terms similarity (or dissimilarity) matrix. Depending on each dataspace-method combination, we have used Euclidean distance or "cosine" distance (i.e. 1-cosine, which weights half of the squared chord distance).

2.5 *Choice of Evaluation Measures*

We chose the four most usual indices encountered in the evaluation literature, i.e. first, Normalized Mutual Information (NMI) and Adjusted Rand Index (ARI), which compute independently from the number and labels of clusters; and second, mean local class-vs.-cluster F-scores (F) and global Purity score (i.e. 1-global error rate) which need the same number of clusters and classes, and same labels. We have aligned the k classes and the k most "analogue" clusters, in the sense of local F-scores, by means of the ranking issued from the leading factor of the Correspondence Analysis of the classes-by-clusters F-score matrix.

2.6 *Code Implementation and Computer Efficiency*

As computer efficiency is out of our goals, we implemented the data transformations, method code, and post-processing code in an Octave environment, on an Intel 6-core I7, 3.33GHz, 48Go RAM computer. Method codes were derived from existing Matlab® codes (links to the original pieces of code are available in the supplementary material). Their degree of computing time optimization varies considerably: e.g. in the case of the 19 000 documents Ng20 test set, from 2 min for 20 elementary runs of the standard "litekmeans.m" code (itself implementing 20 "replicates") to 6 h for one run of Louvain method, and 24h for one run of Hierarchical average-link clustering.

3 Evaluation Results

Our goal was to carry out, for each corpus, the test of each combination between the various data transformations and clustering methods. This was not always possible, due to constraints such as computing time or resources devoted to systematically poor results. Our reference site (Lelu and Cadot 2019) displays the entirety of the results in 29 figures. Figure 1 is one example.

Let us focus now on the measurement tools: we can observe that in the case of the two "balanced" corpora, the four evaluation indices behave in a parallel and orderly manner. In contrast, this parallelism and regular ranking deteriorate in the Reuters' unbalanced corpus, and to a lesser extent when hierarchical methods are used. A thorough investigation could perhaps explain these interesting discrepancies, but is clearly out of our present goals. We have thus chosen the most stable NMI index as a reference measure for ranking each corpus' runs (ACM: 246 runs; Re8: 237 runs; Ng20: 109 runs, summing up to 592 runs).

The best runs of Fig. 2 clearly depend on the corpora. A large variety of dataspace transformations (truncated or not vocabulary and cosine measures, Salton's, Okapi's, or raw dataspace, kernel or Laplacian spectral space) and methods (HC-Ward, K-Means, and NMF) are present. It can be noted that only four out of nine methods can be considered as classical clustering methods. These are K-Means on Salton's

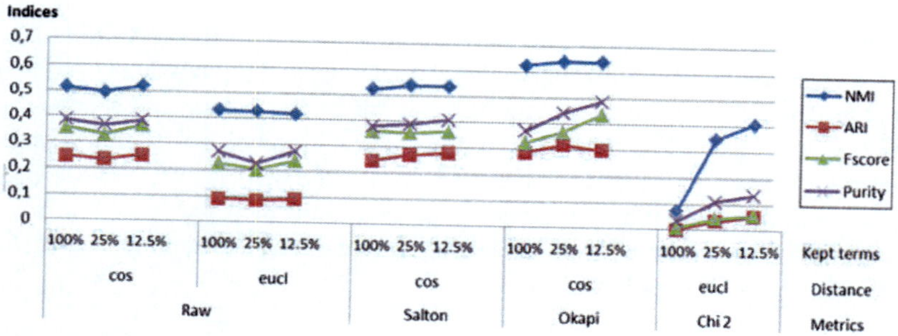

Fig. 1 K-means on ACM corpus

	ACM k=40, 3493 docs, 60768 terms			Reuters'8 k=8, 7674 docs, 8901 terms			20 NewsGroups k=20, 18808 docs, 45434 terms		
Top methods	**1)HC-Ward**	**2)HC-Ward**	**3)K-Means**	**1)K-Means**	**2)HC-ave.**	**3)HC-Ward**	**1)NMF**	**2)HC-Ward**	**3)K-Means**
Dataspace type	Spectral Lapl.	Standard	Spect. Lapl.	Standard	Spect. CA	Kernel	Std	Spectral CA	Standard
#factor or order	40; 80 (=k; 2k)	40 (=k)	80 (=2k)	8 (=k)	8 (=k)	Polynom. 2	20 (=k)	40 (=2k)	20 (=k)
% vocabulary	100%; 12.5%	100.%	12.5%	100%; 12.5%	12.5%	12.5%	100%	100%	12.5-100-25
Metrics	Salton	Okapi	Okapi	Okapi; Salton	Raw	Raw	Okapi	Raw	Salton
NMI	.6980-.6714	.6739	.6712	.6461-.6314	.629	.625	.6252	.6220	.621
Comput. time	77s	75s	6s	10s	24h	800s	107s	4h	150s

Fig. 2 "Top three" methods (NMI criterion) for each corpus. No thresholding of cosine distances needed

or Okapi's dataspace, standard Hierarchical Clustering, and Non-negative Matrix Factorization.

Examining the "top 50" runs of each corpus (ranked by decreasing NMI values), a few common behaviors emerge as follows.

3.1 *Partial Commonalities:*

Concerning ACM corpus: the spectral HC-Ward method in all the variants of the Laplacian Okapi-weighted space dominates massively; the spectral K-Means in the same spaces appears in the top-50 list nine times, standard HC-Ward and standard NMF appear two times and three times, respectively.

Concerning Re8 corpus: Okapi-weighted or Salton-weighted standard K-Means dominate, followed by spectral HC-average in the CA factor space, then by spectral K-Means in the CA (sometimes Laplacian) factor space; three kernel HC-Ward appear in the list, as well as six standard NMF and three Louvain methods in the Salton space.

Concerning Ng20 corpus: standard K-Means comes out on top, together with NMF Okapi in the CA factor space and spectral HC-Ward in the same space. Next come Salton-weighted (sometimes Okapi-weighted) spectral K-Means, and spectral HC-Ward in CA space. Kernel HC-Ward ranks last.

3.2 *Global Commonalities*

As far as inter-corpora comparisons are concerned, we constructed a relative performance indicator by dividing the NMI of a given "data space + method" combination by the maximum NMI observed on this corpus. With three values per combination, we can provide a heuristic view of the overall performance of a particular combination by calculating the average and the maximum range for those three values. The following lines summarize this process for the three combinations which to our view achieve the best compromise between performance and independence from the corpus (embodied in low range values):

1. Standard NMF on Okapi-weighted data, with non-truncated vocabulary: relative NMI: 95.4%, range: 7.9%;
2. Spectral hierarchical clustering (with Ward link) in the space of the $2k$ leading CA factors (k is the number of required clusters), with a strong truncation of the vocabulary (12.5% of the original vocabulary): relative NMI: 92.0%, range: 9.1%;
3. Standard K-Means on Okapi-weighted data, and vocabulary also truncated at 12.5%: relative NMI: 91.1%, range: 18.2%.

The main problem for one to follow recommendation 2 is to build the spectral space for big real-life data. In many computer languages indeed, efficient sparse Singular Value Decomposition procedures exist, appropriate when the problem is to draw a limited number of main eigenvalues and eigenvectors from huge data tables, which is the case in the present study. Otherwise, parallel graphics co-processors may be dedicated to this task.

4 Conclusions and Perspectives

We hope we have brought some clarification to the problem of evaluating text clustering procedures, by considering separately the algorithms and the dataspaces in which they operate. We have achieved some 600 runs of a dozen algorithms and variants, in a few tens various dataspaces, on three prototypical and public access test corpora. We have brought to light an unexpected variety of optimal combinations of methods and dataspaces, from which we have derived three cautious recommendations. The variety of possible transformations and parameters requires a considerable continuation effort for improving our understanding and mastery of artificial versus human categorization processes. We hope that this empirical survey will contribute to such an issue. In a modest first step, we will explore the influence of linguistic preprocessing: choice or elimination of word categories, comparison between taking into account multi-word expressions and kernel expansion of uniterms.

References

Apté, C., Damerau, F., Weiss, S.M.: Automated learning of decision rules for text categorization. ACM Trans. Inf. Syst. **12**(3), 233–251 (1994)

Benzécri, J.: L'analyse des correspondances. l'analyse des données, vol. 2, Dunod. Paris (1973)

Girolami, M.: Mercer kernel-based clustering in feature space. IEEE T Neural Networ. **13**(3), 780–784 (2002)

LABIC stemmer. http://sites.labic.icmc.usp.br/tpt/. Accessed 30 Jan 2020 (2020)

LABIC data. http://sites.labic.icmc.usp.br/text_collections/. Accessed 30 Jan 2020 (2020)

Legendre, P., Gallagher, E.D.: Ecologically meaningful transformations for ordination of species data. Oecologia **129**(2), 271–280 (2001)

Lelu, A., Cadot, M.: Evaluation of text clustering methods and their dataspace embeddings: an exploration. In: IFCS 2019 - 16th International of the Federation of Classification Societies, Thessaloniki, Greece, August 2019. https://hal.archives-ouvertes.fr/hal-02116493

Lewis, D.D., Yang, Y., Rose, T.G., Li, F.: Rcv1: a new benchmark collection for text categorization research. J. Mach. Learn. Res. **5**, 361–397 (2004)

Murtagh, F.: Complexities of hierarchic clustering algorithms: state of the art. Comput. Stat. Quart. **1**(2), 101–113 (1984)

Robertson, S., Walker, S., Jones, S., Hancock-Beaulieu, M., Gatford, M.: Okapi at TREC-3. In: Proceedings of the Third Text REtrieval Conference (TREC). Gaithersburg, USA (1994)

Rossi, R.G., Marcacini, R.M., Rezende, S.O. et al.: Benchmarking text collections for classification and clustering tasks. Technical report 395, Institute of Mathematics and Computer Sciences – University of Sao Paulo. http://repositorio.icmc.usp.br/bitstream/handle/RIICMC/6641/Relat%C3%B3rios%20T%C3%A9cnicas_395_2013.pdf?sequence=1 (2013)

Van Mechelen, I., Boulesteix, A.-L., Dangl, R., Dean, N., Guyon, I., Hennig, C., Leisch, F., Steinley, D. Benchmarking in cluster analysis: a white paper. arXiv preprint arXiv:1809.10496. Accessed 6 Nov 2020 (2018)

Von Luxburg, U.: A tutorial on spectral clustering. Stat. Comput. **17**(4), 395–416 (2007)

Zitt, M., Lelu, A., Cadot, M., Cabanac, G.: Bibliometric delineation of scientific fields. In: Glänzel, W., Moed, H.F., Schmoch, U., Thelwall, M. (eds.) Handbook of Science and Technology Indicators, Springer International Publishing (2018)

Specification of Basis Spacing for Process Convolution Gaussian Process Models

Waley W. J. Liang and Herbert K. H. Lee

Abstract Gaussian process (GP) models have been widely used for statistical modeling of point-referenced data in many scientific applications, including regression, classification, and clustering problems. Standard specification of GP models is computationally inefficient for applications with a large sample size. One solution is to construct the GP by convolving a smoothing kernel with a discretized White noise process, which requires choosing the number of bases. The distance between adjacent bases plays a key role in model accuracy. In this paper, we perform a series of simulations to find a general rule for the basis spacing required for an accurate representation of a discrete process convolution GP model. Under certain common conditions, we find that using a basis spacing of one-quarter the practical range of the process works well in practice.

Keywords Gaussian processes · Process convolutions · Spatial modeling

1 Introduction

A common approach in spatial modeling is to represent the process of interest as a univariate spatial Gaussian process (GP) $\{z(\mathbf{s}) : \mathbf{s} \in \mathbb{R}^d\}$, which is a collection of random variables indexed by points $\mathbf{s}$ in space $\mathbb{R}^d$. Any finite collection of these random variables $\{z(\mathbf{s}_1), \cdots, z(\mathbf{s}_n)\}$ is distributed as multivariate Gaussian with a certain mean function and covariance matrix $\boldsymbol{\Sigma}$. In this paper, we limit our study to isotropic GPs where the correlation between two points depends only on their distance. Common correlation functions include the Gaussian, Exponential, Matérn, and the Spherical class. More details on GP and associated correlation functions can be found in Cressie (1991), Stein (1999), and Paciorek and Schervish (2006).

W. W. J. Liang · H. K. H. Lee (✉)
University of California, Santa Cruz, USA
e-mail: herbie@soe.ucsc.edu

W. W. J. Liang
e-mail: wliang@soe.ucsc.edu

T. Chadjipadelis et al. (eds.), *Data Analysis and Rationality in a Complex World*, Studies in Classification, Data Analysis, and Knowledge Organization,
https://doi.org/10.1007/978-3-030-60104-1_16

Parameter inference under this standard specification often requires an inversion of the covariance matrix whose computational complexity grows at $O(n^3)$, where n denotes the data sample size. This makes a standard GP impractical for applications with a large n, which is now common. This issue becomes even more undesirable for Bayesian inference with Markov chain Monte Carlo (MCMC) which requires such matrix inversion at each of thousands or more of MCMC iterations. Some methods can ameliorate this issue, such as partitioning the spatial domain and fitting a separate GP in each partition (Gramacy and Lee 2008), or reducing the occurrence of such $O(n^3)$ computations for a GP with a single-parameter correlation function (Yang et al. 2014). Another approach for reducing computational cost due to large n is Discrete Process Convolutions (DPC), Higdon (1998), which formulates the GP $z(\mathbf{s})$ by convolving a symmetric kernel $k(\mathbf{u} - \mathbf{s}; \mathbf{Q})$ with a discretized latent process $x(\mathbf{u})$ indexed at a grid of bases $\{\mathbf{u}_j\}_{j=1}^m$:

$$z(\mathbf{s}) = \sum_{j=1}^{m} k(\mathbf{u}_j - \mathbf{s}; \mathbf{Q}) x(\mathbf{u}_j), \quad \mathbf{s} \in \mathscr{S} \subseteq \mathbb{R}^d.$$

Here, $\mathbf{Q}^{-1}$ denotes the covariance matrix of the kernel, $x(\mathbf{u})$ is a White noise process, and m denotes the number of bases. Note that DPC arises as an approximation to continuous process convolutions: $z(\mathbf{s}) = \int_{\mathbb{R}^d} k(\mathbf{u} - \mathbf{s}; \mathbf{Q}) x(\mathbf{u}) d\mathbf{u}$, where the correlation function for z is given by the convolution of the kernel with itself. For instance, when the kernel is a Gaussian density (a common choice in practice), it results in a Gaussian correlation function. We may safely assume that given enough bases, a GP constructed by DPC has a correlation structure that is a close approximation to its counterpart in continuous process convolutions. Figure 1 provides a one-dimensional (1-D) example of a GP constructed by DPC, where the left panel shows a set of six 1-D Gaussian kernels centered at evenly spaced bases on the interval $[-5, 5]$, and the right panel shows the resulting GP obtained as the sum of these kernels. In a

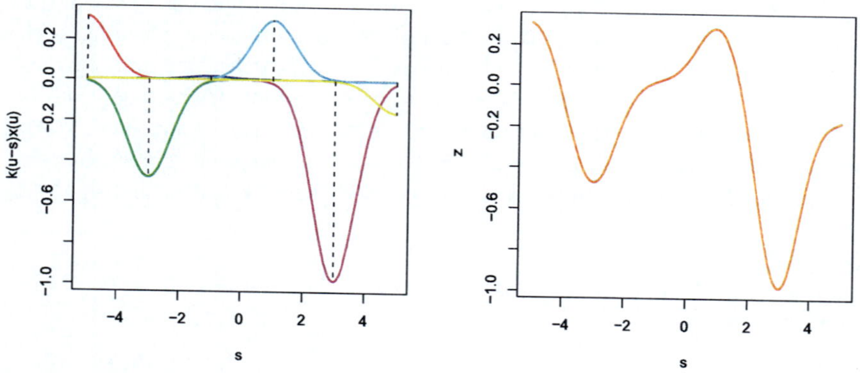

Fig. 1 Left panel shows 1-D Gaussian kernels centered at six evenly spaced bases on the interval $[-5, 5]$, and right panel shows resulting GP obtained as the sum of these kernels

two-dimensional (2-D) domain, bases are specified on a rectangular grid with even spacing in the same dimension, and there can be more bases in one dimension than the other. The computational complexity of DPC grows at $O(m^3)$, where m denotes the number of bases. Hence, DPC is computationally efficient for applications in the low dimension where m is much smaller than n. For this reason, DPC is useful for many environmental applications such as geology and climatology whose spatial domain is naturally 2-D. Given a bounded domain, having a smaller distance between adjacent bases (thus resulting in more bases) allows DPC to better describe local features in the process of interest. Bases are selected by the user before modeling and the optimal basis spacing depends on the application. This paper presents a simulation study to establish a rule-of-thumb for choosing the basis spacing for a GP formulated by DPC. The simulation setup is discussed in detail in Sect. 2, simulation results are presented in Sect. 3, and a summary of our research is given in Sect. 4. In the rest of this paper, the term DPCGP is used to refer to a GP formulated by DPC. A longer version of this paper is available as a University of California, Santa Cruz technical report at https://www.soe.ucsc.edu/research/technical-reports/UCSC-SOE-19-11.

2 Simulation Setup

As DPCGP is computationally efficient only in low dimension, our study focuses on 1-D and 2-D domains. Without loss of generality, the 1-D domain is specified as the unit interval $[0, 1]$ and the 2-D domain is specified as the unit square $[0, 1]^2$. Our study develops a series of DPCGP models on simulated data with bases spaced evenly along each dimension, where the number of bases ranges from 5 to 40 in a step size of 1 for each dimension. For example, if the number of bases is 10 in each dimension of the 2-D domain, then there are a total of $10 \times 10 = 100$ bases. Previous experience with DPCGP shows that using more bases (smaller basis spacing) tends to improve model accuracy, but this effect diminishes as the number of bases is beyond a certain threshold. This threshold varies with respect to the dependence range in the unknown process of interest that DPCGP tries to model. A short range means that a sample point is correlated closely with its neighbors and less with points far away, which requires more bases than a longer range. Our study develops each DPCGP model on simulated data under different combinations of correlation functions and practical ranges (PR). As a property of the correlation function, PR is typically defined as the distance from the origin at which the correlation is 0.05. PR can be thought of as the maximum distance between two points having non-negligible correlation. Our goal is to obtain a rule-of-thumb for basis spacing associated with this threshold with respect to PR.

2.1 Data Generation

Data is simulated from three different correlation functions: Gaussian, Matérn with smoothness parameter $\kappa = 4$, and Exponential, which represent real scenarios where the process of interest is very smooth, somewhat smooth, and non-smooth, respectively. For each correlation function, three PR's, 0.1, 0.2, and 0.3 are considered for response generation. A smaller PR results in a response with more local features, which can be better modeled by DPCGP with a finer basis spacing, up to a certain limit. Two sets of responses are simulated for each combination: one over the 1-D unit interval [0, 1] and the other over the 2-D unit area $[0, 1]^2$. Each response is a random draw of n samples from a multivariate Gaussian distribution with mean zero and covariance matrix equal to the marginal variance times the correlation matrix determined from the correlation function. Here, we let $n = 1000$ for 1-D and $n = 2000$ for 2-D. In all cases, the marginal variance is given a value of 0.25, which gives a marginal standard deviation (SD) of $\sqrt{0.25} = 0.5$. Data are generated by adding zero-mean Gaussian error with SD $= 0.05$ to the response. This level of noise is 10% of the response marginal SD. While only one noise level is considered here, results obtained from this noise level should remain useful for cases with a lower noise level as fewer number of bases are needed for less noisy data in general. Studying a higher noise level is less meaningful as the model can become inaccurate.

2.2 Model Specification

The correlation structure and smoothness of DPCGP mainly depends on the kernel type. In the literature, the Gaussian density is a common kernel choice which induces a Gaussian correlation structure in continuous process convolutions. The resulting GP is infinitely mean-square differentiable, which is a very smooth process. DPC is an approximation of its continuous counterpart, therefore, the resulting correlation structure is also approximately Gaussian given sufficiently fine basis spacing. When the unknown process of interest is less smooth, it requires a kernel supporting adjustment of Lemos and Sansó (2009) defined as follows:

$$k(\mathbf{u} - \mathbf{s}; \mathbf{Q}) = \begin{cases} (1 - D_M^2)^\kappa & \text{if } D_M < 1 \\ 0 & \text{otherwise,} \end{cases} \tag{1}$$

where $D_M = \sqrt{(\mathbf{u} - \mathbf{s})^\top \mathbf{Q}(\mathbf{u} - \mathbf{s})}$ denotes the Mahalanobis distance with covariance matrix $\mathbf{Q}^{-1}$. Since our study concerns isotropic processes only, $\mathbf{Q}^{-1}$ is specified as a diagonal matrix with identical diagonal elements. This isotropic form is called the Bézier kernel whose compact support has a radius equal to the square root of the diagonal elements of $\mathbf{Q}^{-1}$. Increasing κ for a Bézier kernel increases the smoothness of the resulting GP. According to Brenning (2001), the resulting GP is $\lfloor \kappa \rfloor$ times mean-square differentiable. Our study evaluates both the Gaussian and Bézier kernels with a diagonal $\mathbf{Q}^{-1}$ having the same values on the diagonal. For each simulated dataset, a different kernel standard deviation is used such that the kernel induced correlation function has a PR equal to the PR of the correlation function in data generation. This kernel standard deviation is termed practical kernel size (PKS) in the rest of this paper. For example, when PR=0.1, 0.2, and 0.3, the corresponding PKS for a Gaussian kernel is 0.029, 0.058, and 0.087, respectively.

We run a full factorial design with choice of dimension (1-D or 2-D), true correlation function (Gaussian, Matérn ($\kappa = 4$), Exponential), practical range (0.1, 0.2, 0.3), DPCGP kernel (Gaussian or Bézier ($\kappa = 3$)), and number of bases m $(5, 6, \cdots, 39, 40)$. Note that the basis spacing is $1/(m-1)$, which corresponds to $(0.25, 0.2, \cdots, 0.0263, 0.0256)$. Each dataset is randomly divided into a training set and a validation set of equal size. That is, there are 500 samples for each set in 1-D and 1000 samples for each set in 2-D. Each DPCGP model is developed on the training set and evaluated on the validation set. Training is performed using the *lme()* function from the *nlme* R package, which treats DPCGP as a Linear Mixed-Effects Model. RMSE (root-mean-squared-error) against the true response is computed on the validation set to assess model performance. The number of bases at which RMSE starts to saturate is recorded. Sensitivity to kernel size is analyzed by evaluating two (correlation function, kernel) combinations: (Gaussian, Gaussian) and (Matern($\kappa = 4$), Bézier($\kappa = 3$)) with kernel sizes at 80%, 90%, 100%, 110%, and 120% of PKS.

3 Simulation Results

Simulation results are shown in Fig. 2, where each curve corresponds to a specific (correlation function, kernel) combination as indicated in the legend, and it is formed by a series of DPCGP models whose number of bases are shown on the x-axis and the resulting validation RMSE on the y-axis. In the 2-D study (bottom row), number of bases on the x-axis is for a single dimension; total number of bases is the square of this value. The left, center, and right panels correspond to PR of 0.1, 0.2, and 0.3, respectively. In general, validation RMSE starts to saturate when the number of bases reaches a certain threshold. A larger threshold (more bases) is associated with a smaller PR, which is expected since a smaller basis spacing is needed to better describe local features. Within each PR, threshold changes across different (correlation function, kernel) combinations. In general, fewer bases are needed when the kernel induced correlation function is the same or close to that of the data, e.g., (Gaussian, Gaussian) and (Gaussian, Bézier ($\kappa = 3$)). More bases are needed when the kernel induced and data correlation functions are moderately different, e.g., (Matern ($\kappa = 4$), Gaussian) and (Matern ($\kappa = 4$), Bézier ($\kappa = 3$)). Finally, when the difference in correlation structure is large, e.g., (Exponential, Gaussian) and (Exponential, Bézier ($\kappa = 3$)), the model completely misses the true response as indicated by the large validation RMSE which is much larger than the noise SD. This case is excluded from further analysis since the associated models are inaccurate. Table 1 summarizes the near-optimal number of bases and the corresponding basis

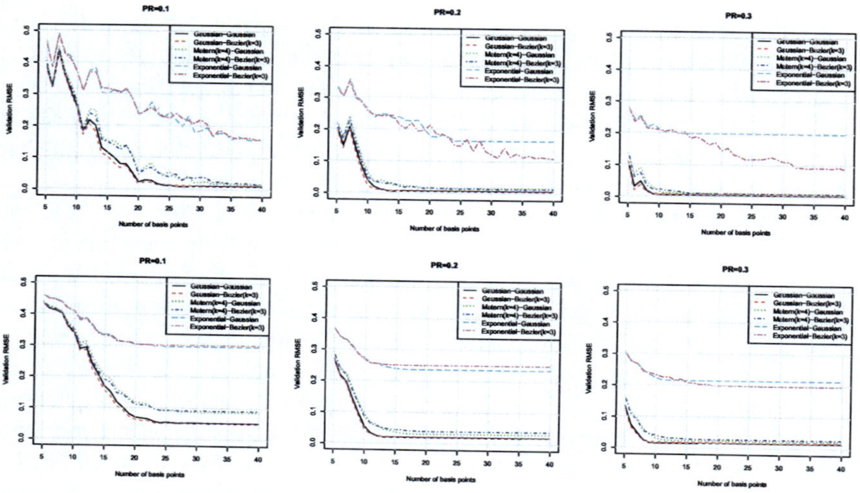

Fig. 2 Validation RMSE versus number of bases for each (correlation function, kernel) combination. Top row shows results for 1-D and bottom row shows results for 2-D

Table 1 Near-optimal number of bases and the corresponding basis spacing for each experiment

Dimension	Correlation function	DPCGP Kernel	Near-optimal (Number of bases, spacing)		
			PR = 0.1	PR = 0.2	PR = 0.3
1-D	Gaussian	Gaussian	(30, 0.0345)	(15, 0.0714)	(10, 0.1111)
	Gaussian	Bézier ($\kappa = 3$)	(30, 0.0345)	(15, 0.0714)	(10, 0.1111)
	Matérn ($\kappa = 4$)	Gaussian	(30, 0.0345)	(20, 0.0526)	(15, 0.0714)
	Matérn ($\kappa = 4$)	Bézier ($\kappa = 3$)	(35, 0.0294)	(20, 0.0526)	(15, 0.0714)
2-D	Gaussian	Gaussian	(25, 0.0417)	(15, 0.0714)	(10, 0.1111)
	Gaussian	Bézier ($\kappa = 3$)	(25, 0.0417)	(15, 0.0714)	(10, 0.1111)
	Matérn ($\kappa = 4$)	Gaussian	(30, 0.0345)	(20, 0.0526)	(15, 0.0714)
	Matérn ($\kappa = 4$)	Bézier ($\kappa = 3$)	(30, 0.0345)	(20, 0.0526)	(15, 0.0714)

spacing under each scenario. These numbers are based on visual inspection from Fig. 2, which may or not may be the exact optimal, but close enough for us to establish a rule-of-thumb for choosing the basis spacing. Results are roughly the same between 1-D and 2-D except for PR $= 0.1$, where DPCGP has good 1-D model fit but struggles to fit the 2-D true response resulting in validation RMSE being slightly higher than noise SD.

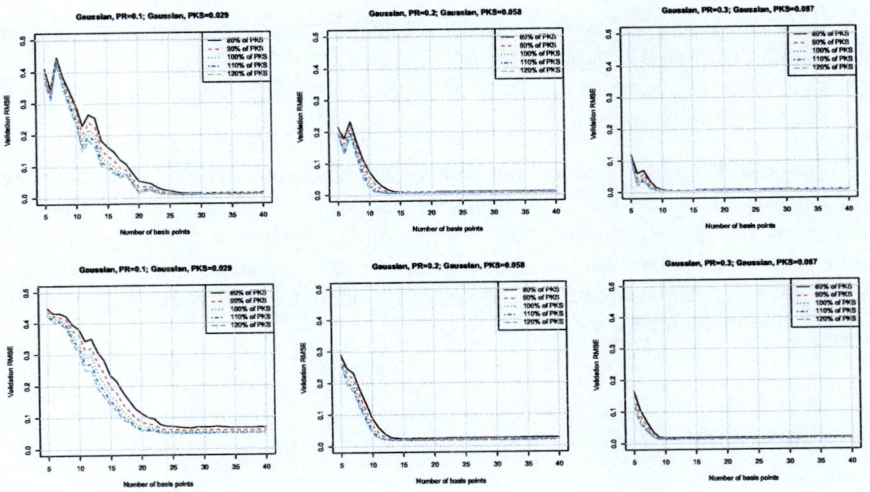

Fig. 3 Sensitivity to kernel size under Gaussian correlation function and Gaussian kernel (1-D at the top and 2-D at the bottom). Kernel SD is varied at 80%, 90%, 100%, 110%, and 120% of PKS

Figure 3 illustrates the effect of kernel size on model performance for two cases when the correlation function and the kernel are both Gaussian. Similar results are obtained for a Matern ($\kappa = 4$) correlation and Bézier ($\kappa = 3$) kernel. Here, kernel size is varied at 80%, 90%, 100%, 110%, and 120% of PKS. These results show that kernel size has limited effect on model performance given enough bases (small enough basis spacing), and the saturation point in each case roughly stays the same. In practice, DPCGP is mostly used for 2-D applications where the kernel induced correlation function is unlikely to be a perfect match to the true correlation structure. A rule-of-thumb for choosing the basis spacing should be robust enough to work in practice. We consider establishing the rule-of-thumb based on the 2-D case where the correlation function is Matern ($\kappa = 4$) and the kernel is Gaussian or Bézier ($\kappa = 3$) due to reasons described above. According to Table 1, the near-optimal basis spacing in this case is 0.0345 for PR = 0.1, 0.0526 for PR = 0.2, and 0.0714 for PR = 0.3. The relationship between basis spacing and PR appears to be linear: basis spacing $= \alpha \times$PR for PR> 0.1, where the least squares estimate of α is found to be around 0.2528. Hence, a reasonable rule-of-thumb for basis spacing can be established as PR/4 for PR≥ 0.1. Note that PR< 0.1 is excluded because DPCGP tends to be unreliable in this region for 2-D. Given this rule-of-thumb, the general steps for setting up a DPCGP model proceed as follows:

1. Scale the spatial domain to [0, 1] or $[0, 1]^2$ and estimate the dependence range from data.
2. Select a kernel whose induced correlation function best describes the process of interest.

3. Specify a kernel size such that the PR of the kernel induced correlation function equals the estimated dependence range in step 1.
4. Calculate basis spacing as estimated dependence range/4, and obtain the corresponding number of bases.

The estimated dependence range may be inaccurate and lead to an over/underestimated kernel size. This is acceptable provided that inaccuracy is not too large, because sensitivity analysis shows that model performance is not very sensitive to kernel size. Overestimation of dependence range can lead to an overestimated basis spacing. This issue can be alleviated by further reducing the basis spacing given by the rule-of-thumb as appropriate.

4 Summary

Research presented in this paper aims to obtain a rule-of-thumb for choosing the basis spacing for DPCGP models. A series of experiments is performed on simulated data based on Gaussian, Matérn ($\kappa = 4$), and Exponential correlation functions. Three different PR's (0.1, 0.2, and 0.3) are evaluated for each case. DPCGP models under the Gaussian and Bézier ($\kappa = 3$) kernels are developed on the training data using different number of bases ranging from 5 to 40. In each case, the kernel size is specified to match the dependence range of the resulting GP to the PR in data generation. Model performance is assessed via RMSE on validation data, and the near-optimal basis spacing is obtained for each PR. A rule-of-thumb for basis spacing is established as PR/4.

References

Brenning, A.: Geostatistics without stationarity assumptions within geographical information systems. Freiberg Online Geosci. **6**, 1–108 (2001)

Cressie, N.: Statistics for Spatial Data. Wiley, New York (1991)

Gramacy, R.B., Lee, H.K.H.: Bayesian treed Gaussian process models with an application to computer modeling. J. Am. Stat. Assoc. **103**, 1119–1130 (2008)

Higdon, D.: A process-convolution approach to modeling temperatures in the North Atlantic Ocean. J. Environ. Ecol. Stat. **5**(2), 173–190 (1998)

Lemos, R.T., Sansó, B.: Spatio-temporal model for mean, anomaly and trend fields of north Atlantic sea surface temperature. J. Am. Stat. Assoc. **104**, 5–18 (2009)

Paciorek, C.J., Schervish, M.J.: Spatial modelling using a new class of nonstationary covariance functions. Environmetrics **17**, 483–506 (2006)

Stein, M.L.: Interpolation of Spatial Data. Springer, New York (1999)

Yang, H., Liu, F., Ji, C., Dunson, D.: Adaptive sampling for Bayesian geospatial models. Stat. Comp. **24**, 1101–1110 (2014)

Estimation of Classification Rules From Partially Classified Data

Geoffrey McLachlan and Daniel Ahfock

Abstract We consider the situation where the observed sample contains some observations whose class of origin is known (that is, they are classified with respect to the g underlying classes of interest), and where the remaining observations in the sample are unclassified (that is, their class labels are unknown). For class-conditional distributions taken to be known up to a vector of unknown parameters, the aim is to estimate the Bayes' rule of allocation for the allocation of subsequent unclassified observations. Estimation on the basis of both the classified and unclassified data can be undertaken in a straightforward manner by fitting a g-component mixture model by maximum likelihood (ML) via the EM algorithm in the situation where the observed data can be assumed to be an observed random sample from the adopted mixture distribution. This assumption applies if the missing-data mechanism is ignorable in the terminology pioneered by Rubin (1976). An initial likelihood approach was to use the so-called classification ML approach whereby the missing labels are taken to be parameters to be estimated along with the parameters of the class-conditional distributions. However, as it can lead to inconsistent estimates, the focus of attention switched to the mixture ML approach after the appearance of the EM algorithm (Dempster et al. 1977). Particular attention is given here to the asymptotic relative efficiency (ARE) of the Bayes' rule estimated from a partially classified sample. Lastly, we consider briefly some recent results in situations where the missing label pattern is non-ignorable for the purposes of ML estimation for the mixture model.

Keywords Bayes' rule · Partially classified data · Semi-supervised learning

G. McLachlan (✉) · D. Ahfock
University of Queensland, Brisbane, QLD, Australia
e-mail: g.mclachlan@uq.edu.au

D. Ahfock
e-mail: d.ahfock@uq.edu.au

T. Chadjipadelis et al. (eds.), *Data Analysis and Rationality in a Complex World*, Studies in Classification, Data Analysis, and Knowledge Organization,
https://doi.org/10.1007/978-3-030-60104-1_17

1 Introduction

We consider the estimation of a classifier from a sample that is not completely classified with respect to the predefined classes. This problem goes back at least to the mid-seventies (McLachlan 1975), and it received a boost shortly afterwards with the advent of the EM algorithm (Dempster et al. 1977) which could be applied to carry out the maximum likelihood (ML) estimation for a partially classified sample. There is now a wide literature on the formation of classifiers on the basis of a partially classified sample or semi-supervised learning (SSL) as it is referred to in the machine learning literature. In the sequel, it is assumed that the features with known class labels are correctly classified, containing no misclassified features as, for example, in McLachnan (1972) and, more recently, Cannings et al. (2020).

More specifically, we focus on the case of $g = 2$ classes C_1 and C_2 in which the p-dimensional feature vector $\boldsymbol{Y}$ measured on an entity is distributed as

$$\boldsymbol{Y} \sim N(\boldsymbol{\mu}_i, \boldsymbol{\Sigma}) \quad \text{in} \quad C_i \quad (i = 1, 2). \tag{1}$$

We let $\boldsymbol{\theta}$ contain the $1 + 2p + \frac{1}{2}p(p+1)$ unknown parameters, consisting of the mixing proportion π_1, the elements of the class means $\boldsymbol{\mu}_1$ and $\boldsymbol{\mu}_2$, and the distinct elements of the common class covariance matrix $\boldsymbol{\Sigma}$. The Bayes' rule of allocation $R(\boldsymbol{y})$ in this case assigns an entity with observed feature vector $\boldsymbol{y}$ to either C_1 or C_2, accordingly as

$$d(\boldsymbol{y}) = \beta_0 + \boldsymbol{\beta}^T \boldsymbol{y}$$

is greater or less than zero, where

$$\beta_0 = -\tfrac{1}{2}(\boldsymbol{\mu}_1 + \boldsymbol{\mu}_2)^T \boldsymbol{\Sigma}^{-1}(\boldsymbol{\mu}_1 - \boldsymbol{\mu}_2) + \log(\pi_1/\pi_2),$$
$$\boldsymbol{\beta} = \boldsymbol{\Sigma}^{-1}(\boldsymbol{\mu}_1 - \boldsymbol{\mu}_2),$$

and where π_i denotes the prior probability of membership of C_i $(i = 1, 2)$; see, for example, McLachlan (1992).

2 History of SSL in Statistics

In his discussion of the paper read to the Royal Statistical Society by Hill (1966), Smith (1966) suggested that in the case of a completely unclassified sample which exhibits bimodality on some feature, a classifier be formed from the unclassified observations on the feature as follows: "One then arbitrarily divides them at the antimode, ... On the basis of this division, we calculate a suitable allocation rule; and, by using this allocation rule, get an improved division, and so on. As far as I know, there is no theoretical research into the effect of 'lifting oneself by one's own bootstraps' in this way."

This led McLachlan (1975) to consider this approach as suggested by Smith (1966) under the normal homoscedastic model (1). Under the latter assumption, the procedure is equivalent to treating the labels of the unclassified features as unknown parameters to be estimated along with $\boldsymbol{\theta}$. This approach became subsequently known as the classification maximum likelihood (CML) approach as considered by Hartley and Rao (1968) among others; see McLachlan and Basford (1988), Sect. 1.12. The CML approach gives an inconsistent estimate of $\boldsymbol{\theta}$ except in special cases like $\pi_1 = \pi_2$.

In order to make the problem analytically tractable for the calculation of the expected error rate of the estimated Bayes' rule, McLachlan (1975) assumed that there were also a limited number n_{ic} of classified features available from C_i in addition to the number of $n_u = n - n_c$ unclassified features, where n denotes the total size of the now partially classified sample and $n_c = n_{1c} + n_{2c}$.

In the sequel, we let $\boldsymbol{x}_{\text{CC}} = (\boldsymbol{x}_1^T, \ldots, \boldsymbol{x}_n^T)^T$ denote n independent realizations of $\boldsymbol{X} = (\boldsymbol{Y}^T, Z)^T$ as the completely classified training data, where Z denotes the class membership of $\boldsymbol{Y}$, being equal to 1 if $\boldsymbol{Y}$ belongs to C_1, and zero otherwise. We let m_j be the missing-label indicator being equal to 1 if z_j is missing and zero if it is available ($j = 1, \ldots, n$). Accordingly, the unclassified sample $\boldsymbol{x}_{\text{PC}}$ is given by those members $\boldsymbol{x}_j$ in $\boldsymbol{x}_{\text{CC}}$ for which $m_j = 0$ and only the feature vectors $\boldsymbol{y}_j$ without their class labels z_j for those members in $\boldsymbol{x}_{\text{CC}}$ for which $m_j = 1$.

3 Asymptotic Expected Error Rate of CML Approach

In practice, $\boldsymbol{\beta}$ has to be estimated from available training data. It can be calculated iteratively as described in the previous section. More formally, it can be obtained iteratively by applying the expectation–maximization (EM) algorithm of Dempster (1977) with the following modification (McLachlan 1982). Namely, the E-step is executed using outright (hard) rather than fractional (soft) assignment of each unclassified feature to a component of the mixture as with the standard application of the EM algorithm. We let $\hat{\boldsymbol{\beta}}_{\text{PC}}^{(k)}$ denote the estimator after the kth iteration of the vector $\boldsymbol{\beta} = (\beta_0, \boldsymbol{\beta}^T)^T$ of discriminant function coefficients obtained by the classification ML approach applied to the partially classified sample $\boldsymbol{x}_{\text{PC}}$. The estimated Bayes' rule using $\hat{\boldsymbol{\beta}}_{\text{PC}}^{(k)}$ for $\boldsymbol{\beta}$ in the Bayes' rule $R(\boldsymbol{\beta})$ is denoted by $\hat{R}_{\text{PC}}^{(k)}$. The (overall) conditional error rate of $\hat{R}_{\text{PC}}^{(k)}$ is denoted by $\text{err}(\hat{\boldsymbol{\beta}}_{\text{PC}}^{(k)}; \boldsymbol{\theta})$.

Then the expected excess error rate of the estimated Bayes' rule $\hat{R}_{\text{PCC}}^{(k)}$ is defined after the kth iteration by $E\{\text{err}(\hat{\boldsymbol{\beta}}_{\text{PC}}^{(k)}; \boldsymbol{\theta})\} - \text{err}(\boldsymbol{\theta})$, where $\text{err}(\boldsymbol{\theta})$ is the optimal error rate.

In the present SSL context, McLachlan (1975) showed in the case of equal, known prior probabilities that the overall expected error rate of this classifier after the kth iteration is given as $n_u \to \infty$, by

$$E\{\mathrm{err}(\hat{\boldsymbol{\beta}}_{\mathrm{PC}}^{(k)}; \boldsymbol{\theta})\} = \Phi(-\tfrac{1}{2}\Delta) + \{\phi(\tfrac{1}{2}\Delta)/4\}\, a_1^{(k)} + O(n_c^{-2}), \tag{2}$$

where

$$\begin{aligned} a_1^{(k)} &= h_1^{2k}\frac{\Delta}{4} + h_2^{2k}\frac{p-1}{\Delta}\left(\frac{1}{n_{1c}} + \frac{1}{n_{2c}}\right) + h_2^{2k}\frac{(p-1)\Delta}{n_c - 2}, \\ h_1 &= \phi(\tfrac{1}{2})[4\phi(\tfrac{1}{2}) + \Delta\{1 - 2\Phi(-\tfrac{1}{2})\}], \\ h_2 &= \{\phi(\tfrac{1}{2})\}^2\,(4 + \Delta^2)/h_1, \end{aligned}$$

and where $\Delta = \{(\boldsymbol{\mu}_1 - \boldsymbol{\mu}_2)^T \Sigma^{-1}(\boldsymbol{\mu}_1 - \boldsymbol{\mu}_2)\}^{1/2}$ is the Mahalanobis distance between the class-conditional distributions and $\mathrm{err}(\boldsymbol{\theta}) = \Phi(-\frac{1}{2}\Delta)$.

As it can be shown that both $|h_1|$ and $|h_2|$ are always less than one, and it follows from (2) that the expected error rate of $\hat{R}_{\mathrm{PC}}^{(k)}$ decreases after each iteration and converges to the optimal error rate as $k \to \infty$.

4 Asymptotic Relative Efficiency of ML Approach

The construction of classifiers from partially classified data can be undertaken also by the fitting of finite mixture models. The ML estimate of the vector of parameters $\boldsymbol{\theta}$ can be obtained via the EM algorithm of Dempster (1977). As noted in McLachlan (2000), it was the publication of this seminal paper that greatly stimulated interest in the use of finite mixture models.

We let

$$\log L_{\mathrm{C}}(\boldsymbol{\theta}) = \sum_{j=1}^{n}(1 - m_j)[z_j \log\{\pi_1\phi(\boldsymbol{y}_j; \boldsymbol{\mu}_1, \boldsymbol{\Sigma})\} + (1 - z_j)\log\{\pi_2\phi(\boldsymbol{y}_j; \boldsymbol{\mu}_2, \boldsymbol{\Sigma})\}] \tag{3}$$

$$\log L_{\mathrm{UC}}(\boldsymbol{\theta}) = \sum_{j=1}^{n} m_j \log \sum_{i=1}^{2} \pi_i \phi(\boldsymbol{y}_j; \boldsymbol{\mu}_i, \Sigma), \tag{4}$$

$$\log L_{\mathrm{PC}}^{(\mathrm{ig})}(\boldsymbol{\theta}) = \log L_{\mathrm{C}}(\boldsymbol{\theta}) + \log L_{\mathrm{UC}}(\boldsymbol{\theta}), \tag{5}$$

where $\phi(\boldsymbol{y}_j; \boldsymbol{\mu}, \boldsymbol{\Sigma})$ denotes the multivariate normal density with mean $\boldsymbol{\mu}$ and covariance matrix $\boldsymbol{\Sigma}$. In situations where one proceeds by ignoring the "missingness" of the class labels, $L_{\mathrm{C}}(\boldsymbol{\theta})$ and $L_{\mathrm{UC}}(\boldsymbol{\theta})$ denote the likelihood function formed from the classified data and the unclassified data, respectively, and $L_{\mathrm{PC}}^{(\mathrm{ig})}(\boldsymbol{\theta})$ is the likelihood function formed from the partially classified sample $\boldsymbol{x}_{\mathrm{PC}}$. The log of the likelihood $L_{\mathrm{CC}}(\boldsymbol{\theta})$ for the completely classified sample $\boldsymbol{x}_{\mathrm{CC}}$ is given by (3) with all $m_j = 0$.

Situations in the present context where it is appropriate to ignore the missing-data mechanism in carrying out likelihood inference are where the missing labels are missing at random in the framework for missing data pioneered by Rubin (1976).

Table 1 Asymptotic relative efficiency of $\hat{R}_{\mathrm{PC}}^{(\mathrm{ig})}$ compared to $\hat{R}_{\mathrm{CC}}$

π_1	$\Delta = 1$	$\Delta = 2$	$\Delta = 3$	$\Delta = 4$
0.1	0.0036	0.0591	0.2540	0.5585
0.2	0.0025	0.0668	0.2972	0.6068
0.3	0.0027	0.0800	0.3289	0.6352
0.4	0.0038	0.0941	0.3509	0.6522
0.5	0.0051	0.1008	0.3592	0.6580

This will be the case in the present context if the missingness of the labels does not depend on the features nor the labels (missing completely at random) or if the missingness depends only on the features (missing at random), as in McLachlan and Basford (1988).

We let $\hat{\boldsymbol{\theta}}_{\mathrm{CC}}$ and $\hat{\boldsymbol{\theta}}_{\mathrm{PC}}$ be the estimate of $\boldsymbol{\theta}$ formed by consideration of $L_{\mathrm{CC}}(\boldsymbol{\theta})$ and $L_{\mathrm{PC}}^{(\mathrm{ig})}(\boldsymbol{\theta})$, respectively, and we let $\hat{\boldsymbol{\beta}}_{\mathrm{CC}}$ and $\hat{\boldsymbol{\beta}}_{\mathrm{PC}}^{(\mathrm{ig})}$ be the estimates of $\boldsymbol{\beta}$ formed from the elements of $\hat{\boldsymbol{\theta}}_{\mathrm{CC}}$ and $\hat{\boldsymbol{\theta}}_{\mathrm{PC}}^{(\mathrm{ig})}$, respectively. The relative efficiency of the estimated Bayes' rule $R(\hat{\boldsymbol{\beta}}_{\mathrm{PC}}^{(\mathrm{ig})})$ compared to the rule $\hat{R}_{\mathrm{CC}}$ using $\hat{\boldsymbol{\beta}}_{\mathrm{CC}}$ for $\boldsymbol{\beta}$ based on the completely classified sample $\boldsymbol{x}_{\mathrm{CC}}$ is defined by

$$\mathrm{ARE}(R_{\mathrm{PC}}^{(\mathrm{ig})}) = \frac{E\{\mathrm{err}(\hat{\boldsymbol{\beta}}_{\mathrm{CC}}; \boldsymbol{\theta})\} - \mathrm{err}(\boldsymbol{\theta})}{E\{\mathrm{err}(\hat{\boldsymbol{\beta}}_{\mathrm{PC}}^{(\mathrm{ig})}; \boldsymbol{\beta})\} - \mathrm{err}(\boldsymbol{\theta})}, \tag{6}$$

where the expectation in the numerator and denominator of the right-hand side of (6) is taken over the distribution of the estimators of $\boldsymbol{\beta}$ and is expanded up to terms of the first order.

Under the assumption that the class labels are missing always completely at random, (that is, the missingness of the labels does not depend on the data), Ganesalingam and McLachlan (1978) derived the ARE of $\hat{R}_{\mathrm{PC}}^{(\mathrm{ig})}$ compared to $\hat{R}_{\mathrm{CC}}$ in the case of a completely unclassified sample ($\gamma = n_u/n = 1$) for univariate features ($p = 1$). Their results are listed in Table 1 for $\Delta = 1, 2, 3$, and 4. O'Neill (1978) extended their result to multivariate features and for arbitrary γ using the result of Efron (1975) for the information matrix of $\boldsymbol{\beta}$ in applying logistic regression. His results showed that this ARE was not sensitive to the values of p and does not vary with p for equal class prior probabilities. Not surprisingly, it can be seen from Table 1 that the ARE of $\hat{R}_{\mathrm{PC}}^{(\mathrm{ig})}$ for a totally unclassified sample is low, particularly for classes weakly separated as represented by $\Delta = 1$ in Table 1.

In other work on the ARE of $\hat{R}_{\mathrm{PC}}^{(\mathrm{ig})}$ compared to $\hat{R}_{\mathrm{CC}}$, McLachlan (1995) evaluated it where the unclassified univariate features had labels missing always at random due to truncation of the features.

5 Modelling Missingness for Unobserved Class Labels

In many practical applications, class labels are assigned by experts. Manual annotation of the dataset can induce a systematic missingness mechanism. This led Ahfock and McLachlan (2019a, b) to pursue the idea that the probability that a particular feature is unlabelled is related to the difficulty of determining its true class label. As an example, suppose medical professionals are asked to classify each image from a set of MRI scans into three groups, tumour present, no tumour present, or unknown. It seems reasonable to expect that the unassigned images will correspond to those that do not present clear evidence for the presence or absence of a tumour. The unlabelled images will exist in regions of the feature space where there is class overlap. In these situations, the unlabelled features carry additional information that can be used to improve the efficiency of parameter estimation.

The missing-data mechanism of Rubin (1976) is specified in the present context by the conditional distribution

$$\text{pr}\{M_j = m_j \mid \boldsymbol{y}_j,\, z_j; \boldsymbol{\kappa}\} \quad (j = 1,\, \ldots,\, n), \tag{7}$$

where $\boldsymbol{\kappa}$ is a vector of parameters. Ahfock and McLachlan (2019a, b) proposed that

$$\begin{aligned} \text{pr}\{M_j = 1 \mid \boldsymbol{y}_j, z_j\} &= \text{pr}\{M_j = 1 \mid \boldsymbol{y}_j\} \\ &= q(\boldsymbol{y}_j; \boldsymbol{\theta}, \boldsymbol{\xi}), \end{aligned} \tag{8}$$

where the parameter $\boldsymbol{\xi}$ is distinct from $\boldsymbol{\theta}$. On putting $\boldsymbol{\Psi} = (\boldsymbol{\theta}^T, \boldsymbol{\xi}^T)^T$, an obvious choice for the function $q(\boldsymbol{y}_j; \boldsymbol{\Psi})$ is the logistic model

$$q(\boldsymbol{y}_j; \boldsymbol{\Psi}) = \frac{\exp\{\xi_0 + \xi_1 e_j\}}{1 + \exp\{\xi_0 + \xi_1 e_j\}}, \tag{9}$$

where

$$e_j = -\sum_{i=1}^{2} \tau_i(\boldsymbol{y}_j; \boldsymbol{\theta}) \log \tau_i(\boldsymbol{y}_j; \boldsymbol{\theta}) \tag{10}$$

denotes the entropy for $\boldsymbol{y}_j$, and $\tau_i(\boldsymbol{y}_j; \boldsymbol{\theta})$ is the posterior probability that the jth entity with observed feature $\boldsymbol{y}_j$ belongs to Class C_i $(1 = 1, 2)$.

The log of the full likelihood function for $\boldsymbol{\Psi}$ is given by

$$\log L_{\text{PC}}^{(\text{full})}(\boldsymbol{\Psi}) = \log L_{\text{PC}}^{(\text{ig})}(\boldsymbol{\theta}) + \log L_{\text{PC}}^{(\text{miss})}(\boldsymbol{\Psi}), \tag{11}$$

where

$$\log L_{\text{PC}}^{(\text{miss})}(\boldsymbol{\Psi}) = \sum_{j=1}^{n} [(1 - m_j) \log\{1 - q(\boldsymbol{y}_j; \boldsymbol{\Psi})\} + m_j \log q(\boldsymbol{y}_j; \boldsymbol{\Psi})] \tag{12}$$

is the log likelihood function for $\boldsymbol{\Psi}$ formed on the basis of the missing-label indicators m_j $(j = 1, \ldots, n)$.

6 Fractionally Supervised Classification

In this section, we make use of the model (9) to examine the potential usefulness of fractionally supervised classification (FSC) as proposed by Vrbik and McNicholas (2015) and considered further by Gallaugher and McNicholas (2019). With this approach, the parameter $\boldsymbol{\theta}$ is estimated by consideration of the objective function $L_{\text{PC}}^{(\alpha)}(\boldsymbol{\theta})$ defined for a given α in [0,1] by

$$\log L_{\text{PC}}^{(\alpha)}(\boldsymbol{\theta}) = \alpha \log L_{\text{C}}(\boldsymbol{\theta}) + (1 - \alpha) \log L_{\text{UC}}(\boldsymbol{\theta}).$$

One suggestion for the choice of α in practice is to use BIC (Schwarz 1978).

We report here a simulation experiment undertaken by Ahfock and McLachlan (2019a) in which a partially classified sample of size $n = 500$ was generated on each of $N = 100$ replications. Bivariate features were generated from a mixture of two normal bivariate distributions in equal proportions ($\pi_1 = \pi_2 = 0.5$) with unequal covariance matrices, where the two components correspond to $g = 2$ classes. The component means were given by $\boldsymbol{\mu}_1 = (0, 0)^T$ and $\boldsymbol{\mu}_2 = (0, 3)^T$ with the component-covariance matrices having unit variances for both variables with correlation 0.7 in the first component and zero correlation in the second component. The conditional distribution of the missing-label indicators M_j was specified by the model (9) with $\xi_0 = -5$ and $\xi_1 = 100$. For each partially classified sample $\boldsymbol{x}_{\text{PC}}$ generated, the estimate $\hat{\boldsymbol{\theta}}_{\text{PC}}^{(\alpha)}$ of $\boldsymbol{\theta}$ was calculated via maximization of the objective function $L_{\text{PC}}^{(\alpha)}$ for a grid of values of α, along with the estimate $\hat{\boldsymbol{\theta}}_{\text{PC}}^{(\text{full})}$ using the full likelihood function $L_{\text{PC}}^{(\text{full})}(\boldsymbol{\Psi})$. On each replication, the adjusted Rand index (ARI) for the estimated Bayes' rule was obtained by applying it to 2,000 data points in a test set. The average values of these ARI's are displayed in Fig. 1. They show that as α moves away from a small neighbourhood of $\alpha = 0.5$, the performance of the rule using the fractionally supervised estimate falls dramatically. The horizontal line in Fig. 1 is the simulated value of the ARI for the use of $\hat{\boldsymbol{\theta}}_{\text{PC}}^{(\text{full})}$.

The advantage of this simulation experiment is that it provides a framework for examining the behaviour of a proposed fractionally supervised approach to the clustering of a partially classified sample. A realistic model is used to generate the unclassified data according to the degree of difficulty in their being classified correctly.

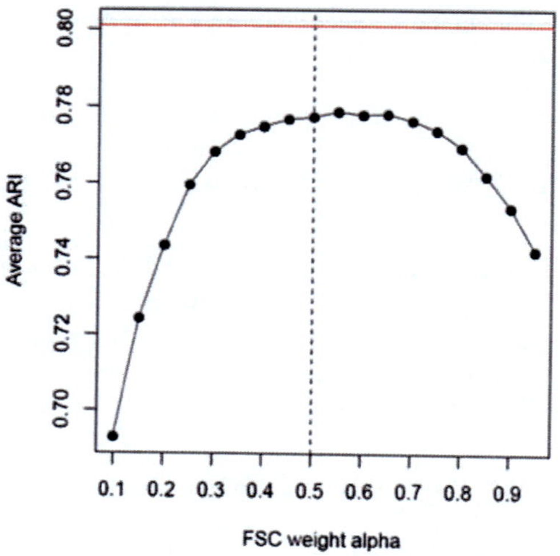

Fig. 1 Plot of simulated ARI for various values of α

References

Ahfock, D., McLachlan, G.J.: On missing data patterns in semi-supervised learning. ePreprint arXiv:1904.02883 (2019a)

Ahfock, D., McLachlan, G.J.: An apparent paradox: a classifier trained from a partially classified sample may have smaller expected error rate than that if the sample were completely classified. ePreprint arXiv:1910.09189v2 (2019b)

Cannings, T.I., Fan, Y., Samworth, R.J.: Classification with imperfect training labels. Biometrika **107**, 311–330 (2020)

Dempster, A.P., Laird, N.M., Rubin, D.B.: Maximum likelihood from incomplete data via the EM algorithm (with discussion). J. R. Statist. Soc. B **39**, 1–22 (1977)

Efron, B.: The efficiency of logistic regression compared to normal discriminant analysis. J. Am. Stat. Assoc. **70**, 892–898 (1975)

Gallaugher, M., McNicholas, P.D.: On fractionally-supervised classification: weight selection and extension to the multivariate t-distribution. J. Classif. **36**, 232–265 (2019)

Ganesalingam, S., McLachlan, G.J.: The efficiency of a linear discriminant function based on unclassified initial samples. Biometrika **65**, 658–665 (1978)

Hartley, H.O., Rao, J.N.K.: Classification and estimation in analysis of variance problems. Int. Stat. Rev. **36**, 141–147 (1968)

Hills, M.: Allocation rules and their error rates (with discussion). J. R. Statist. Soc. B **28**, 1–31 (1966)

McLachlan, G.J.: Asymptotic results for discriminant analysis when the initial samples are misclassified. Technometrics **14**, 415–422 (1972)

McLachlan, G.J.: The classification and mixture maximum likelihood approaches to cluster analysis. In: Krishnaiah, P.A. Kanal, L. (eds.) Handbook of Statistics Vol. 2, North-Holland, pp. 199–208. Amsterdam (1982)

McLachlan, G.J.: Discriminant Analysis and Statistical Pattern Recognition. Wiley, New York (1992)

McLachlan, G.J.: Iterative reclassification procedure for constructing and asymptotically optimal rule of allocation in discriminant analysis. J. Am. Stat. Assoc. **70**, 365–369 (1975)

McLachlan, G.J., Basford, K.E.: Mixture Models: Inference and Applications to Clustering. Marcel Dekker, New York (1988)

McLachlan, G.J., Gordon, R.D.: Mixture models for partially unclassified data: a case study of renal venous renin levels in essential hypertension. Stat. Med. **8**, 1291–1300 (1989)
McLachlan, G.J., Peel, D.: Finite Mixture Models. Wiley, New York (2000)
McLachlan, G.J., Scot, D.: On the asymptotic relative efficiency of the linear discriminant function under partial nonrandom classification of the training data. Stat. Comput. Simul. **52**, 452–456 (1995)
O'Neill, T.J.: Normal discrimination with unclassified observations. J. Am. Stat. Assoc. **73**, 821–826 (1978)
Rubin, D.B.: Inference and missing data. Biometrika **63**, 581–592 (1976)
Schwarz, G.: Estimating the dimension of a model. Ann. Stat. **6**, 461–464 (1978)
Smith, C.A.B.: Contribution to the discussion of paper by M. Hills. J. R. Stat. Soc. **28**, 21 (1966)
Vrbik, I., McNicholas, P.D.: Fractionally-supervised classification. J. Classif. **32**, 359–381 (2015)

Correspondence Analysis and Kriging: Projection of Quantitative Information on the Factorial Maps

George Menexes and Thomas Koutsos

Abstract In this study, a methodological scheme is proposed for the combined use of Analyse Factorielle des Correspondances—AFC (or Correspondence Analysis) and the Ordinary Kriging method to display values of quantitative variables as supplementary points (or as "supplementary data") onto the factorial maps (or planes) resulting from the application of AFC to a contingency table of two categorical variables. The proposed method can also be generalized in the case of Multiple Correspondence Analysis (Analyse des Correspondances Multiples). The kriging method is widely used as one of the most effective spatial interpolation techniques. The proposed methodological scheme is demonstrated using hypothetical data from a 5 × 4 contingency table (sites×crops). Fertilizer mean costs will be used as supplementary points or "supplementary data". Also, a specific data coding scheme is proposed aiming at a better presentation and interpretation of the graphical results.

Keywords Supplementary points · Spatial interpolation · Factorial planes · Ordinary kriging

1 Introduction

Analyse Factorielle des Correspondances (AFC) (in French, or Correspondence Analysis, in English) is a non-linear multidimensional data analysis method suitable for graphically exploring the association between two or more categorical variables. The main goal of AFC is to highlight and graphically represent visible and/or hidden relations in the data structure. The method has found numerous applications in the social sciences (De Nooy 2003; Torres and Greenacre 2002) and biometry (ter Braak

G. Menexes · T. Koutsos (✉)
School of Agriculture Faculty of Agriculture Forestry and Natural Environment Hellas, Aristotle University of Thessaloniki, 54124 Thessaloniki, Greece
e-mail: tkoutsos@agro.auth.gr

G. Menexes
e-mail: gmenexes@agro.auth.gr

T. Chadjipadelis et al. (eds.), *Data Analysis and Rationality in a Complex World*, Studies in Classification, Data Analysis, and Knowledge Organization,
https://doi.org/10.1007/978-3-030-60104-1_18

et al. 2002). Although AFC is intended for the analysis of two-way contingency tables, it can also handle many different types of data matrices. In practice, AFC is applied to a transformed two-way matrix (not necessarily a contingency table), constructed from the initial data. The only input requirement of the method is a matrix with homogeneous non-negative entries, which has at least one non-zero entry to every column and every row.

A conceptual presentation of AFC can be provided in a great variety of ways (Benzécri 1992; Greenacre 2017). That is probably the main reason why this method "surfaced" on several occasions and with various names during the twentieth century. As a descriptive technique, it is used to examine the relationships between categorical variables by mapping the initial data onto factorial axes, in a similar way to Principal Components Analysis. Both the theoretical and practical aspects of AFC have been covered in depth in a large collection of introductory and advanced textbooks (Benzécri 1992; Greenacre 1984, 2017; Lebart et al. 1984).

One advantage of AFC is that it can handle additional rows and/or columns of data that are not of primary research and exploration interest, but they are very useful for interpreting the results that were obtained from the analysis of the original data (Benzécri 1992; Greenacre 1984, 2017). These additional data can be projected (or suppressed) afterwards on the factorial maps obtained from the application of AFC on the original data, without affecting the relationship between the variables, the relative positions of the projected rows and columns, and, consequently, having no influence on the construction of the factorial axes. The only restriction is that these additional data, called "supplementary", should be comparable to the existing original categorical data and on the same scale. The supplementary points could be considered as "supplementary data" ("loanword", term traditionally used in libraries), since they can provide additional information and insights about one or more features of the primary data and/or summarize basic information which can make interpretation of the originally analysed data easier. An issue arises when the additional supplementary data is not categorical but quantitative (measured on interval or ratio scale). As far as we know, there are no published studies addressing this issue. In the present study, a methodological scheme is proposed for the combined use of AFC and the kriging method (Matheron 1963) to display values of quantitative variables as supplementary points (or as "supplementary data") on the factorial planes resulting from the application of AFC on a contingency table of two categorical variables.

The paper is organized as follows: Sects. 2 and 3 are devoted to a conceptual presentation of the bivariate and multivariate version of AFC, respectively. We will not provide a detailed mathematical description of the method since it is well known and grounded. In Sect. 4, a short presentation of the kriging method is given. In Sect. 5, the application of the proposed methodological scheme is presented using hypothetical data from a 5×4 contingency table (sites$\times$crops). Fertilizer costs (mean values) will be the supplementary points or the "supplementary data". In addition, a specific data coding scheme is proposed aiming at a better presentation and interpretation of the results. The paper concludes in Sect. 5.

2 Simple and Multiple Correspondence Analysis

In the bivariate case, the input data for AFC (or Simple Correspondence Analysis-SCA) is a $k \times m$ contingency table of frequencies. AFC can be viewed as a method that quantifies qualitative data, with a simultaneous dimensionality reduction. Technically, AFC assigns weights and scores to the rows and columns of the contingency table in such a way that many interesting properties are optimized (Greenacre 1984). Multiple Correspondence Analysis—MCA (Analyse des Correspondances Multiples, in French) also known as Homogeneity Analysis or Dual Scaling is an extension of SCA suitable for analysing more than two categorical variables. MCA may be described as an SCA of the indicator or design matrix, **Z**, with binary or dummy coding (0–1). The rows of **Z** are defined by the observations (subjects or objects), and the columns by the total number of variable categories or modalities. Each cell matrix of **Z** contains either "0" or "1". Since only one modality of a variable can be assigned, each row contains p "1s" and $q - p$ "0s" (disjunctive form), where p is the total number of variables and q is the total number of categories of the p variables.

3 The Kriging Method

3.1 Introduction to Kriging

The problem of predicting values at non-sampled points based on existing observed data is of great interest in many scientific disciplines. Therefore, methods for delivering accurate predictions, such as the best linear unbiased prediction (BLUP), are highly appreciated. Kriging is a geostatistical interpolation gridding method and has been proven useful and popular in many fields, producing visually appealing maps from irregularly spaced data (Krige 1951). Kriging also incorporates anisotropy and underlying trends, suggested in existing data, in an efficient and natural manner, producing new accurate grids of data. Originally, the idea of "optimal linear prediction" was presented by Kolmogorov (1991).

3.2 Kriging Types and Modelling Tools for Interpolation

Irrespective of the kriging type, the final aim is always to achieve the best linear unbiased prediction based on the calculation of prediction variance as the variance of the difference of the linear predictor and the measured data. Therefore, using kriging we can predict the value of the random function $Z = Z(x)$ at any arbitrary location of interest x_0, i.e. the value $Z(x_0)$, based on the nearest measured observations $z(x_i)$ of $Z(x)$ at the $n \in \mathbb{N}$ sample points x_i. In fact, kriging uses a weighted average of these observations at the sample points xi to reveal the spatial structure of data. These weights depend on the assumptions on the $\mu\,(x)$, as well as on the variogram or covariance function of $Z(x)$. As this prediction variance minimizes, the accuracy of the linear predictor increases, and the best linear prediction is achieved (Kolmogorov 1991; Matheron 1963).

There are several types of kriging but all rely on the same concept of predicting values at not sampled locations, Webster and Oliver (2007): (1) *Simple Kriging* is the simplest case of kriging prediction used for predicting Z(x) at any arbitrary point x_0; (2) *Ordinary Kriging* is the most frequently used interpolation kriging type in practice, and (3) *Universal Kriging* is the most general considered model compared to the previous ones. Predicting the mean value of Z(x) over the spatial data domain D is called Kriging the mean. For the needs of this work, Ordinary Kriging was used.

4 Application of the Proposed Method

Combining multiple correspondence analysis with factorial kriging analysis has been used for geochemical mapping (Reis et al. 2004). The combined use of these two methods was also effective in the case of separating the chemical elements that are related to natural sources from the anthropogenic sources (Patinha et al. 2008). The Kriging estimation procedure has also been tested in the case of the construction of designs for test-control field experiments (Wiens and Zhou 2008) but not in combination with AFC, according to the knowledge of the authors. In this section, the application of the proposed methodological scheme is presented using, as an example, hypothetical data from a 5 × 4 contingency table of absolute frequencies (sites × crops). Fertilizer mean costs will be used as the supplementary points or the "supplementary data" (Table 1).

Plots of sites' (S) and crops' (C) profiles are presented separately on AFC factorial maps 1×2 in Fig. 1a and b, respectively, along with the projection of the spatial distribution of the corresponding mean cost of the fertilizer used calculated by using Ordinary Kriging. The AFC was performed using the CHIC software ver. 1.1 (Markos

Table 1 Distribution of four different plant species (C) (C1: rice; C2: winter cereals; C3: corn; and C4: cotton) at five different geographical sites (S), along with the corresponding Mean Cost (in Euros) of the fertilizer used

Crops sites	C1	C2	C3	C4	Total	Mean cost
S1	23	10	3	11	47	14,064
S2	61	15	7	15	98	19,286
S3	26	8	4	7	45	28,378
S4	13	8	7	12	40	25,225
S5	8	1	0	5	14	35,929
Totals	131	42	22	50	245	21,885
Mean cost	21,954	18,095	28,381	22,160	21,885	

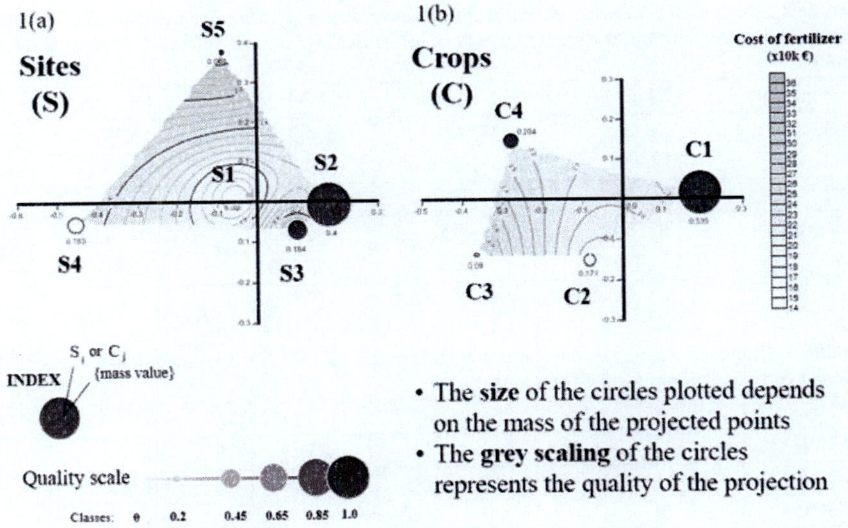

Fig. 1 Plots of sites' (S) and crops' (C) profiles separately, on (1a) and (1b), respectively, along with the projection of spatial distribution of the corresponding mean cost of fertilizer used, estimated by Ordinary Kriging. The size of the circles plotted depends on the mass of the projected points, while the grey colour scaling of the circles represents the quality of the projection (where C1: rice; C2: winter cereals; C3: corn; and C4: cotton)

et al. 2010). The coordinates of the sites' and crops' points on the AFC factorial axes along with the corresponding fertilizer mean costs from Table 1 were entered in Surfer ®*R* v.13.4.553, which was used for the spatial interpolation. Site S1 is mainly associated with a cost of about 14,000 Euros concerning the cost of fertilizer used, site S2 with about 19,000 Euros, site S3 with about 28,000 Euros, site S4 with about 25,000 Euros, and site S5 with about 36,000 Euros, respectively (Fig. 1a). The graphical results presented in Fig. 1b, that is the crops' profiles, can be interpreted in the same way. For a specialist in agriculture, the biological interpretation of the differentiation between the sites and between the crops could be enriched with the additional "supplementary data" relative to the corresponding mean cost of fertilizer used.

The problem, now, is how the data presented in Table 2 (individual mean costs of fertilizer used per site and crop) could be utilized to enhance the presentation and interpretation of the association (or interaction) between the sites and the crops. For this reason, we propose the following specific data coding scheme (Table 3): (a) Let **Z** be the 245×9 design matrix with "logical" (binary) coding (0–1), as in the case of Multiple Correspondence Analysis for two categorical variables (5 sites and 4 crops), (b) identify each one of the 245 rows (objects) by the corresponding combination of the site by crop categories (5×4=20 combinations), (c) aggregate the **Z** matrix according to the Distributional Equivalence Theorem (Benzécri, 1992) in order to form the 20×9 contingency table, (d) apply the AFC method on the aggregated matrix of the previous step, (e) enter the estimated coordinates of the combined

Table 2 Mean Cost (in Euros) of fertilizer used individually noted at five different geographical sites (S), and for four different plant species (C) (C1: rice; C2: winter cereals; C3: corn; and C4: cotton)

Crops sites	C1	C2	C3	C4
S1	13,957	17,200	4,333	14,091
S2	18,738	16,867	29,857	19,000
S3	28,808	25,625	30,250	28,857
S4	30,154	15,375	36,143	20,083
S5	33,875	7,000	45,000	35,929

Table 3 Proposed coding and data table combining sites (S1, S2, S3, S4) and crops (C1: rice; C2: winter cereals; C3: corn; and C4: cotton)

	S1	S2	S3	S4	S5	C1	C2	C3	C4
S1C1	23	0	0	0	0	23	0	0	0
S2C1	0	61	0	0	0	61	0	0	0
S3C1	0	0	26	0	0	26	0	0	0
S4C1	0	0	0	13	0	13	0	0	0
S5C1	0	0	0	0	8	8	0	0	0
S1C2	10	0	0	0	0	0	10	0	0
S2C2	0	15	0	0	0	0	15	0	0
S3C2	0	0	8	0	0	0	8	0	0
S4C2	0	0	0	8	0	0	8	0	0
S5C2	0	0	0	0	1	0	0	0	0
S1C3	3	0	0	0	0	0	0	3	0
S2C3	0	7	0	0	0	0	0	7	0
S3C3	0	0	4	0	0	0	0	4	0
S4C3	0	0	0	7	0	0	0	7	0
S5C3	0	0	0	0	1	0	0	1	0
S1C4	11	0	0	0	0	0	0	0	11
S2C4	0	15	0	0	0	0	0	0	15
S3C4	0	0	7	0	0	0	0	0	7
S4C4	0	0	0	12	0	0	0	0	12
S5C4	0	0	0	0	5	0	0	0	5

profiles (combinations of sites by crops categories) on the AFC factorial axes along with the data presented in Table 2 in the Ordinary Kriging method, and (f) plot the data. Figure 2 presents the plot of the sites' (S) and crops' (C) combined profiles (20 combinations) along with the projected spatial distribution of the corresponding mean costs of the fertilizer used per site and crop (from Table 2).

Finally, in both Figs. 1 and 2, the size of the circles (markers) plotted depends on the corresponding points' mass values, while the greyscale colour of the circles

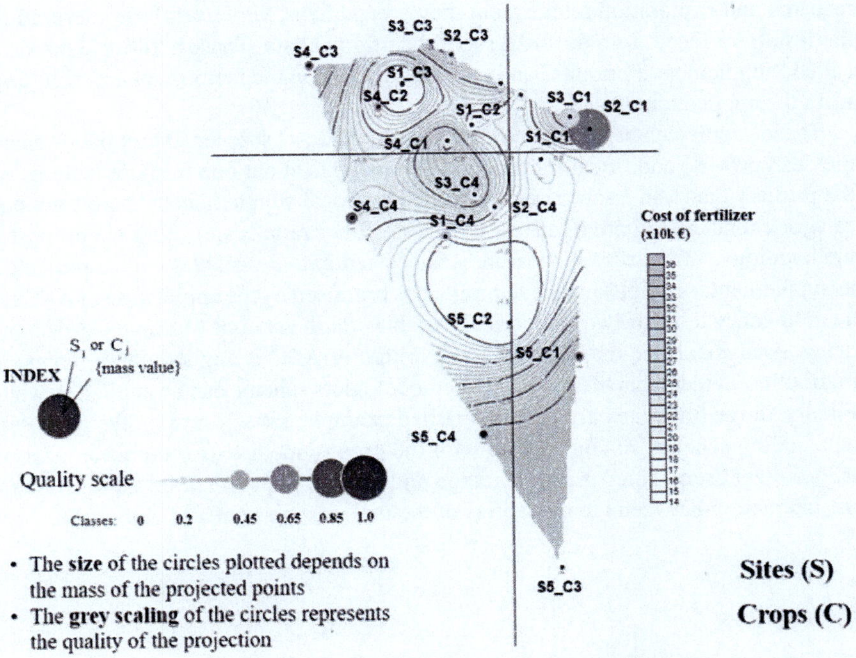

Fig. 2 Plotting the sites' (S) and crops' (C) profiles along with the spatial distribution of mean cost (in Euros) of the fertilizer used, estimated by Ordinary Kriging. The size of the circles plotted depends on the mass of the projected points, while the grey scaling colour of the circles represents the quality of the projection (C1: rice; C2: winter cereals; C3: corn; and C4: cotton)

depends on the quality of the corresponding projected points. Now, based on the information presented in Fig. 2, it can be noted, for example, that the combination S5–C3 (that is, corn in site S5) is associated with very high costs of the fertilizer used, the combination S5–C4 (that is, cotton in site S5) is associated with high costs, the combination S2–C4 (that is, cotton in site S2) is associated with medium costs, and the combination S5–C2 (that is winter cereals in site S5) is associated with very low costs of fertilizer used.

5 Conclusions

The kriging method is widely used as one of the most effective spatial interpolation techniques, mainly in geosciences. The AFC (Correspondence Analysis) method is a non-linear multidimensional data analytic method suitable for graphically exploring the association between two or more categorical variables, highlighting and graphically representing visible and/or hidden relations in the data structure. One advantage of the AFC is that it can handle additional categorical data that are not of primary research and exploration interest, but they are, perhaps, very useful for interpreting the results surfaced from the analysis of the original data. These additional data are called "supplementary points" and can be projected on the factorial planes resulting from the application of AFC.

These supplementary points could be considered as "supplementary data", since they can provide additional information and insights about one or more features of the primary data. An issue arises when the additional supplementary points are not categorical but quantitative (measured on interval or ratio scale). With the proposed methodology, values or statistical indices of quantitative variables can be projected as supplementary points on the factorial maps produced by the application of AFC on a contingency table of two categorical variables. In the case of Multiple Correspondence Analysis, since the same basic algorithm of AFC is applied on an indicator matrix, it is evident that the proposed methodological scheme can be applied as well. Finally, in the bivariate case (in the reported example, sites × crops), the proposed data coding scheme, in conjunction with the kriging method (for the mean cost of the fertilizer used), aided the presentation and the interpretation of the association or the interaction between the modalities of the two variables entered in the AFC.

References

Benzécri, J.-P.: Correspondence Analysis Handbook, CRC Press LLC (1992)

De Nooy, W.: Fields and networks: correspondence analysis and social network analysis in the framework of field theory. Poetics **31**(5–6), 305–327 (2003)

Greenacre, M.J.: Theory and Applications of Correspondence Analysis. Academic Press, London (1984)

Greenacre, M.J.: Correspondence Analysis in Practice. CRC Press (2017)

Kolmogorov, A.N.: The local structure of turbulence in incompressible viscous fluid for very large Reynolds numbers. Proc. Math. Phys. Eng. Sci. **434**, 9–13 (1991)

Krige, D.G.: A statistical approach to some basic mine valuation problems on the Witwatersrand. J. South. Afr. Inst. Min. Metall. **52**, 119–139 (1951)

Lebart, L., Morineau, A., Warwick, K.M.: Multivariate Descriptive Statistical Analysis: Correspondence Analysis and Related Techniques for Large Matrices. Wiley, New York (1984)

Markos, A., Menexes, G., Papadimitriou, I.: The CHIC analysis software v1.0. In: Loracek-Junge, H., Weihs, C. (eds.) Classification as a Tool for Research, pp. 409–416. Springer, Berlin (2010)

Matheron, G.: Principles of geostatistics. Econ. Geol. **58**, 1246–1266 (1963)

Patinha, C., Correia, E., Ferreira da Silva, E., Simoes, A., Reis, P., Morgado, F., Cardoso Fonseca, E.: Definition of geochemical patterns on the soil of Paul de Arzila using correspondence analysis. J. Geochem. Explor. **98**, 34–42 (2008)

Reis, A.P., Sousa, A.J., Ferreira de Silva, E., Patinha, C., Fonseca, E.C.: Combining multiple correspondence analysis with factorial kriging analysis for geochemical mapping of the gold–silver deposit at Marrancos (Portugal). Appl. Geochem. **19**, 623–631 (2004)

ter Braak, C.: Ordination. In: Jongman, R., ter Braak, C., van Tongeren O. (eds.) Data Analysis in Community and Landscape Ecology, pp. 91–173. Cambridge University Press (2002)

Torres, A., Greenacre, M.: Dual scaling and correspondence analysis of preferences, paired comparisons and ratings. Int. J. Res. Mark. **19**, 401–405 (2002)

Webster, R., Oliver, M.A.: Geostatistics for Environmental Scientists, 2nd edn. Wiley (2007)

Wiens, D.P., Zhou, Z.: Robust estimators and designs for field experiments. J. Stat. Plan Inference. **138**, 93–104 (2008)

Intertemporal Exploratory Analysis of E-Commerce From Greek Households from Official Statistics Data

Stratos Moschidis and Athanasios Thanopoulos

Abstract E-commerce worldwide is transforming the economy at the macro level while also affecting the consumption habits of households. This paper aims to map the effects of these changes through exploratory data analysis. For this purpose, data from Official Statistics were analyzed, as they were collected via sample surveys of Greek Statistical Authority (ELSTAT) from 2009 to 2018. This work is of multiple interest, as not only the phenomenon under study is an area of general scientific interest, but also the period of data collection includes the time horizon of the beginning of the economic crisis in Greece.

Keywords Official Statistics · Exploratory analysis · Clustering · e-commerce

1 Introduction

E-commerce as a form of commerce continues to dynamically reshape the global economy, Thiebaut (2019), with significant effects on the domestic economies, Ho et al. (2007); Laudon and Traver (2019). National Statistical Services (NSS) carry out census surveys on household use of e-commerce. Similarly, the Greek Statistical Authority (ELSTAT) records household use of e-commerce on a yearly basis as part of the European Statistical System (ESS). This recording started in 2003 and has been systematically organized since 2008. It should be mentioned that the use of e-commerce by households is a subset of the annual survey on the use of information and communication technologies by households (ICT).

It is well known that e-commerce can be a growth driver for an economy, Chaffey et al. (2019); Laudon and Traver (2019). It is also widely accepted that the last decade (2009–2018) has been important for Greece, both politically and financially.

S. Moschidis (✉) · A. Thanopoulos
Hellenic Statistical Authority, 46 Pireos St. Eponiton St., 185 10 Piraeus, Greece
e-mail: smos@statistics.gr

A. Thanopoulos
e-mail: a.thanopoulos@statistics.gr

T. Chadjipadelis et al. (eds.), *Data Analysis and Rationality in a Complex World*, Studies in Classification, Data Analysis, and Knowledge Organization,
https://doi.org/10.1007/978-3-030-60104-1_19

Events such as the economic crisis, financial surveillance, Savage and Verdun (2015), the recourse to International Monetary Fund (IMF), Kickert and Ongaro (2019); Visvizi (2012), and continuous political changes have been milestones of this period. The purpose of this work is to investigate the use of e-commerce services by Greek households via descriptive and exploratory data analysis. Although implied, a cause-effect type relationship is not examined in the present work. However, descriptive and exploratory analysis of the use of e-commerce data, based on official statistics, can give us valuable insights into the way individuals and their characteristics have interacted over time, Kitsios et al. (2013). The findings can be used as a basis for discussion on the adoption of policies related to economic development and administrative reforms. In addition, they can be a useful tool for stakeholders in developing entrepreneurship in the field of e-commerce and the entrepreneurial ecosystem that surrounds it.

2 Methodology

The research design is a survey study, based on three-dimensional stratified sampling, following the established standards of Eurostat. The survey data were obtained by Greek Statistical Authority for the period between 2009 and 2018. Data collection was carried out via a questionnaire that was created by the authority for this purpose. During the process of questionnaire design and training, the requirement of data comparability imposes on all Member States to take into account the guidelines and the proposed content by Eurostat, once it is adapted to the particularities of each country, without however being able to change the questions from which the e-Europe indicators arise. A total of 9, 460 questionnaires were received.

The following variables were used in the study:

- SEX: sex1 (male), sex2 (female)
- EMPST (employment status): empst1 (Employed or self-employed (including family-owned workers), empst2 (Unemployed), empst3 (Student), empst4 (Not financially active (retired, soldier, inactive, etc.).))
- HHIQ (Household income): hhiq1 (Lowest quartile), hhiq2 (second lowest quartile), hhiq3 (second highest quartile), hhiq4 (highest quartile)
- AGE from continuous variable transformed into 4 categories: age1 (16–18), age2 (18–34), age3 (35–50), age4(50–)
- FOOD (food purchase): food0 (no purchase), food1 (purchase)
- MED (medicines purchase): med0 (no purchase), med1 (purchase)
- CLOTH (clothing purchase): cloth0 (no purchase), cloth1 (purchase)
- HOL (holiday purchase): hol0 (no purchase), cloth1 (purchase)
- TIC (ticket purchase): tic0 (no purchase), tic1 (purchase)
- YEAR (year of purchase, 2009–2018): year9, year10, year11, ..., year18.

Data analysis was based on Multiple Correspondence Analysis and Ascending Hierarchical Clustering, applied in the framework of the French School of Data

Analysis, Benzécri (1992). Descriptive analysis was performed using IBM SPSS Statistics version 23 and exploratory methods were applied with the software package M.A.D, Karapistolis (2002).

3 Results

One of the most interesting findings of descriptive analysis was a notable increase (250%) in e-commerce food purchases during the last year of the survey (2018). A steady increase was also observed in the participation of women in e-commerce, which for the first time surpassed that of men in 2015, while the same was observed in 2018. We should also underline a significant increase in the participation of people belonging to the older age group.

Multiple Correspondence Analysis (MCA) was applied to the most highlight dominant trends from 2009 to 2018, Moschidis (2009), in relation to the e-commerce respondents' choices (see also Papaioannou et al. 2015). Data input was a slice of the Burt table or a Burt subtable, as shown in Table 1 (see also Moschidis 2015). Years are in the rows of the data table and the remaining variables are in the columns, in order to give a temporal overview of the trends under investigation.

Table 2 shows the absolute and relative inertia of each axis. We observe that the addition of the fourth axis contributes a very small percentage to the total inertia and therefore we will consider only the first three axes for interpretation. These three axes account for about 92% of the total inertia.

Table 3 presents the most important points of the row-column cloud forming the first factorial axis (F1). As most important were considered the points that had a CTR value higher than the average contribution for this axis. The average contribution (cut-off value) for the row points is given by the ratio $\frac{1000}{\text{number of rows}}$ and the average contribution for the column points is given by $\frac{1000}{\text{number of columns}}$.

The first axis contrasts the years 2013 and 2018. The year 2018 is characterized by individuals who purchase food, clothes ,or medical products online and belong to the highest income quartile. The year 2013 is associated with individuals who do no buy clothes online and belong to the second lowest income quartile.

The second axis contrasts the years 2009–2011 with the year 2016. Years 2009–2011 are associated with individuals who belong to the highest income quartile, buy holiday packages online and do not buy clothes. The year 2016 is associated with individuals who buy clothing and medical products online and belong to the second lowest income quartile.

Finally, the third axis contrasts years 2012 and 2013 with years 2014 and 2015. The former group of years is associated with individuals from the lowest income quartile, who buy holiday products online. The latter year group is associated with individuals that belong to the second highest income quartile and buy tickets online.

The first factorial plane, shown in Fig. 1, can be considered important because the first and the second factorial axes explain together almost 76% of the total variance

Table 1 Burt subtable with years in the rows and the remaining variables in the columns

IND	sex1	sex2	empst1	empst2	empst3	empst4	hhiq1	...	hhiq2	cloth1	hol0	hol1	tic0	tic1
Year9	233	152	288	15	44	38	89	...	68	84	297	88	309	76
Year10	289	184	343	26	58	46	124	...	107	113	358	115	380	93
Year11	396	287	444	63	95	81	77	...	105	206	437	246	546	137
Year12	356	263	403	89	71	56	186	...	286	250	456	163	463	156
Year13	435	374	514	109	91	95	308	...	397	288	637	172	638	171
Year14	452	520	567	127	124	154	132	...	347	484	794	178	816	156
Year15	692	608	823	166	144	167	188	...	467	576	1131	169	1137	163
Year16	622	591	759	164	120	170	243	...	559	552	1024	189	1054	159
Year17	754	690	913	198	137	196	354	...	363	797	1092	352	1168	276
Year18	794	768	1004	196	150	212	378	...	387	905	1095	467	1212	350

Table 2 MCA inertia table

Axis	Inertia	%inertia	Sum
1	0.017	42.30	42.30
2	0.013	33.23	75.52
3	0.006	16.53	92.05
4	0.0013	3.20	95.25
5	0.0010	2.48	97.73
6	0.0003	0.92	98.65
7	0.0003	0.82	99.47
8	0.0001	0.36	99.83
9	0.00006	0.17	100

Table 3 Coordinates, contribution (CTR), and correlation (COR) of the points on the first factorial axis

Category	F1	COR	CTR
Year18	−246	942	566
Year13	150	376	108
Food1	−671	826	254
Med1	−471	734	150
hhiq4	−397	315	142
Cloth1	−175	537	85
hhiq2	190	333	73
Cloth0	142	537	70

(inertia). On this map, we observe the formation of three year groups. The first group consists of the years 2017 and 2018, the second consists of the years 2012–2016, and the third consists of the years 2009–2011. The appearance of consecutive years in the same group is interesting because from 2009, Greece faced significant economic and political changes.

Next, Ascending Hierarchical Clustering (CAH) was applied to the so-called indicator matrix in order to validate the groups obtained via MCA and further identify the most important attributes of each group (see also Moschidis 2017, for a similar treatment). The chi-square distance was used as the distance metric and Ward's criterion as the preferred linkage criterion. This set of options is also consistent with the MCA algorithm.

The results of CAH are shown in Table 4. Column *d* corresponds to the intra-class inertia (Moschidis 2009) that was used as a criterion for choosing the number of clusters. A great increase in *d* is a sign of disruption of clustering. In our case, this happens on the transition from three to four clusters. Therefore, three groups were selected, corresponding to nodes 15, 16, and 17. The first cluster (node 15) consists of the years 2009–2011 (16% of the total observations). The second cluster (node 16)

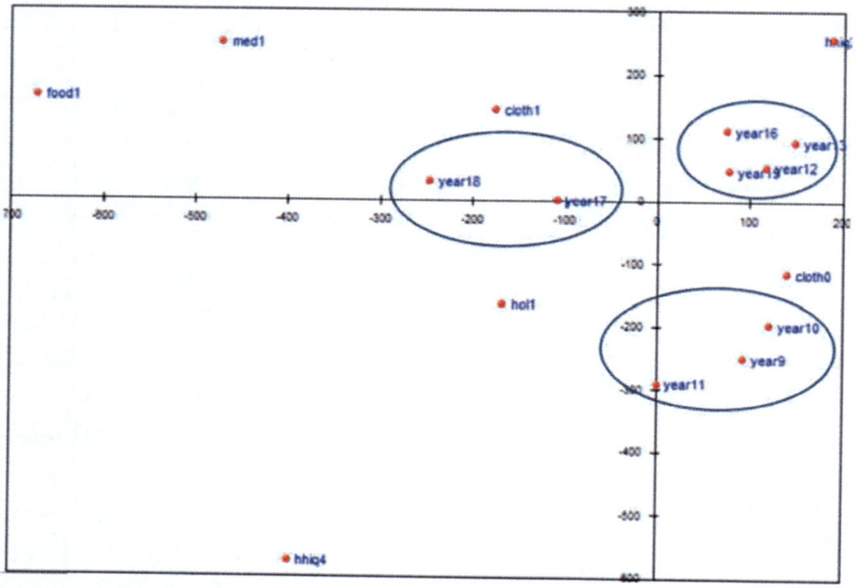

Fig. 1 First factorial plane (axes 1 and 2)

Table 4 Results of ascending hierarchical clustering

Node	A(i)	B(i)	Mass	d	id	ie	lr
11	1	2	0.09	0.00023	0.00023	0.04138	0.99447
12	4	5	0.15	0.00051	0.00074	0.04087	0.98221
13	6	7	0.24	0.00065	0.00139	0.04022	0.96659
14	13	8	0.36	0.00126	0.00265	0.03896	0.93631
15	11	3	0.16	0.00156	0.00421	0.0374	0.89882
16	9	10	0.31	0.00262	0.00683	0.03478	0.83586
17	12	14	0.51	0.00623	0.01306	0.2855	0.68615
18	15	17	0.68	0.01318	0.02624	0.01537	0.36942
19	18	16	1	0.01535	0.04159	0.00002	0.00054

consists of the years 2017 and 2018 (32%), and the third cluster (node 17) consists of the years 2012–2016 (52%).

The first cluster is characterized by individuals that belong to the highest income quartile, buy holiday packages or tickets, do not buy medicine, food or clothes online, and are women, self-employed, employed, or students. The second cluster consists of individuals who buy holiday packages, tickets, food, medicine or clothes, belong to the highest income category, and are over 35. Finally, the third cluster consists of individuals who belong to the second lowest income category, are students or unemployed, and do not buy food and holiday packages online.

4 Conclusions

Overall, results showed an increase in the number of people using e-commerce in Greece over the last decade. With regard to gender, men were more active users than women during the first years of the period under study, but this trend eventually disappeared or even reversed with women being more active than men, thus shaping a new landscape. A significant increase in the presence of economically inactive people in e-commerce should also be emphasized. Familiarity with new technologies and the exploration of opportunities could be reasonable explanations.

Exploratory data analysis revealed three distinct clusters: the first cluster consisting of years 2009–2011, the second cluster consisting of years 2012–2016, and the third cluster of years 2017–2018. This grouping is interesting since significant economic and political changes took place in Greece within each period. Relating these clusters with additional economic as well as sociopolitical characteristics can give a more insightful look at the reasons behind the shaping of e-commerce over the last decade in Greece.

References

Benzécri, J.-P.: Correspondence Analysis Handbook. CRC Press (1992)

Chaffey, D., Hemphill, T., Edmundson-Bird, D.: Digital Business and E-Commerce Management. Pearson UK (2019)

Ho, S., Kauffman, R., Liang, T.: A growth theory perspective on B2C e-commerce growth in Europe: an exploratory study. Electron Commer. R A. **6**(3), 237–259 (2007)

Karapistolis, D.: The software MAD. Dat. Anal. Bull. **2**, 133–147 (2002)

Kickert, W., Ongaro, E.: Influence of EU (and IMF) on domestic consolidation and reform: introduction. Public Manag. Rev. **21**(9), 1261–1264 (2019)

Kitsios, F., Moschidis, O., Livanis, E.: Service innovation strategies in Greek hotel sector: an exploratory study using the statistical method of multidimensional analysis. Int. J. Data Anal. Tech. Strateg. **5**(1), 49–62 (2013)

Laudon, K., Traver, C.: E-Commerce 2018. Pearson Education Limited, Upper Saddle River (2019)

Moschidis, O.: A different approach to multiple correspondence analysis (MCA) than that of specific MCA. Mathématiques et sciences humaines **5**(1), 77–88 (2009)

Moschidis, O.: A method for transforming ordinal variables. In: Palumbo, F., Montanari, A., Vichi, M. (eds.) Data Science. Studies in Classification, Data Analysis, and Knowledge Organization, pp. 285–294. Springer, Cham (2017)

Moschidis, O.: Unified coding of qualitative and quantitative variables and their analysis with ascendant hierarchical classification. Int. J. Data Anal. Tech. Strateg. **7**(2), 114–128 (2015)

Papaioannou, E., Georgiadis, C., Moshidis, O., Manitsaris, A.: Factors affecting customers' perceptions and firms' decisions concerning online fast food ordering. Int. J. Agric. Environ. Inf. Syst. **6**(1), 48–78 (2015)

Savage, J., Verdun, A.: Strengthening the European Commission's budgetary and economic surveillance capacity since Greece and the euro area crisis: a study of five Directorates-General. J. Eur. Public Policy **23**(1), 101–118 (2015)

Thiebaut, R.: AI revolution: how data can identify and shape consumer behavior in ecommerce. In: Sergi, B.S., Scanlon, C.C. (eds.) Entrepreneurship and Development in the 21st Century, pp 191–229. Emerald Publishing Limited (2019)
Visvizi, A.: The crisis in Greece and the EU-IMF rescue package: Determinants and pitfalls. Acta Oeconomica **62**(1), 15–39 (2012)

Benchmarking in Cluster Analysis: A Study on Spectral Clustering, DBSCAN, and K-Means

Nivedha Murugesan, Irene Cho, and Cristina Tortora

Abstract We perform a benchmarking study to identify the advantages and the drawbacks of Spectral Clustering and Density-Based Spatial Clustering of Applications with Noise (DBSCAN). We compare the two methods with the classic K-means clustering. The methods are performed on five simulated and three real data sets. The obtained clustering results are compared using external and internal indices, as well as run times. Although there is not one method that performs best on all types of data sets, we find that DBSCAN should generally be reserved for non-convex data with well-separated clusters or for data with many outliers. Spectral Clustering has better overall performance but with higher instability of the results compared to K-means, and longer run time.

Keywords Spectral clustering · DBSCAN · K-means

1 Introduction

When performing cluster analysis, there are many decisions that must be made including data preprocessing, selecting the number of clusters, choosing appropriate clustering techniques for a given data set, etc. For each of these choices, the researchers need to decide from several options, with many new options constantly being introduced in the statistical literature. This has emphasized the importance of extensively and carefully comparing new techniques with existing ones, i.e., benchmarking in cluster analysis (Mechelen et al. 2018). Clustering techniques can be broadly divided into

N. Murugesan (✉) · I. Cho · C. Tortora
Department of Mathematics and Statistics, San José State University, One Washington Square, San José, CA 95192, USA
e-mail: nivedha.murugesan@sjsu.edu

I. Cho
e-mail: sisang.cho@sjsu.edu

C. Tortora
e-mail: cristina.tortora@sjsu.edu

T. Chadjipadelis et al. (eds.), *Data Analysis and Rationality in a Complex World*, Studies in Classification, Data Analysis, and Knowledge Organization,
https://doi.org/10.1007/978-3-030-60104-1_20

two categories: hierarchical and partitioning. Partitioning techniques include model-based clustering (e.g., Gaussian Mixture models McLachlan and Basford 1988), partition-based clustering (e.g., K-means MacQueen 1967), graph-based clustering (e.g., Spectral Clustering Shi and Malik 2000), and density-based clustering (e.g., Density-Based Spatial Clustering of Applications with Noise (DBSCAN) Ester et al. 1996). One of the advantages of graph-based and density-based clustering is that they can detect clusters of arbitrary shapes, such as non-convex clusters. Furthermore, they do not make assumptions about the distributions of the clusters. Among graph-based techniques, Spectral Clustering (SC) (Shi and Malik 2000) is one of the most used, (e.g., see Fitch et al. 2019; Paccanaro et al. 2006). In SC, the first step is to create a similarity matrix W between all the data points. The eigenvalues of the Laplacian matrix of W are found in order to perform dimensionality reduction. Finally, the observations are clustered in fewer dimensions using K-means clustering. SC works well on non-convex data sets, however, one disadvantage of SC is its high computational cost, due to which the method is generally applied to smaller data sets. There is ongoing research focusing on the expansion to higher dimensional data (e.g., Peng et al. 2018; Wang et al. 2015; Wu et al. 2014). Among density-based clustering techniques, one of the most commonly used is DBSCAN (Ester et al. 1996), (e.g., see Emadi and Mazinani 2018; Francis et al. 2011). It focuses on partitioning observations into clusters based on the surrounding density. If a point does not have enough nearest neighbors, it is labeled as a noise point. This allows the algorithm to find clusters of various shapes and to filter out the outliers. DBSCAN can be used on large data sets with a relatively small computational cost and is primarily used on data sets characterized by outliers. An advantage is that the number of clusters does not need to be specified, however, DBSCAN does not perform too well on clusters of varying densities, often returning either too few or too many clusters depending on how closely the observations are located to each other. In this benchmark study, we investigate how SC and DBSCAN perform on simulated and real data sets with different characteristics. Moreover, we compare their performances with one of the most common and simple clustering techniques, K-means (MacQueen 1967). Given the number of clusters G, K-means randomly initializes the means of the clusters and each observation is assigned to the cluster with the nearest mean based on the Euclidean distance. The cluster means are recalculated and the observations are reallocated to the nearest cluster. This process is repeated either until convergence or for a specified number of iterations. Two main drawbacks of K-means are its sensitivity to outliers and impossibility of detecting non-convex clusters. SC and DBSCAN have been compared to other techniques in the recent statistical literature (for e.g., Berhane 2020; McInnes et al. 2020), the goal of this paper is to compare the performances of the three techniques on a variety of simulated and real data sets to emphasize the conditions under which each method performs the best.

2 Methodology

The benchmark study is performed using R software (Core 2017). The function *specc()* from the package "kernlab" (Karatzoglou et al. 2004), implements SC, the function *dbscan()* from the package "dbscan" (Hahsler and Piekenbrock 2017), implements DBSCAN, and the function *kmeans()* provided in the R package "stats" (Core 2017), implements K-means. The code used for this paper is available on GitHub (Murugesan et al. 2020).

K-means and SC require the user to specify the number of clusters. For simulated data sets, the number of clusters is already known and is specified accordingly in each scenario. For real data sets, the number of clusters is chosen based on internal evaluation criteria listed in Sect. 2.1. Since the clustering outcome for K-means highly depends on the initial starting values, we specify $nstart = 50$ to reduce the effect of the starting values on the results. We leave the other two input parameters as the default, i.e., Hartigan-Wong algorithm and the number of maximum iterations as 10, since the algorithm converges before it reaches the maximum on all the data sets. For SC, we use the Gaussian kernel, the width parameter is automatically determined, and the number of iterations is kept as the default value of 200. DBSCAN requires the choice of two parameters *minPts* and *eps*. For *minPts*, Schubert et al. (2017), recommends setting $minPts = 4$, while (Sander et al. 1998), recommends $minPts = 2{\times}dim$ where dim represents the number of dimensions in the data set. Since we want to choose the optimal parameter value for each data set, we tried both $minPts = 4$ and $minPts = 2{\times}dim$ and selected the value that produced the best results based on the internal evaluation criteria listed in Sect. 2.1. Ultimately, this procedure resulted in setting $minPts = 4$ for all data sets except for the forest fire data set in the empirical data section. Next, we determine the optimal *eps* value through the use of the *kNN()* function (Hahsler and Piekenbrock 2017). For each data set, the *kNN()* function takes user input for *minPts* and computes the smallest *eps* value for each observation such that each observation has the specified number of *minPts* within a radius of *eps*. The index of each observation is plotted against its corresponding *eps* value that was calculated using *kNN()*. Using this plot, we select the elbow point as the optimal *eps* value for that data set. If there is no obvious elbow point, we try a few values of *eps* and select the value that produces the best results based on internal criteria (Sect. 2.1).

2.1 *Benchmarking Evaluation Criteria*

The true partitions for the simulated data sets are known, and therefore, the accuracy of each algorithm is evaluated by using three external indices that compare clustering labels to the true labels. The first is the Adjusted Rand Index (ARI), which uses pairwise agreements to compare two partitions (Hubert and Arabie 1985). The second is the Jaccard Index (JAC), which measures the similarity between two partitions by dividing the intersection size by the union size (Desgraupes 2018). The final index is Fowlkes-Mallows Index (FM), which computes the geometric mean of the precision (positive predictive rate) and sensitivity (true positive rate) proportions (Desgraupes 2018). The respective R functions used are *ARI()* from the R package "MixGHD" (Tortora et al. 2017), and *extCriteria()* from the R package "clusterCrit" (Desgraupes 2018), for the JAC and FM indices. For all three indices, the values range between 0 and 1, where a value of 1 indicates a perfect classification agreement between the two compared partitions. The use of three indices produces a more comprehensive assessment of the algorithms in various scenarios.

For the empirical data sets, internal indices are used to select the number of clusters and to analyze results for each algorithm. The first index is the S-Dbw index, which uses the density of clusters to measure inter-cluster separation (Liu et al. 2010). Smaller values of the S-Dbw index indicate a better outcome. Since the S-Dbw index works based on the density of clusters, it can evaluate clustering results only for larger data sets. For smaller empirical data sets (e.g., the bike data set), the Calinski-Harabasz (CH) index is used. This index is based on average between-cluster and within-cluster sum of squares (Liu et al. 2010); a larger CH index indicates a better outcome. The last index used is the silhouette value, which measures the similarity of an observation to its own cluster (Rousseeuw 1987). Silhouette values range from -1 to 1, where a value closer to 1 means the observation is closer to its assigned cluster and a value closer to -1 means the observation is closer to a neighboring cluster and is, therefore, poorly classified. The average silhouette width of a partition is used to assess cluster validity. It can also indicate if the chosen number of clusters is appropriate (Rousseeuw 1987). The respective R functions used for the internal indices are *intCriteria()* from the R package "clusterCrit" (Desgraupes 2018), for the S-Dbw and CH indices and *silhouette()* from the R package "cluster" (Maechler et al. 2019), for the silhouette values.

3 Simulated Data Sets

We use five different simulated scenarios. For each scenario (aside from the first data set), we generate 20 simulations and compute the average ARI, JAC, and FM indices with the respective standard deviations for each clustering algorithm. One data set per scenario is shown in Fig. 1, with the true partition (leftmost), clusters labeled by K-means, DBSCAN, and SC (rightmost). Each row in the figure pertains to each of the five simulated data sets used.

The first simulated data set is the "multishapes" data set from the R package "factoextra" (Kassambara and Mundt 2017), characterized by 1100 observations and three variables. In the true partition, there are five distinct clusters with the noise points grouped as a sixth cluster. For the second simulated scenario, the function *mlbench.spirals()* from the R package "mlbench" (Leisch and mlbench 2010), was used to create two entangled spirals with some noise. The true partition has 400 observations, three variables, and two clusters. The third simulated scenario consists

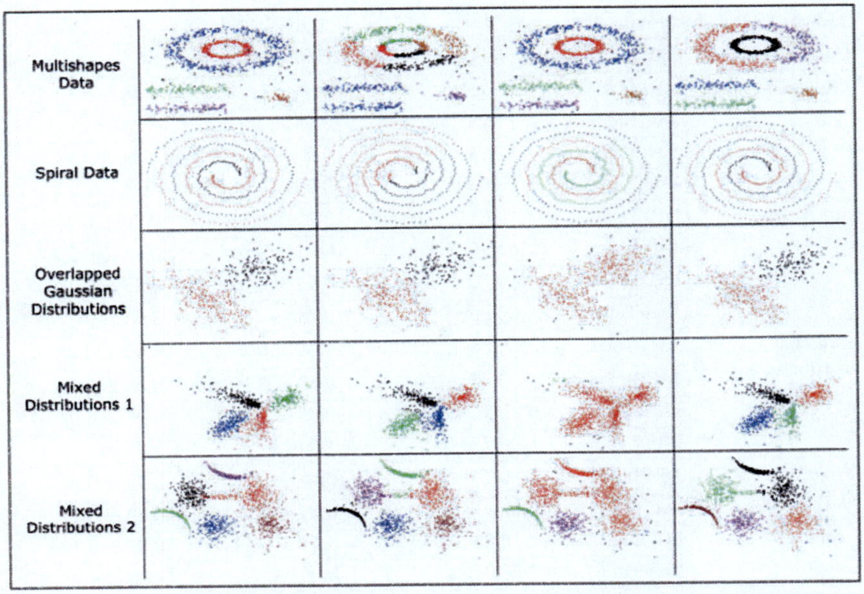

Fig. 1 Simulated data sets, clustering results in each row represent (in order from left to right): true partition, K-means, DBSCAN, and SC

of two overlapping clusters generated from two separate Gaussian distributions, 150 observations from each cluster, for a total of 300 observations. The fourth simulated scenario contains four overlapping clusters generated from four different distributions: a generalized hyperbolic distribution (GHD) (Barndorff-Nielsen 1977), a multiple-scaled generalized hyperbolic distribution (MSGHD) (Tortora et al. 2019), a contaminated normal distribution (Tukey 1960), and a Gaussian distribution. We generated 200 observations from each distribution, for a total of 800 observations. The last simulated scenario contains six clusters from six distributions: two separate Gaussian distributions, two separate von Mises-Fisher distributions, and two separate GHDs. Uniform noise connects the two Gaussian distributions. We generated 200 observations from each cluster and 40 noise observations, for a total of 1240 observations. The functions used to generated the data were *rmvnorm()* from the R package "mvtnorm" (Genz et al. 2019), *rGHD()*, *rMSGHD()* from the R package "MixGHD" (Tortora et al. 2017), *rCN()* from the R package "ContaminatedMixt" (Punzo et al. 2018), *rmovMF()* from the R package "movMF" (Hornik and Grün 2014), and *runif()* from the base R package "stats" (Core 2017). For details on the parameters used see Murugesan et al. (2020). We generated 20 data sets for the last four simulated scenarios and reported the means and standard deviations of the external indices. Furthermore, the run time[1] of one simulation for each clustering method was measured. These results have been summarized in Table 1. For non-convex data

[1]Performed on a MacBook Pro 2017 - Processor: 2.3 GHz Intel Core i5; Memory: 8GB.

Table 1 Results for simulated data sets

Data set	Index	K-Means	DBSCAN	SC
Multishapes data	ARI Index	0.25	0.96	0.78
	JAC Index	0.27	0.95	0.72
	FM Index	0.44	0.97	0.84
	Run Time	0.030 s	0.003 s	40.933 s
Spiral data	ARI Index	0.01 ± 0.00	0.33 ± 0.17	0.78 ± 0.38
	JAC Index	0.34 ± 0.00	0.45 ± 0.05	0.84 ± 0.25
	FM Index	0.50 ± 0.00	0.63 ± 0.06	0.89 ± 0.19
	Run Time	0.009 s	0.003 s	2.359 s
Gaussian distributions	ARI Index	0.86 ± 0.05	0.08 ± 0.24	0.80 ± 0.28
	JAC Index	0.87 ± 0.04	0.50 ± 0.10	0.86 ± 0.13
	FM Index	0.93 ± 0.03	0.68 ± 0.07	0.92 ± 0.08
	Run Time	0.717 s	1.046 s	1.576 s
Mixed distributions 1	ARI Index	0.79 ± 0.11	0.46 ± 0.25	0.73 ± 0.21
	JAC Index	0.74 ± 0.12	0.47 ± 0.15	0.69 ± 0.17
	FM Index	0.85 ± 0.08	0.65 ± 0.12	0.82 ± 0.12
	Run Time	0.016 s	0.003 s	14.441 s
Mixed distributions 2	ARI Index	0.85 ± 0.01	0.40 ± 0.21	0.80 ± 0.09
	JAC Index	0.78 ± 0.02	0.39 ± 0.15	0.72 ± 0.11
	FM Index	0.88 ± 0.01	0.59 ± 0.12	0.84 ± 0.07
	Run Time	0.034 s	0.006 s	60.062 s

with obvious patterns, it is evident that K-means generally fails in accuracy. In contrast, DBSCAN and SC are able to produce clusters that more closely resemble the intended clusters in the Multishapes and spiral data simulations. However, when it comes to the simulations produced using mixed distributions, K-means and SC are rather effective in identifying the separate clusters while DBSCAN is not. Although SC performs well overall on all five scenarios, the computational cost is one of the biggest disadvantages. While K-means and DBSCAN take less than one second to cluster 1240 observations in the final scenario, SC takes close to a minute. Furthermore, SC generally has the highest index standard deviations, indicating lack of stability over multiple simulations. Looking at the index values, run times, and overall performance, K-means performs best on convex data sets, DBSCAN performs best on non-convex high density or well-separated clusters, and SC performs well on either type of data set but with less stability and a high run time. However, run times can be affected by many factors including the way the algorithms are coded. Therefore, the run time should be considered as simply a rough estimation of the method's speed.

4 Empirical Data Sets

For the empirical data sets, the results of all internal indices are summarized in Table 1, in Appendix.[2] The bike data set, which was obtained from GitHub repository of user "mdancho84", is the first of the three empirical ones. It contains 97 different bike models and 35 variables including the type of bike, frame material, price range, and 30 bike shops. For each bike shop, the value for each observation i is the proportion of sales made from selling bike model i. The goal is to cluster the 30 bike shops into appropriate groups based on the scaled quantities of bike model orders. Starting with K-means, the appropriate number of clusters is selected by comparing the Calinski-Harabasz indices and average silhouette widths for different values of G (number of clusters). Recall that higher values of CH and silhouette indices indicate a better choice of G. For K-means, the indices are maximized for $G = 2$. However, for both $G = 2$ and $G = 3$, about two-thirds of the observations are grouped in one cluster. Therefore, $G = 4$ is selected, since the observations are more spread out over the four clusters. There are 13 shops in cluster 1, six in cluster 2, eight in cluster 3, and three in cluster 4. Table 2 summarizes the mode of each variable in the clusters obtained using K-means with parameter $centers = 4$. Each cluster indeed highlights a different specific feature in each variable, indicating that K-means has grouped together bike shops that primarily sell similar models. Therefore, it appears that K-means has performed well. For DBSCAN, the parameters used are $eps = 0.09$ and $minPts = 4$. The method returns three clusters for which the average silhouette width is 0.16 and the CH index is 3.67. The silhouette widths of all observations in cluster 1 are negative, indicating they are all incorrectly classified. Since it appears that K-means has produced an appropriate partition, ARI and JAC indices can be used to compare the partitions obtained from K-means and DBSCAN. The ARI between K-means and DBSCAN is 0.26 and the JAC index is 0.34. Furthermore, the average silhouette width and CH index are lower for the DBSCAN results when compared to the K-means results. Therefore, K-means has performed better on this data set. Finally, SC is conducted for different values of G. The silhouette and CH indices are highest for $G = 2$ and $G = 4$. However, for $G = 2$, a majority of observations are again grouped into one cluster. For better spread, the value $G = 4$ is selected. The ARI value between K-means and SC is 1, i.e., perfect classification agreement, meaning either method can be utilized for this data set. However, if we have a similar data sets with more variables, K-means may be preferred over SC due to speed.

Next, the forest fire data set contains information about burned areas in the north-east region of Portugal. This data set is from the UCI Machine Learning Repository. There are 517 observations and the variables used in the clustering are temperature, relative humidity, wind, and rain. For K-means, average silhouette widths and S-Dbw indices are used to select an appropriate number of clusters. Recall that the S-Dbw index should be minimized for better results. Based on the table, the optimal value for G is 2. However, after considering silhouette plots, $G = 3$ is selected since the observations are well distributed over the clusters and the two internal indices

[2]Link: https://irene1014.github.io/benchmarking-study-cluster-analysis/.

Table 2 Clustering results from K-means on bike data

Cluster	Category 1	Category 2	Frame	Price
Cluster 1	Mountain	Sport	Aluminum/Carbon	[415, 3500)
Cluster 2	Road	Endurance/Elite	Carbon	[3500, 12790)
Cluster 3	Road	Endurance/Elite	Aluminum/Carbon	[415, 3500)
Cluster 4	Mountain	Over mountain	Carbon	[3500, 12790)

are still relatively optimal. Based on the data visualization (Murugesan et al. 2020), the observations seem to be well separated and appropriately grouped together. For DBSCAN, the parameters used are $eps = 5.2$ and $minPts = 8$. The algorithm returns two clusters for which the average silhouette width is 0.47 and the S-Dbw index is 2.91. Although these indices might imply that DBSCAN has comparatively performed well, it is important to look closer at each cluster. Out of 517 observations, 489 are grouped into one cluster, which is highly undesired. Since this data set contains too many observations that are located close to each other, DBSCAN tends to group many observations together. Once again, it is determined that K-means is a more appropriate method for this data set compared to DBSCAN. For SC, based on the internal indices, and the distribution of observations over cluster, $G = 3$ seems optimal. However, even the best partition for SC has many observations in one cluster and has more negative silhouette values compared to the results obtained from K-means. Furthermore, the cluster sizes are more balanced for K-means. Based on these results, K-means performs better on the forest fire data set when compared to the other two methods.

Finally, "Inside Airbnb" is an independent, noncommercial site that allows people to view how Airbnb, a well-established website for room or home rental, is being used around the world. We utilize data about Airbnb listings located in Santa Clara County, California. There are 16 variables, of which we use the eight continuous variables. The data set contains a total of 5854 observations, from which a random subset of 500 observations is sampled to keep the computational cost to a minimum. As before, G for K-means is selected based on two internal indices. The average silhouette width does not differ much for each value of G, but the S-Dbw index increases as the value of G increases. Thus, $G = 2$ or $G = 3$ may be the optimal choice. For $G = 3$, the observations are well-spread out so this value is selected for K-means. The mean values for price and minimum nights and the modes for neighborhood and room type in each cluster are summarized in Table 3. The 500 Airbnbs grouped by K-means, DBSCA, and SC are displayed in a geographical plot based on latitude and longitude in Fig. 1, in Appendix (the plot was created using R packages "maps" Becker et al. 2018, "mapdata" Becker and Wilks 2018, "ggmap" Kahle and Wickham 2013, and "ggplot2" Wickham 2016). For DBSCAN, the specified parameter values are $eps = 0.03$ and $minPts = 4$. The results for DBSCAN are summarized in Table 3. Almost all the observations are grouped in the same cluster, which is significantly different from the K-means results. For SC, based on the internal indices, $G = 3$ is selected for which

Table 3 Clustering results from K-means and DBSCAN on airbnb data

Method	Cluster	Neighborhood	Room type	Price	Nights
K-means	Cluster 1	San José	Private room	$167.04	5.71
	Cluster 2	Mtn view	Private room	$163.80	6.15
	Cluster 3	Sunnyvale	Private room	$122.08	15.64
DBSCAN	Cluster 1	Independent areas	Entire home/Apt	$421.90	84.09
	Cluster 2	San José	Private room	$142.51	7.98
	Cluster 3	San José	Private room	$172	2.12

the observations are well-spread out, the index values are comparatively optimal, and there are fewer negative silhouette values. Comparing the results of K-means and SC using the geographical plot (Fig. 1, in Appendix), it appears that Airbnb listings in the same neighborhood tend to have similar rental characteristics. The JAC value between K-means and SC is 0.67, which aligns well with the similarity between the geographical plots. Once again, K-means or SC is preferred over DBSCAN because the clusters are more distinguishable. Although K-means and SC produce similar results for this data set, we keep in mind that a random subset of data was sampled to keep the computational cost of SC to a minimum.

5 Conclusion

It comes as no surprise that there is not one single method that performs the best on all types of data sets. The goal of the analysis should be considered when choosing an appropriate clustering algorithm for a given data set. If the goal is to identify space with high density data or well-separated clusters by space with low density data, then DBSCAN may be preferred. DBSCAN can detect non-convex clusters and is not impacted by outliers that should be classified as noise points. For example, on the multishapes data, DBSCAN consistently performed better than the other two methods. Moreover, DBSCAN has the advantage that the number of clusters does not need to be specified by the user, which can be a challenge for K-means and SC as observed in the empirical data applications. However, when it comes to data without well-separated clusters, K-means and SC achieve better clustering results, even if some of the clusters are non-convex. SC is preferred over K-means on non-convex data and it is the most flexible methods among the three. In fact, the lowest ARI obtained by SC on the simulated data sets is 0.73, indicating that SC performed well on all five simulated scenarios. However, although the overall performance of SC in terms of average ARI is better than K-means on simulated data, the solutions can be unstable and the computational cost of SC is still a challenge for larger data sets.

References

Barndorff-Nielsen, O.E.: Exponentially decreasing distributions for the logarithm of particle size. Proc. R. Soc. Lond. A. **353**(1674), 401–419 (1977)

Becker, R.A., Wilks, A.R.: (Orig. S code), R. Brownrigg (R Version), T.P. Minka and A. Deckmyn (Enhancements), maps: Draw Geographical Maps. R package v. 3.3.0 (2018)

Becker, R.A., Wilks A.R.: (Original S code), R. Brownrigg (R Version), mapdata: Extra Map Databases. R package version 2.3.0. (2018)

Berhane, F.: Data distributions where Kmeans clustering fails: Can DBSCAN be a solution? https://datascience-enthusiast.com/Python/DBSCAN_Kmeans.html (2020)

Desgraupes, B.: clusterCrit: Clustering Indices. R package version 1.2.8. (2018)

Emadi, H.S., Mazinani, S.M.: A novel anomaly detection algorithm using DBSCAN and SVM in wireless sensor networks. Wirel. Pers Comm. **98**, 2025–2035 (2018)

Ester, M., Kriegel, H., Sander, J., Xu, X.: A density-based algorithm for discovering clusters in large spatial databases with noise. In: KDD-96 Proceedings, pp. 226-231 (1996)

Fitch, J.P., Khan, N.A.M., Tortora, C.: Back pain: a spectral clustering approach. Arch. Data Sci. Ser. B **1**(1) (online first) (2019)

Francis, Z., Villagrasa, C., Clairand, I.: Simulation of DNA damage clustering after proton irradiation using an adapted DBSCAN algorithm. Comput. Methods Programs Biomed. **101**(3), 265–270 (2011)

Genz, A., Bretz, F., Miwa, T., Mi, X., Leisch, F., Scheipl, F., Hothorn, T.: mvtnorm: Multivariate Normal and t Distributions, R package version 1.0-11 (2019)

Hahsler, M., Piekenbrock, M.: dbscan: Density Based Clustering of Applications with Noise (DBSCAN) and Related Algorithms, R package version 1.1-1 (2017)

Hornik, K., Grün, B.: movMF: An R package for fitting mixtures of von mises-fisher distributions. J. Stat. Soft. **58**(10), 1–31 (2014)

Hubert, L., Arabie, P.: Comparing partitions. J. Classif. **2**(1), 193–218 (1985)

Kahle, D., Wickham, H.: ggmap: spatial visualization with ggplot2. R J. **5**(1), 144–161 (2013)

Karatzoglou, A., Smola, A., Hornik, K., Zeileis, A.: kernlab—An S4 Package for kernel methods in R. J. Stat. Soft. **11**(9), 1–20 (2004)

Kassambara, A., Mundt, F.: factoextra: Extract and Visualize the Results of Multivariate Data Analyses, R package version 1.0.5 (2017)

Leisch F., Dimitriadou, E.: mlbench: Machine Learning Benchmark Problems, R package version 2.1-1 (2010)

Liu, Y., Li, Z., Xiong, H., Gao, X. Wu, J.: Understanding of internal clustering validation measures. In: Proceedings of the 2010 IEEE International Conference on Data Mining, pp. 911-916 (2010)

MacQueen, J.: Some methods for classification and analysis of multivariate observations. In: Proceedings of the Fifth Berkeley Symposium on Mathematical Statistics and Probability, vol. 1: Statistics, pp. 281-297. University of California Press, Berkeley, California (1967)

Maechler, M., Rousseeuw, P., Struyf, A., Hubert, M., Hornik, K.: cluster: Cluster Analysis Basics and Extensions, R package version 2.1.0 (2019)

McInnes, L., Healy, J., Astels, S.: Comparing python clustering algorithms. https://hdbscan.readthedocs.io/en/latest/comparing_clustering_algorithms.html

McLachlan, G.J., Basford, K.E.: Mixture Models: Inference and Applications to Clustering, Statistics, Textbooks and Monographs, vol. 84 (1988)

Mechelen, I.V., Boulesteix, A., Dangl, R., Dean, N., Guyon, I., Hennig, C., Leisch, F., Steinley, D.: Benchmarking in cluster analysis: A white paper (2018). arXiv:1809.10496v2

Murugesan, N., Cho, I., Tortora, C.: Benchmarking in cluster analysis: a study on spectral clustering, DBSCAN and K-means (github repository) https://irene1014.github.io/benchmarking-study-cluster-analysis (2020)

Paccanaro, A., Casbon, J.A., Saqi, M.A.S.: Spectral clustering of protein sequences. Nucleic. Acid Res. **34**(5), 1571–1580 (2006)

Peng, H., Pavlidis, N., Eckley, I., Tsalamanis, I.: Subspace clustering of very sparse high-dimensional data. In: 2018 IEEE International Conference on Big Data, pp. 3780–3783 (2018)
Punzo, A., Mazza, A., McNicholas, P.D.: ContaminatedMixt: An R package for fitting parsimonious mixtures of multivariate contaminated normal distributions. J. Stat. Softw. **85**(10), 1–25 (2018)
R Core Team: R: A language and environment for statistical computing, R Foundation for Statistical Computing, Vienna, Austria (2017)
Rousseeuw, P.J.: Silhouettes: a graphical aid to the interpretation and validation of cluster analysis. J. Comput. Appl. Math. **20**, 53–65 (1987)
Sander, J., Ester, M., Kriegel, H., Xu, X.: Density-based clustering in spatial databases: The algorithm GDBSCAN and its applications. Data Min. Knowl. Disc. **2**, 169–194 (1998)
Schubert, E., Sander, J., Ester, M., Kriegel, H., Xu, X.: DBSCAN revisited, revisited: why and how you should (Still) use DBSCAN. ACM T Database Syst. **42**(3), 19 (2017)
Shi, J., Malik, J.: Normalized cuts and image segmentation. IEEE Trans. Pattern Anal. Mach. Intell. **228**, 888–905 (2000)
Tortora, C., ElSherbiny, A., Browne, R.P., Franczak B.C., McNicholas, P.D.: MixGHD: Model Based Clustering, Classification and Discriminant Analysis Using the Mixture of Generalized Hyperbolic Distributions, R package version 2.1 (2017)
Tortora, C., Franczak, B.C., Browne, R.P., McNicholas, P.D.: A mixture of coalesced generalized hyperbolic distributions. J. Classif. **36**, 26–57 (2019)
Tukey, J.W.: A survey of sampling from contaminated distributions. In: Olkin, I., Ghurye, S.G., Hoeffding, W., Madow, W.G., Mann, H.B. (eds.) Contributions to Probability and Statistics: Essays in Honor of Harold Hotelling, pp 448–495 (1960)
Wang, S., Chen, F., Feng, J.: Spectral clustering of high-dimensional data via nonnegative matrix factorization. In: 2015 International Joint Conference on Neural Networks (IJCNN), pp 1–8 (2015)
Wickham, H.: ggplot2: Elegant Graphics for Data Analysis. Springer, NY (2016)
Wu, S., Feng, X., Zhou, W.: Spectral clustering of high-dimensional data exploiting sparse representation vectors. Neurocomputing **135**, 229–239 (2014)

Detection of Topics and Time Series Variation in Consumer Web Communication Data

Atsuho Nakayama

Abstract Consumers' personal interests are influenced by new product strategies, such as marketing communication schemes, and these can change over time. Thus, it is important to consider temporal variation in trending consumer interests. We aimed to detect temporal variations in consumer web communication data using weight coefficients between entries and topics obtained from nonnegative matrix factorization. The weight coefficient, which indicates the strength between an entry and a topic, was modeled with a Bayesian network to capture changes in the topic over time. Bayesian networks, commonly used in a wide range of studies such as anomaly detection, reasoning, and time series prediction, build models from data using Bayesian inference for probability computations. The causations can be modeled by representing conditional dependence based on the edges in a directed graph of the Bayesian network.

Keywords Bayesian networks · Text mining · Time series variation

1 Introduction

As the number and distribution of posts on the internet continue to increase, the ability to identify market trends based on consumer web communication data has become increasingly important. Uploading of data is a major part of daily life, with the sharing of information via texts, tweets, photographs, and videos on social media. Thus, uploading habits tend to represent the activities, interests, and opinions of individuals. The purpose of this study was to detect trending topics in web communications' data among consumers, with a focus on the associated temporal variation.

We examined the variation in topics related to a new product launch by classifying Twitter text into clusters based on the co-occurrence of words. A single tweet is sufficient to express an idea and even to write a short story in Japanese, given that just

A. Nakayama (✉)
Tokyo Metropolitan University, 1-1 Minami-Ohsawa, Hachioji-shi 192-0397, Japan
e-mail: atsuho@tmu.ac.jp

T. Chadjipadelis et al. (eds.), *Data Analysis and Rationality in a Complex World*, Studies in Classification, Data Analysis, and Knowledge Organization,
https://doi.org/10.1007/978-3-030-60104-1_21

a few characters can convey a considerable amount of information. Here, we chose keywords representing various topics from Twitter entries and tracked the weekly variation in these topics. We then classified the topics and performed a dimensionality reduction of the extracted text using nonnegative matrix factorization (NMF) (Lee and Seung 2000). Based on the NMF results, the relationships between entries and topics over time were modeled using Bayesian networks. Specifically, we examined the weight coefficients of the matrix H obtained from NMF analysis to determine the strengths of the associations between entries and topics.

Personal concerns are influenced by new product strategies, such as marketing communications, and these can change over time. Thus, it is important to consider the temporal variation of the topics in consumers' web communication data. Here, we aimed to detect temporal variation by applying the weight coefficients of matrix H between entries and topics obtained from NMF to Bayesian networks. A Bayesian network is a type of probabilistic graphical model based on Bayesian inference from probability computations. The causations can be modeled by representing conditional dependence as edges in a directed graph of the Bayesian network. As such, Bayesian networks are used in a wide range of applications for anomaly detection, reasoning, and time series prediction.

Nakayama (2017) investigated the ability to detect trending topics and temporal variation in social media communications among consumers related to the introduction of new products. Specifically, words extracted from Twitter data were classified using NMF to identify weekly variations in topics. Aggregate data represented in matrix form as weekly averages of matrix H were analyzed using multidimensional preference scaling (Carroll 1972). The weight coefficients of H indicate the entries' contributions to topics; these coefficients were averaged on a weekly basis. The results showed that the personal concerns and tweet contents of Twitter users were influenced by new product strategies (e.g., marketing strategies), and these strategies changed over time. For more details on the results of the analysis, see Nakayama (2017), who revealed that personal concerns are influenced by new product schemes but were unable to elucidate the time series variation associated with the topics. In the current study, we aimed to reveal the temporal variation in trending consumer topics related to a new beverage launch and why this change occurs. As such, matrix H was analyzed without aggregation using Bayesian networks.

The rest of this paper is organized as follows. Bayesian networks are introduced in Sect. 2. Data preparation is described in Sect. 3. In Sect. 4, we explain our analysis of topic variation patterns in consumer web communication data using Bayesian networks. Finally, in Sect. 5, we present our conclusions.

2 Bayesian Networks

Bayesian networks are used for modeling various phenomena, and many practical applications have been reported (e.g., in academia; biology; business and finance; capital equipment; causal learning; computer games, vision, hardware, and software;

data mining; medicine; natural language processing and speed recognition; planning; psychology; reliability analysis; scheduling; and vehicle technology). Bayesian networks are also used in control and malfunction diagnoses and in weather forecasting (Neapolitan 2003).

In this study, two artificial intelligence technologies were applied: text mining and Bayesian networks; this approach has been applied extensively in the modeling of text data. Hearst (1999) pointed out that text data are a rich source of information regarding causal relationships that can be exploited, independent of the number of relationships to take into account. In the study, data mining, information access, and corpus-based computational linguistics were described and discussed with respect to their relationships with text data mining. The author also included examples of real text data mining efforts and briefly outlined the means to extract and analyze exploratory data from text.

Sanchez-Graillet and Poesio (2004) examined domain-independent methods for acquiring information from text causal knowledge encoded as Bayesian networks. They performed a subjective evaluation of networks revealed in a comparison between the network structure generated by the system and that created manually by observing causal patterns in the text. The system was evaluated using text from five domains: medical diagnosis, health care information, software failure diagnostics, engine tip forums, and social forums, all obtained from the web. The structure generated was similar to the one created manually. Sanchez-Graillet and Poesio (2004) also extensively investigated the automatic extraction of Bayesian networks based on textual properties that describe the types of relationships by which the concepts associated with the nodes are linked.

Blanco et al. (2008) performed research on the automated extraction of causal relations among semantic objects such as events, people, entities, and factual data. They proposed a system for the detection of marked and explicit causations between a verb phrase and a subordinate clause. From this, they presented a supervised method for the detection and extraction of causal relations from open domain text. This simple approach demonstrated high performance.

Trovati et al. (2017) introduced a systematic method to extract and populate fragments of Bayesian networks from textual sources, based on grammar and lexical properties and the topological features of networks extracted subsequently. The aim of their study was to provide an agile yet accurate method to identify and assess probabilistic relationships among concepts. They pointed out that typically this is a complex problem, especially when addressing large volumes of unstructured data sets, i.e., big data. They showed that embedding the knowledge in suitable Bayesian networks allows for efficient knowledge extraction.

In this study, we analyzed Japanese Twitter data related to a new beverage launch. Notably, the data preparation of Japanese text takes considerably more effort than that of its English language counterpart. One of the most difficult natural language processing problems in Japanese is tokenization. This is referred to as the "wakachi-gaki" problem. In most Western languages, words are delimited by spaces and punctuation. In Japanese, words are not separated by spaces. Consider the following sentence, "新商品はとても美味しい" (shinshouhinhatotemooishii). The English translation of this sentence is: "The new product is very appealing." In contrast, there are no spaces or separation symbols between Japanese words. We used morphological analyses, such as tokenization, stemming, and part-of-speech tagging, to separate the words as follows. The Japanese morphological analyzer ChaSen was used to separate words in passages and distinguish all nouns, verbs, and adjectives. ChaSen (http://chasen.naist.jp/) is a fast, customizable Japanese morphological analyzer that takes the form of a hierarchical structure. It is designed for generic use and can be applied to a variety of language processing tasks. A detailed discussion of ChaSen can be found in Kudo et al. (2004).

新商品	は	とても	美味しい
shinshouhin	ha	totemo	oishii
noun	Japanese particle	adverb	adjective

3 Data Preparation

We searched for Twitter entries expressing interest in a new, inexpensive, beer-like beverage called "金のオフ" (Kin no Off) produced by Sapporo Breweries, Ltd. (Tokyo, Japan). The text data from Japanese Twitter entries containing the product name were collected at 5 min intervals. In total, 4,622 tweets were acquired over the period from September 2, 2011 to May 18, 2012. These are the same data as those used by Nakayama (2017). Data were then tokenized using the Japanese morphological analyzer ChaSen. The data from Nakayama (2017), comprised 4,232 entries of 358 words. As the data revealed co-occurrences between 358 words in the selected entries, we decided to use the complementary similarity measure (CSM) (Sawaki and Hagita 1996), in this study as well. CSM allows words that appear during a specific period to be extracted, even if their frequency is low. Words related to the launch of the new product, TV commercials about the new product, and the purchase and consumption of the product were extracted as feature words. The entry × word matrix obtained from Twitter entries was sparse and high-dimensional; thus, it was necessary to perform a dimensionality reduction analysis. Methods used for this type of analysis of sparse matrices include latent semantic analysis (LSA) or latent semantic indexing (LSI) (Deerwester et al. 1990), and probabilistic latent semantic analysis (PLSA) or probabilistic latent semantic indexing (PLSI) (Hofmann 1999). Nakayama (2017) employed NMF (Lee and Seung 2000), to reduce the dimensionality. Ding and Peng (2006) showed that both NMF and PLSI (PLSA) optimize the same objective function, ensuring that the use of NMF and PLSI are equivalent. In Lee and Seung (2000), NMF dimensionality reduction was represented using (1)

$$V \approx W \times H \tag{1}$$

where matrix V is a nonnegative $m \times n$ matrix, such as that in an entry $\times$ word matrix. Matrix W contains nonnegative basis vectors and indicates the strength of associations between words and topics. Matrix H comprises nonnegative coefficients and represents the strength of associations between entries and topics. Nakayama (2017) classified words extracted from Twitter data related to a new, inexpensive, beer-like beverage named "金のオフ" (Kin no Off) produced by Sapporo Breweries, Ltd.; the maximum number of topics was 10, and the minimum number was 4. Nakayama (2017) discussed eight topics for interpretation. Table 3 of Nakayama (2017), lists the eight topics and the top 10 heavily weighted words in the basis vector W. Nakayama (2017) also characterized these topics based on Japanese word interpretation and divided the eight topics into three groups. The first group included the review topics, which consisted of Topics 1, 3, 7, and 8. Tweets classified as Topic 1 contained a link to an external website and product images. Topic 3 was related to reviews of the purchasing behavior, as well as information about the new product. Topic 7 was related to the brewery's release of the new product. Topic 8 was related to experiences of actually consuming the product. The second group was made up of advertising topics, Topics 2, 5, and 6. Topic 2 was about advertisements that appeared on trains. Topic 5 dealt with TV commercials. Topic 6 was concerned with performers in TV commercials. The third group consisted only of Topic 4. Topic 4 was not associated with inexpensive beer-like beverages, and the product name was used as a keyword to extract Twitter entries that occurred in a different context.

4 Detection of Topics and Time Series Variation

We modeled the temporal variation in trending topics related to the release of a new beverage product using the weight coefficients in H representing the strength of the associations between entries and topics. The relationships between entries and topics were analyzed using Bayesian networks to capture the changes over time in trending topics related to the release of the new beverage product. Bayesian networks are a type of probabilistic graphical model that builds models from data using Bayesian inference for probability computations (Neapolitan 2003). The causations are modeled by representing conditional dependence based on edges in a directed graph of the Bayesian networks. This approach is used in a wide range of studies related to anomaly detection, reasoning, and time series prediction, in which the relationship between random variables is expressed using nodes and directed links. Bayesian networks can handle nonlinearity, non-Gaussian relationships, and interactions between random variables, for flexible modeling in theoretical and practical applications such as those found in the industry. Recent advances in computer and information communication technologies have involved the use of Bayesian network analysis to clarify the phenomena behind big data on human behavior and natural science. The Bayesian network represents stochastic knowledge and events as stochastic models through a graphical representation of a finite number of random variables (each random variable or set thereof) and their relationships as nodes and

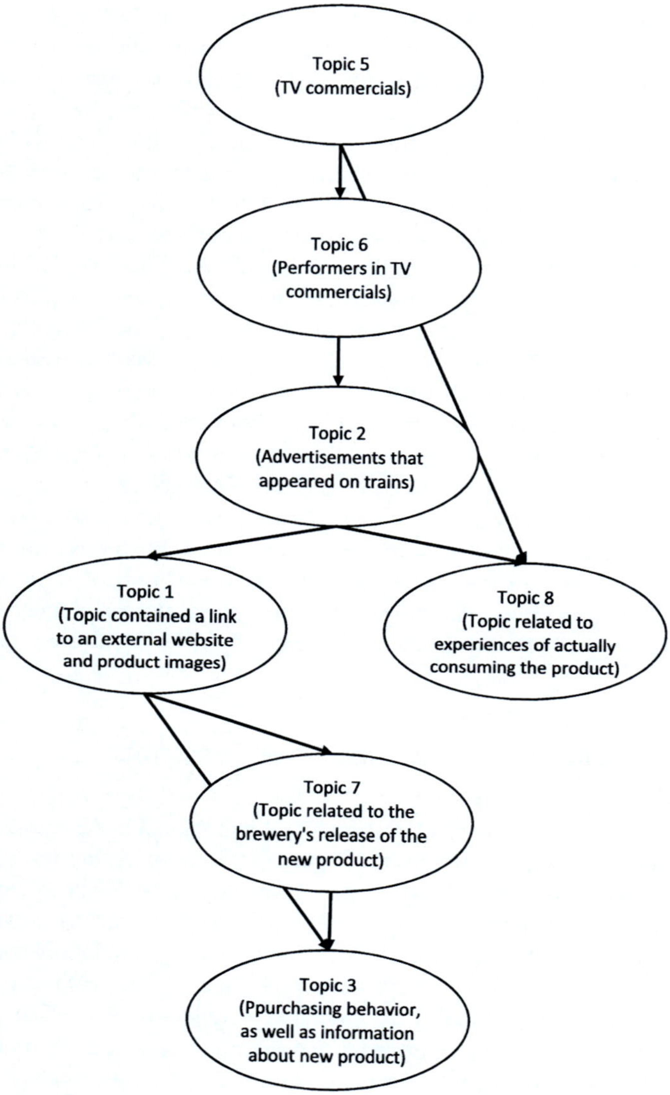

Fig. 1 Results of the detection of topics and time series variation using Bayesian network analysis

directed links. Let the set of random variables be $x_1,\ldots, x_N$ and the joint probability distribution be $p(x_1, \ldots, x_N)$. Let $p(x_i|x_j)$ be the conditional probability of random variable x_i given a given random variable x_j, and let $pa(x_i)$ be the set of variables with a directed link to x_i. Then, the joint probability distribution of the model is defined as (2).

$$p(x_1, \ldots, x_N) = \prod_{i=1}^{N} p(x_i | pa(x_i)) \quad (2)$$

The convenience of Bayesian networks is the ability to decompose the joint probability distribution of all random variables into a product of conditional probabilities between certain local variables. Efficient calculation is possible for marginalization operations of the joint probability distribution. Moreover, a stochastic inference can be carried out efficiently based on the random variable state, such as the assignment of a state value based on observation data and knowledge. The Bayesian network representation also facilitates visual understanding of the dependencies between variables.

Here the topic variation included two groups: one group consisting of topics 1, 3, 7, and 8 corresponding to new product reviews and the second that included topics 2, 5, and 6 related to advertising (Fig. 1). Topics 2, 5, and 6 are the topics associated with advertising tweets. At first, consumers tweeted their impressions of the TV commercials. Then, they tweeted about the performers in the commercials in more detail. After that, people tended to tweet about advertisements other than TV commercials. This was followed by a post of their impressions of the new product. Topics 1, 3, 7, and 8 were related to new product reviews, in the form of impressions after product consumption that commonly included images. Reviews related to product use were posted immediately after watching TV commercials about the product. This was followed by informational posts regarding the launch of the new product and their impressions of, in this case, the taste of the new beverage. Finally, tweets concerning purchasing behavior and additional product information were posted (Fig. 1).

5 Conclusion

In this study, we aimed to model trends in consumer web communication data of new products based on text mining data obtained from Twitter feeds using Bayesian network analysis. Topics on the introduction of a new beverage were extracted by applying NMF as a dimensionality reduction tool to analyze Twitter entries. Specifically, temporal variation was detected using the weight coefficients between entries and topics obtained from NMF. The weight coefficient, which indicates the strength between an entry and a topic, was modeled using a Bayesian network to capture changes in the topic over time.

This modeling demonstrated the process of topic occurrence and disappearance and revealed the variation in topics since the new product release.

Early in the launch of a new product, topics related to the TV commercials for that product appeared upon its release. The topics initially included comments on the commercial itself that eventually transitioned to comments about the TV commercial's performers, and finally, to topics about advertisements on trains. This transition shows that the changes in the topic were related to the consumer's awareness of the new product. This was closely followed by tweets related to product impressions related to the consumption of the new product and topics related to information about the product with images and external links. Topics related to information about the product with images and external links were replaced by topics related to the brewery's release of the new product and then changed to reviews of purchasing behavior.

As consumers' purchasing behavior changed from a trial purchase to repeat purchases, topics about product information with images and external links transitioned to topics that provided more detailed information about the product. Trends in topics related to the new product were modeled using Bayesian networks. Topic variation patterns were influenced by new product strategies, such as marketing communications, and their variation over time.

Acknowledgements We express our gratitude to the anonymous referees for their valuable reviews. This work was supported by JSPS KAKENHI Grant Number JP20K01963, JP19H01532, JP19K21702.

References

Blanco, E., Castell, N., Moldovan, D.: Causal relation extraction. In: Calzolari, N., Choukri, K., Maegaard, B., Mariani, J., Odijk, J., Piperidis, S., Tapias D. (eds.) Proceedings of the Sixth International Conference on Language Resources and Evaluation (LREC'08), pp. 28–30. European Language Resources Association (ELRA), Marrakech, Morocco (2008)

Carroll, J.D.: Individual differences and multidimensional scaling. In: Shepard, R.N., Romney, A.K., Nerlove S.B. (eds.) Multidimensional Scaling, vol. I Theory, pp. 105–155. Seminar Press, New York (1972)

Deerwester, S., Dumais, S., Furnas, G.W., Landauer, T.K., Harshman, R.: Indexing by latent semantic analysis. J. Am. Soc. Inf. Sci. Tec. **41**(6), 391–407 (1990)

Ding, C., Li, T., Peng, W.: Nonnegative matrix factorization and probabilistic latent semantic indexing: Equivalence, chi-square statistic, and a hybrid method. In: Proceedings of the 21st National Conference on Artificial Intelligence and the 18th Innovative Applications of Artificial Intelligence Conference (AAAI'06), pp. 342–347 (2006)

Hearst, M.A.: Untangling text data mining. In: Proceedings of the 37th Annual Meeting of the Association for Computational Linguistics, pp. 3–10. Association for Computational Linguistics, College Park, Maryland (1999)

Hofmann, T.: Probabilistic latent semantic analysis. In: Proceedings of the 15th Conference on Uncertainty in Artificial Intelligence, pp. 289–296 (1999)

Kudo T., Yamamoto, K., Matsumoto, Y: Applying conditional random fields to Japanese morphological analysis. In: Proceedings of the 2004 Conference on Empirical Methods in Natural Language Processing (EMNLP-2004), pp. 230–237 (2004)

Lee, D.D., Seung, H.S.: Algorithms for non-negative matrix factorization. In: Leen, T., Dietterich, T., Tresp, V. (eds.) Advances in Neural Information Processing Systems, vol. 13, pp. 556–562. MIT Press (2000)

Nakayama, A.: The classification and visualization of Twitter trending topics considering time series variation. In: Palumbo F., Montanari A., Vichi M. (eds.) Data Science. Studies in Classification, Data Analysis, and Knowledge Organization, pp. 161–173. Springer, Cham (2017)

Neapolitan, R.E.: Learning Bayesian Networks. Prentice-Hall Inc, Upper Saddle River, NJ, USA (2003)

Sanchez-Graillet, O, Poesio, M.: Acquiring Bayesian networks from text. In: Proceedings of the Fourth International Conference on Language Resources and Evaluation LREC'04. European Language Resources Association (ELRA), Lisbon, Portugal (2004)
Sawaki, M., Hagita, N.: Recognition of degraded machine-printed characters using a complementary similarity measure and error-correction learning. IEICE T Inf. Syst. **E79-D**(5), 491–497 (1996)
Trovati, M., Hayes, J., Palmieri, F., Bessis, N.: Automated extraction of fragments of Bayesian networks from textual sources. Appl. Soft Comput. **60**, 508–519 (2017)

Classification Through Graphical Models: Evidences From the EU-SILC Data

Federica Nicolussi, Agnese Maria Di Brisco, and Manuela Cazzaro

Abstract The purpose of this work is to evaluate the level of perceived health by studying possible factors such as personal information, economic status, and use of free time. The analysis is carried out on the European Union Statistics on Income and Living Conditions (EU-SILC) survey covering 31 European countries. At this aim, we take advantage of graphical models that are suitable tools to represent complex dependence structures among a set of variables. In particular, we consider a special case of Chain Graph model, known as Chain Graph models of type IV for categorical variables. We implement a Bayesian learning procedure to discover the graph which best represents the dataset. Finally, we perform a classification algorithm based on classification trees to identify clusters.

Keywords Chain regression graph models · Bayesian learning procedure · Perceived health

1 Introduction

The purpose of this study is to take advantage of graphical models to evaluate the dependence relationships between the level of perceived health and possible explanatory variables concerning personal information, economic status, and use of free time. The latter variables are dichotomous or categorical ones and all together they constitute a contingency table. To evaluate the relationship between the level of perceived health and the other variables, we take advantage of graphical models that offer a

F. Nicolussi
University of Milan, via Conservatorio 7, 20122 Milano MI, Italy
e-mail: federica.nicolussi@unimi.it

A. M. Di Brisco · M. Cazzaro (✉)
University of Milano-Bicocca, Via Bicocca Degli Arcimboldi 8, 20126 Milano MI, Italy
e-mail: manuela.cazzaro@unimib.it

A. M. Di Brisco
e-mail: agnese.dibrisco@unimib.it

T. Chadjipadelis et al. (eds.), *Data Analysis and Rationality in a Complex World*,
Studies in Classification, Data Analysis, and Knowledge Organization,
https://doi.org/10.1007/978-3-030-60104-1_22

rigorous statistical instrument of analysis. In particular, we adopt the approach of Marchetti and Lupparelli (2011) to Chain Graph models that simplifies the interpretation of the parameters. Indeed, these particular Chain Graph models, called "of type IV", are suitable to consider the variables as *purely explicative*, *purely response*, or *mixed* variables. To select the model that represents the dataset at best, we consider a Bayesian learning algorithm, see Corander et al. (2008), understood as the posterior distribution over graphical models. The key idea of this learning procedure is to penalize models with too many parameters. Following the same rationale, we propose an alternative learning procedure based on a new score and depending on new stopping rules. Finally, through the estimated probability function of the graphical model, we are able to classify subjects in clusters given by the level of the perceived health. The goodness of this classification is evaluated and compared with standard classification techniques.

The paper is organized as follows. Section 2 illustrates the Chain Graph models and a suitable parametrization to model the dependence relationships. The algorithm adopted to learn the best fitting model is illustrated in Sect. 2.1, while Sect. 2.2, is devoted to describe the classification procedure. Finally, Sect. 3, shows an application on the EU-SILC dataset.

2 Chain Graph Models

Graphical models provide a dependence structure for log-linear models for contingency tables. The (possibly complex) system of dependencies can be easily represented through graphs, where each vertex corresponds to one variable and an (un)directed arc between a couple of vertices indicates a (symmetric) asymmetric dependence. Moreover, complex structures of dependencies can be modeled by including both directed and undirected arcs in the graph. In this paper, we take advantage of the Chain Graph (CG) models the main feature of which is the absence of cycles or semi-cycles, meaning that there is no (semi) directed paths that start and end in the same vertex. In particular, to represent the dependence structure of the variables through these chain graphs, we adopt the Markov property of type IV, see Marchetti and Lupparelli (2011), that are suitable to represent recursive multivariate logistic regression models. A relevant advantage of these models is in the possibility of understanding each independence marginally with respect to the other response variables, thus simplifying the system of multivariate logistic regressions to a great extent. The variables are split in *responses* and *covariates*.

For instance, by looking at Fig. 1, variables X_1 and X_2, represented by the vertices 1 and 2, play the role of covariates that affect, in different ways according to the presence/absence of the arcs, the response variables X_3, X_4 and X_5, represented by the vertices 3, 4, and 5. Thus, variables can be collected into components, T_1 and T_2, according to their nature. To model the dependence structure among a set of categorical variables, we take advantage of a generalization of the classical log-linear models, known as Hierarchical Multinomial Marginal (HMM) models. It is worth

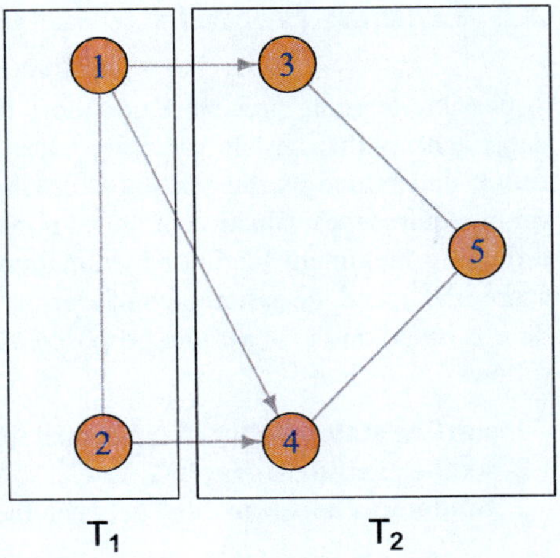

Fig. 1 Example of a chain graph composed by five vertices

noting that the parameters of HMM models, denoted with the symbol η, are suitable to describe the conditional and marginal dependence relationships among a set of variables classified as responses and covariates, see Bartolucci et al. (2007).

The HMM parameters are contrasts of logarithms of probabilities defined on marginal and joint distributions according to certain properties of hierarchy and completeness. Each parameter is characterized by the marginal distribution where it is evaluated, $\mathscr{M}$, by the set of the variables involved by the parameter, $\mathscr{L}$, and by the values of variables in $\mathscr{L}$ to which it refers, $i_{\mathscr{L}}$: $\eta_{\mathscr{L}}^{\mathscr{M}}(i_{\mathscr{L}})$. Please note that classical-log-linear models are a special case of HMM models, such that there is only one marginal set equal to the whole set of variables. When $\mathscr{L}$ is composed of only one variable, the HMM parameter is a logit. Otherwise, if $\mathscr{L}$ is composed of a couple of variables, the HMM parameter becomes a logarithm of odds ratios. As the cardinality of $\mathscr{L}$ increases, the HMM parameters become logarithm of contrasts of odds ratios.

Any missing arc in the graph, i.e., any independence relationship among two or more variables, results in a set of certain null ηs parameters. The remaining unconstrained parameters describe the dependencies among variables.

The inference on the CG models and consequently on HMM models is carried out through the likelihood ratio test G^2 (Marchetti and Lupparelli 2011). Each model to be tested is compared with the HMM saturated model (i.e., the model without constraints on the ηs parameters). Asymptotically, the G^2 statistic follows the χ^2 distribution with k degrees of freedom equal to the number of constrained parameters.

The next section is dedicated to describe and discuss the algorithm used to choose the best fitting CG model.

2.1 Learning Procedure

A possible learning procedure to choose the best fitting CG model, i.e., the set of dependencies that best fit the data, takes advantage of a Bayesian learning algorithm, understood as the posterior distribution over graphical models. The algorithm requires the evaluation of the marginal likelihood, which can be approximated through a maximum likelihood estimation of the Bayesian Information Criterion score (BIC), and the assignment of a prior probability to the graph. The used procedure is based on the algorithm proposed by Corander et al. (2008), and it works as follows:

1 **starting state**: G_0, the graph containing **no edges**
2 **set the candidate graph** $G_t = G_0$
3 **randomly choose an edge** between the nodes δ and γ:

- if the edge is present in G_t, remove it
- if the edge is absent in G_t, add it

4 **calculate** the score of G_t: $score(G_t)$
5 **calculate** $P = min\left[(score(G_t) - score(G_0)), \mathbf{1}\right]$
6 **choose** the graph: set $G_0 = G_t$, with probability P, otherwise set $G_0 = G_0$.
7 **repeat** from step 2
8 arrest rule

Stop the procedure if at least one of the two following conditions is FALSE:

I the number of times we choose, consecutively, the graph G_0 is less than two times the number of possible edges of the model (i.e., 4 variables $\rightarrow$ 6 edges);
II the selected graph has not been tested against all the other possible graphs (less than an edge).

We propose to use the following score:

$$score(G_t) \simeq S(\mathbf{X}|G_t) \cdot 2^{-\lambda_{G_t}},$$

where $S(\mathbf{X}|G_t)$ is the approximation of the marginal likelihood of $\mathbf{X}$, the dataset, using BIC and λ_{G_t} is the penalization term. In Corander et al. (2008), the penalization term is $\lambda_{G_t} = dim(\Theta_{G_t})$ that is the maximum number of *free parameters* in a distribution with the parameter restrictions induced by G_t (penalization for dense graph). In addition, we consider the penalization term equal to the number of edges in the graph: $\lambda_{G_t} = \#E$. This new penalization term is supported from the necessity to be less restrictive in some learning procedures, especially when the number of variables is not so high.

This part of the analysis was carried out with the statistical software R (Core Team 2019), and the package hmmm, Colombi et al. (2014).

2.2 *Classification*

CG models can be adopted as classification tools. Indeed, any CG model provides a fitted joint distribution of a set of variables. When the aim of the analysis is to classify n subjects into k groups (that can be the categories of the variable(s) Y) given a set of factors, X, (the remaining variables used to fit the model), we can derive the conditional distribution of the variable(s) given all the other variables, $p(Y|X)$. Each new subject will be classified in the group y_j, with $j = 1, \ldots, k$, with probability $p(Y = y_j|X = x)$, where x is the vector of the levels that the subject assumes on the X variables.

The literature concerning classification methods for categorical variables given a set of factors, X, is extensive. Here, we compare our model with two well-established classification techniques: the CART (classification tree algorithm based on a generalization of the Gini impurity index) and the *random forest* (a collection of tree structured classifiers), see Breiman (2001).

To assess the best classification technique, we compute the Brier score (BS) that evaluates the accuracy of probabilistic predictions of a categorical variable and it is defined as follows:

$$\mathbf{BS} = \frac{1}{n}\sum_{i=1}^{n}\sum_{j=1}^{k}(p_{ij} - o_{ij})^2$$

where n is the sample size, k is the number of categories of the variable to be predicted, p_{ij} is the predicted probability for category j and subject i and o_{ij} is the observed outcome (1 if subject i has predicted variable level equal to j, 0 otherwise). The greater the BS index the greater the classification error, see Breiman et al. (1984).

3 Application

The EU-SILC survey aims to gather multidimensional data and to monitor the poverty level, social inclusion, and living conditions of European countries. For the purpose of our study, we focus on evaluating the level of perceived health (H), which is quantified into 3 levels from "good" = 1 to "bad" = 3. The sample consists of 345 501 adult subjects from 31 European countries. Possible factors that determine the level of perceived health may include personal information, economic status (status in employment, income bracket), and use of free time (capacity to afford paying for one week annual holiday, regular participation in a leisure activity, and access to a small amount of money each week for personal use). Within the set of personal information, we consider the age (A) categorized in four levels determined from the quartiles, the gender (G) with categories *Male* = 1 and *Female* = 2 and the educational level (E) with levels *Upper secondary education*= 1, *Bachelor degree* = 2, *Master/Doctoral degree* = 3. The variables apt to define the economic status are the working condition (W): *Employee* = 1, *Unemployed* = 2, *Unfit to work* = 3, *Family Worker* = 4; and the Equivalised disposable household income (I) with categories: *Less than poverty threshold* = 1, *Between poverty threshold and Median* = 2, *Over Median* = 3. Finally, we consider the variable Capacity to afford paying for one week annual holiday away from home (J) with levels *Yes* = 1 and *No* = 2. To evaluate the relationship between the level of perceived health and the other variables, we take advantage of graphical models that offer a rigorous statistical instrument of analysis.

In order to implement this part, we took advantage of the R `randomForest` and `rpart` packages, Liaw and Wiener (2002), Therneau and rpart (2019).

3.1 Results

The best fitting CG model, obtained by using both the scores presented in the Sect. 2.1, is described from the graph in Fig. 2. The independence relationships underlying this CG model are $GE \perp H|AWIJ$ and $G \perp IJ|AE$.

Table 1 shows the non-null second order parameters concerning the variable H.

Since in Table 1 (top left), all log-odds ratios referred to variables A and H are positive, we can assess that the perceived health condition gets worse with age—a trend that is particularly evident for higher levels of age. Instead, all log-odds ratios referred to variables I and H, Table 1 (top right), are negative meaning that the perceived health condition gets better as the disposable household income is higher than the poverty threshold.

In the light of the relation between variables J and H, Table 1 (bottom left), we deduce that perceived health improves with the ability to afford a week annual holiday. Finally, the relation between variable W and H, Table 1 (bottom right), suggests that the perceived health condition gets worse in the transition from W = *employee* to W = *family worker* status.

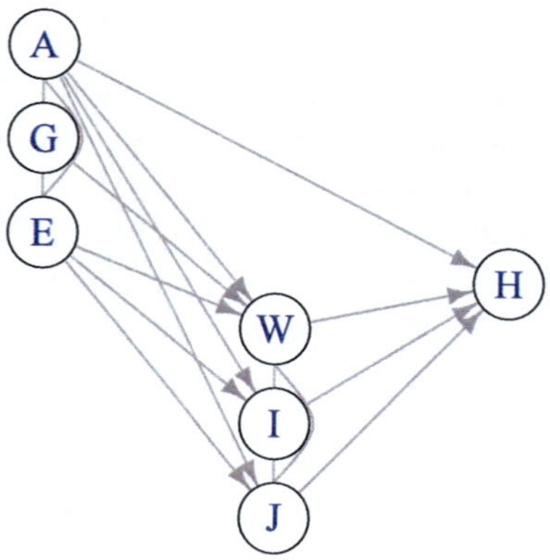

Fig. 2 Chain graph model which best represent the data with $BIC = -1\,599\,073$ and $BS = -1\,599\,513$

Table 1 Log-odds ratio concerning pair of variables

	X_2	H	
X_1	i_{X_1}/i_{X_2}	2	3
A	2	0.652	0.939
	3	1.442	0.670
	4	0.926	**1.958**
	X_2	H	
X_1	i_{X_1}/i_{X_2}	2	3
I	2	**–0.461**	–0.337
	3	–0.380	–0.199
	X_2	H	
X_1	i_{X_1}/i_{X_2}	2	3
J	2	0.166	**0.269**
	X_2	H	
X_1	i_{X_1}/i_{X_2}	2	3
W	2	0.108	2.106
	3	0.238	1.256
	4	**4.666**	0.875

As the last step of the analysis, we consider the strength of our model as a classifier. In particular, we evaluate the BS on the n subjects by using the predicted probabilities obtained from the CG model represented from the graph in Fig. 2. We also carried out the CART and the Random Forest analysis on the same dataset. The Brier score of the CG model is 0.35, the one of the CART is 0.37 and the one of the Random Forest is 0.44. Thus, we can conclude that the CG model provides a satisfactory behavior as a classification technique.

4 Conclusion

In this work, we use the CG model with the HMM model to provide a graphical representation of the system of dependencies among a set of variables, to model the joint probability function of the set of variables involved, and finally to classify a variable according to a set of factors. The choice of the best fitting model is attained by a Bayesian algorithm carried out with two different scores. All these features are discussed and highlighted through an application on EU-SILC data.

Acknowledgements This paper is based on data from Eurostat, EU Statistics on Income and Living Conditions [2016].
The research leading to these results has received funding from the European Union's Horizon 2020 research and innovation programme under grant agreement No. 730998, InGRID-2 Integrating Research Infrastructure for European expertise on Inclusive Growth from data to policy.

References

Bartolucci, F., R. Colombi, Forcina, A.: An extended class of marginal link functions for modelling contingency tables by equality and inequality constraints. Stat. Sinica **17**, 691–711 (2007)

Breiman, L., Friedman, J.H., Olshen, R.A., Stone, C.J.: Classification and Regression Trees. CRC Press (1984)

Breiman, L.: Random forests. Mach. Learn. **45**, 5–32 (2001)

Colombi, R., Giordano, S., Cazzaro, M.: hmmm: An R Package for Hierarchical Multinomial Marginal Models. J. Stat. Soft. **59**(11), 1–25 (2014)

Corander, J., Ekdahl, M., Koski, T.: Parallell interacting MCMC for learning of topologies of graphical models. Data Min. Knowl. Disc. **17**(3), 431–456 (2008)

Liaw, A., Wiener, M.: classification and regression by randomForest. R News **2**(3), 18–22 (2002)

Marchetti, G.M., Lupparelli, M.: Chain graph models of multivariate regression type for categorical data. Bernoulli **17**(3), 827–844 (2011)

R Core Team. R: A Language and Environment for Statistical Computing. R Foundation for Statistical Computing, Vienna, Austria (2019)

Therneau, T., Atkinson, B.: rpart: Recursive Partitioning and Regression Trees. R package version 4.1–15 (2019)

A Simulation Study for the Identification of Missing Data Mechanisms Using Visualisation

Johané Nienkemper-Swanepoel, Niël Le Roux, and Sugnet Gardner-Lubbe

Abstract Understanding the cause of the missingness in data is a science of its own and is of great importance for the application of valid and unbiased analysis techniques for missing data. The distribution of missingness is defined by certain dependencies on either observed or missing values in a data set, and therefore, requires a multivariate visualisation to attempt to identify the missing data mechanism (MDM). Multivariate categorical data sets containing missing data entries can be separated into observed and unobserved (or missing) subsets by creating an additional category level (CL) for each variable with missing responses in the indicator matrix. Subset multiple correspondence analysis (sMCA) can then be applied to the recoded indicator matrix to obtain separate biplots for the observed and missing subsets. The sMCA biplot of missing categories enables the exploration of the missing values which could expose non-response patterns. Partitioning around medoids (pam) clustering is used to determine whether sufficient clustering structures can be identified in the sMCA biplot of missing responses. A simulation study consisting of data sets with different sample sizes are generated from three distributions. Artificial missingness is created by deleting values according to MAR and MCAR MDMs with different percentages of missing values. The influence of the underlying distribution on the outcome of the clustering techniques will be presented. The insight obtained from the simulation results provides guidelines for the identification of the MDM in real data applications.

Keywords Categorical data · Clustering · Missing data · Missing data mechanism · Subset multiple correspondence analysis

J. Nienkemper-Swanepoel (✉) · N. Le Roux · S. Gardner-Lubbe
Stellenbosch University, Stellenbosch, South Africa
e-mail: nienkemperj@sun.ac.za

N. Le Roux
e-mail: njlr@sun.ac.za

S. Gardner-Lubbe
e-mail: slubbe@sun.ac.za

T. Chadjipadelis et al. (eds.), *Data Analysis and Rationality in a Complex World*, Studies in Classification, Data Analysis, and Knowledge Organization,
https://doi.org/10.1007/978-3-030-60104-1_23

1 Introduction

It is crucial to understand the cause of missingness before selecting a missing data handling technique (Buhi et al. 2008; Kowarik and Templ 2016). Visualisations of missing values could expose structures and patterns that are not perceptible in a data table (Templ et al. 2012). The occurrence of missing values can be explained as the result of a random process referred to as the missing data mechanism (MDM) (Van Buuren 2012). Three mechanisms are defined: missing at random (MAR), missing completely at random (MCAR) and missing not at random (MNAR). The MAR MDM occurs when the missing values are dependent on the observed responses, whereas an MCAR MDM occurs when the missing observations are independent of observed and unobserved observations. Missing observations that are unobserved due to the MNAR mechanism are dependent on information that is not captured by the questionnaire, therefore, dependent on other missing values (García-Laencina et al. 2010).

Biplots are utilised for the visualisations in this paper, in which the samples and variables can be displayed simultaneously in a lower (two or three) dimensional configuration. Biplots are not limited to certain variable types, but the displays differ for continuous and categorical variables. Continuous variables are represented as calibrated axes, an axis for each continuous variable. The multiple axes are not orthogonal but similar to two-dimensional scatterplots, sample points are projected perpendicularly to a calibrated axis to obtain the response value. Only multivariate nominal scaled categorical data are considered in this paper. Each sample in the data set, typically the rows of the data matrix, is characterised by the responses to categorical variables that consist of distinct category levels (CLs).

In a biplot for categorical variables, the CLs are illustrated as a simplex of points referred to as the category level points (CLPs), one point for each CL (Gower and Hand 1996). The display of the variables, whether it is an axis or CLP simplex, is considered to provide a 'framework' or 'reference system' of the display (Cox and Cox 2001). The proximity of the sample points and CLPs reveal associations. Dissimilarities are illustrated through points that are not situated in close proximity, whereas closely positioned points reflect high association and similarity (Gower et al. 2011). The sample points are positioned at the vector sum of its observed CLPs (Gower and Hand 1996).

Subset correspondence analysis (sCA) has been used to explore incomplete categorical data consisting of two-way contingency tables (Greenacre 2017; Hendry et al. 2014). The complete correspondence analysis (CA) can be restricted using a chosen subset of a data matrix whilst maintaining the original column and row masses for the calculation of the distances. Therefore, the total variation is partitioned into components associated with the various subsets and no interpretable information is lost. This idea is extended to multiple correspondence analysis (MCA), referred to as subset MCA (sMCA).

The data matrix of a multivariate categorical data set is commonly coded as an indicator matrix of zeros and ones. The columns of the indicator matrix represent the

CLs and the rows correspond to the samples. A particular response will be represented by a one in the column corresponding to the chosen CL and zero elsewhere. In order to expose the missing data structures, the indicator matrix is recoded by adding new CLs for missing responses using single active data handling. Since single active handling only creates one missing CL per variable, it is assumed that all samples with a missing response for a particular variable are similar. This assumption might not be a true reflection of the MDM, and therefore, single active handling could be inappropriate if the data are missing due to an MCAR MDM (Van der Heijden and Escofier 1997). This being said, in the context of this paper, the single active handling is used to distinguish between missing and observed information irrespective of MDM and forms part of a step in the process of treating missing values before obtaining final inferences. Therefore, the single active handling enables the formation of two subsets which are available for separate investigation as opposed to using the actively handled data set as a whole for standard complete data analysis methods without forfeiting or compromising observed information.

Apart from the benefits of missing data analysis, sMCA also provides a visualisation benefit in the sense that it improves visual representations where large data sets become uninterpretable due to cluttered displays. A set of visualisations can be used with different subsets of the data, whilst maintaining the explained variation of the analysis.

Since the aim is to understand the cause of missingness, the emphasis is on the subset of unobserved CLs. Therefore, the missing CLPs and the corresponding samples from the sMCA solution are configured in a biplot. It is hypothesised that if insufficient clustering structures occur between the missing CLs in the sMCA biplot of missing values, this could indicate independence, and therefore, relate to the MCAR MDM. The contrary is hypothesised for sufficient clustering structures, which could indicate missing CLs that are highly associated, and therefore, reflect a MAR MDM.

After experimenting with various clustering techniques, it was found that the partitioning around medoids (pam) clustering technique (Maechler et al. 2017), is suitable and robust for this particular application. The pam approach is flexible and since the number of clusters can be specified, it enables automation of the algorithm over 1000 simulation repetitions. Therefore, pam will be used to determine the number of clusters that can be successfully separated in the incomplete sMCA biplots.

2 Methodology

2.1 Subset Multiple Correspondence Analysis

MCA is applied by performing CA on the indicator matrix. However, for sMCA a subset of the recoded indicator matrix, $\mathbf{G}_{\text{recoded}}$ (single active handling), is isolated for the analysis. The subset indicator matrix, $\mathbf{G}_h$, is transformed by the diagonal matrices of row weights from the full recoded indicator matrix, $\mathbf{G}_{\text{recoded}}$, and column weights, $\mathbf{C}$, from the subset indicator matrix, $\mathbf{G}_h$: $\mathbf{S} = p^{-1/2}\mathbf{G}_{(n\times h)}\mathbf{C}^{-1/2}_{(h\times h)}$, where p is the number of variables and h is the number of CLs in the particular subset.

The singular value decomposition (SVD) of $\mathbf{S}$ results in the following:

$$\mathbf{S}_{(n\times h)} = \mathbf{U}_{(n\times r)}\boldsymbol{\Lambda}_{(r\times r)}\mathbf{V}^T_{(r\times h)},$$

where n is the number of samples and r is the reduced dimension. The singular vectors are represented in $\mathbf{U}$ and $\mathbf{V}$ with the singular values represented in decreasing order in $\boldsymbol{\Lambda}$ (Greenacre 2017; Greenacre and Pardo 2006).

The sMCA biplot is constructed by plotting the samples using the first two columns of the following matrix multiplication (Gower et al. 2011):

$$p^{-1/2}\mathbf{U}_{(n\times 2)}\boldsymbol{\Lambda}_{(2\times 2)}.$$

The CLPs are obtained from using the first two columns of the following matrix multiplication:

$$\mathbf{C}^{-1/2}_{(h\times h)}\mathbf{V}_{(h\times 2)}.$$

The sMCA method is available in the in the R package, *ca* (Nenadić and Greenacre 2007).

2.2 Partitioning Around Medoids

We are interested in determining whether a sufficient clustering structure exists for the CLPs, since this could lead to emphasising the association between missing responses and subsequently identifying the MDM. Since cluster analysis is applied to the reduced dimension sMCA solution, this is regarded as a tandem clustering approach (Mitsuhiro and Yadohisa 2015).

The pam method is repeatedly applied by separating the CLPs into all possible number of medoids between two and one less than the number of missing CLPs in the biplot. This will establish the optimal number of medoids for the identification of discerning patterns to consequently confirm the MDM.

The average silhouette width or silhouette coefficient, $s(i)$, is used to determine whether the number of predetermined medoids successfully discriminates between the clusters.

Guidelines to decide whether a silhouette value reflects substantial clustering structures are not available, but $s(i) > 0.5$ is regarded as an above average measure reflecting the efficient identification of clustering structures. It is advised that a silhouette value below 0.25 is an indication that no notable clusters are present in a data set (Struyf et al. 1997). Carrying forward, a $s(i) \geq 0.5$ will be regarded as an indication of well separated clusters, which illustrates dependency between missing response CLs. A $s(i) < 0.5$ will be indicative of no substantial clustering structures, and therefore, independence of missing response CLs.

The *pam* method is available in the R package, *cluster* (Maechler et al. 2017).

3 Simulated Data

Simulated data play a vital role in the evaluation of missing data techniques. An extensive simulation study has been conducted considering a 1000 repetitions of 162 unique data scenarios with the following parameters: (1) Simulation distribution: Dirichlet, normal and uniform. (2) Sample size: 100, 1000 and 3000. (3) Number of variables: 5, 10 and 15. (4) Percentage of missing values: 10, 30 and 50%. (5) MDM: MCAR and MAR.

The simulated continuous complete data sets are categorised by dividing the continuous responses into a randomly determined number of CLs per variable before creating artificial missingness.

The interested reader may consult the given reference (Nienkemper-Swanepoel 2019), for the complete simulation protocol.

4 Results

A fixed random seed is used to ensure the reproducibility of results. Creating an artificial MCAR MDM requires the missing values to be independent of any observed or missing information, and therefore, represents an unbiased sample from what would have been the complete data set. A 1000 randomly generated seed values are used throughout the simulations, which means that each first simulation (irrespective of data scenario) uses the first seed value, the second simulation uses the second seed value until the final simulation is completed. Utilisation of the same seed values results in similar positions of the missing observations for the same repetition of the simulations, irrespective of the simulated distribution and data scenario. Due to the similarity of the visualisations obtained from the MCAR MDMs, irrespective of the simulated distribution, only one MCAR distribution per missing percentage is visualised along with the sMCA biplots of the missing subsets obtained from the MAR simulations (normal, uniform and Dirichlet) to ease the visual interpretation. First, a selection of sMCA biplots are investigated in Fig. 1. The sample points are depicted by green open circles and the CLPs by black triangles.

Structured sample points are observed in some of the visualisations, which could be a reflection of recurring response patterns due to the recoding of the indicator matrix using single active handling. It is also plausible that the structured sample points indicate the categorical scale of the data, since only qualitative responses are possible. There are visible differences between the MAR and MCAR visualisations. As the percentage of missing values increases, the sample points for the MAR MDM

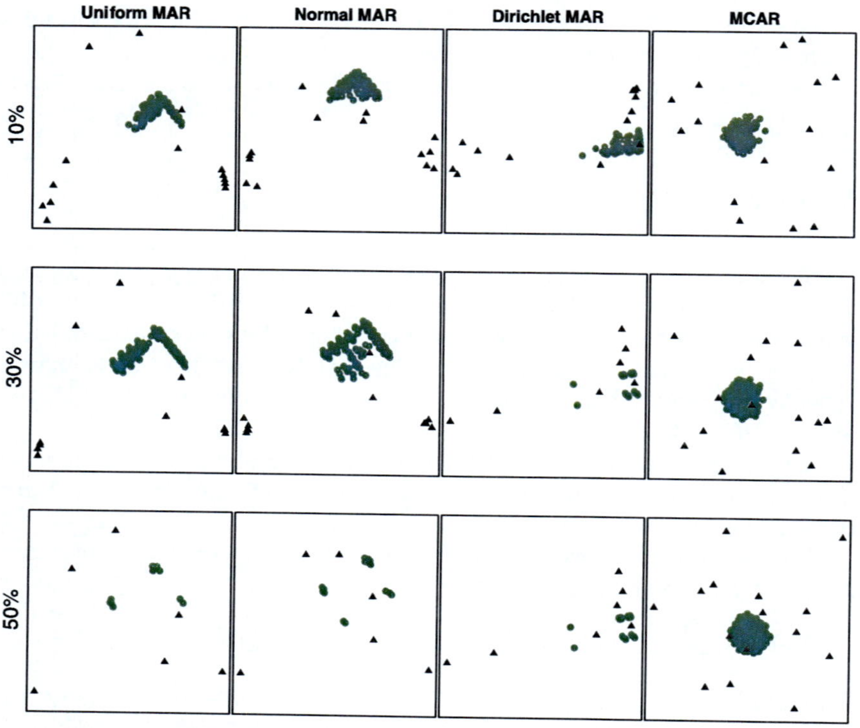

Fig. 1 sMCA biplot of missing subsets of a selected repetition ($n = 1000$, $p = 15$)

overlap. Additional to the sample point behaviour, it is noticeable that the CLPs also show separation with overlapping points in the MAR configurations. This is not the case for the MCAR visualisations, where the samples form a circular cloud of points with approximately equally scattered CLPs which suggest that there is no particular separation. The random cloud of points resembles homogeneous spread which upholds the definition of the MCAR MDM. The visible separation between CLPs in the MAR sMCA biplots is expected to result in higher silhouette coefficients than MCAR sMCA biplots.

A selection of the silhouette coefficients obtained from the clustering of the sMCA biplots presented in Fig. 1, are illustrated in the scatterplots of Fig. 2.

The scatterplots are divided into vertical intervals to distinguish between the silhouette coefficients obtained from the number of clusters that are specified for the 1000 repetitions for each simulated scenario. The MAR MDM is generated by conditioning the missingness on an observed variable, therefore, one variable will be fully observed, and consequently, there is one less missing CLP than the corresponding MCAR MDM simulation. Each scatterplot has the same number of vertical intervals to enhance the comparison between different panels.

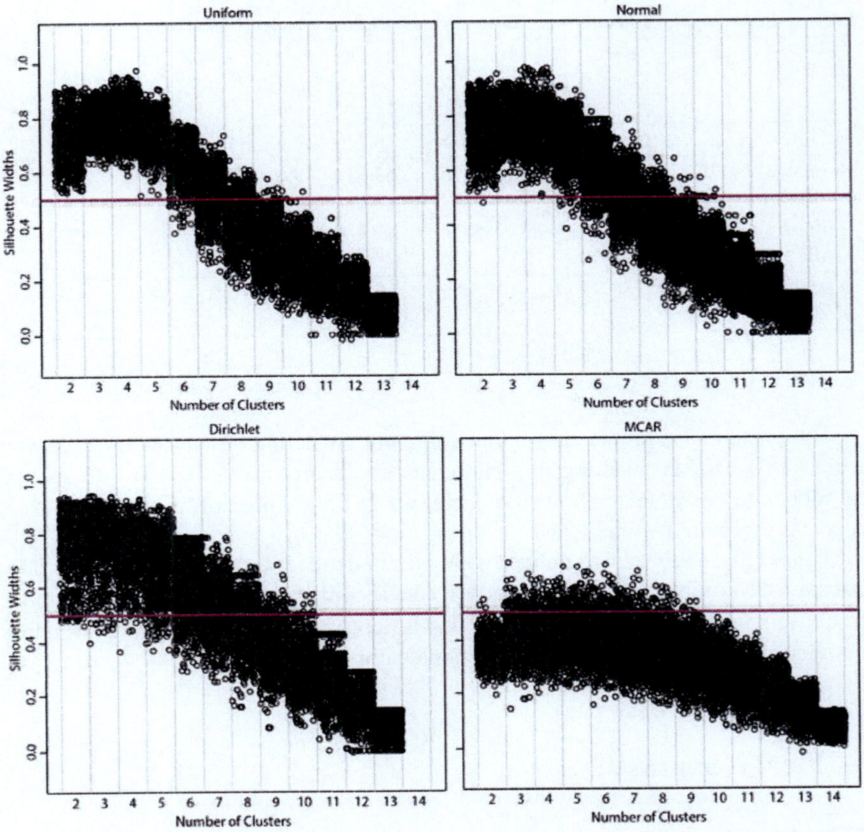

Fig. 2 Silhouette coefficients of simulated data sets ($n = 1000$, $p = 15$) with 30% missing values

A solid horizontal line is visualised where $s(i) = 0.5$ which enables a quick appreciation of the trend of silhouette widths below and above the 0.5 benchmark value. The focus is on identifying the difference in patterns/trends between the MAR and MCAR MDMs.

It is clear that the MCAR MDM results in lower silhouette coefficients with a large proportion of values occurring below the 0.5 threshold which is not observed for the MAR MDMs. Silhouette coefficients decrease as the number of clusters increases, especially when more than five clusters are specified. Similar results are observed for the uniform and normal simulations with slightly lower and more dispersed silhouette coefficients observed from the Dirichlet simulations.

In order to distinguish between the two MDMs based on silhouette coefficients, a stricter benchmark of 0.6 is proposed to suggest a MAR MDM. It is conjectured that the frequency of silhouette widths above 0.6 for MAR MDMs will be considerably more than the frequency for MCAR MDMs. The percentages of silhouette widths above 0.6 presented in Fig. 2, are given in Table 1.

Table 1 Percentage of simulations resulting in a slhouette coefficient above 0.6

	$k = 2$	$k = 3$	$k = 4$	$k = 5$	$k = 6$	$k = 7$	$k = 8$	$k = 9$	$k = 10$ to $k = 14$
MCAR	0	0.5	0.7	0.8	0.7	0.3	0	0	0
Normal MAR	95.1	99.5	97.8	89.5	63.1	19.6	0.7	0.1	0
Uniform MAR	92.8	100.0	99.9	98.4	81.9	25.8	0.4	0	0
Dirichlet MAR	90.7	93.7	92.2	78.5	65.3	36.3	5.2	0	0

The presented example shows that sufficient clustering structures are observed from MAR MDMs resulting in a silhouette coefficient above 0.6. The frequency of silhouette coefficients above 0.6 decreases as the number of clusters increases, especially beyond five medoids.

The results in this section only provide a limited reflection of the simulation study, but from the complete simulation study, it can be conjectured that an optimal number of clusters for the evaluation of the MDM is approximately a third of the available number of missing CLPs in the recoded indicator matrix.

5 Final Remarks

Only selected results are presented in the paper, however, the conclusions reflect the findings of the complete simulation study. It was found that data sets with small dimensions ($n = 100$ and $p = 5$) do not result in discernible differences between MAR and MCAR MDMs. Separation is observed between the CLPs of MAR MDMs with overlapping points in the separate groups with an increase in the size of the data set. The sample points also result in structured patterns which converge to overlapping separated groups as the percentage of missing values increases. To the contrary, MCAR MDM sMCA biplots do not show a clear separation between CLPs and the sample points tend to form a cloud of points, especially in a higher percentage of missing values. Therefore, the visual differences between the MDMs are noticeable. The visual interpretations were confirmed by cluster analysis on the CLPs of the sMCA biplots.

The silhouette coefficients confirmed that the clustering structure of data sets is more evident for larger data sets, both sample size and number of variables, with a large percentage of missing values. It was observed that the data sets simulated from a uniform distribution achieved slightly higher silhouette coefficients than the Dirichlet and normal distributions. However, similar patterns in silhouette coefficients were observed for the uniform and normal distributions.

An optimal number of clusters to use with respect to the parameters of a given data set has been identified. In general, the smallest number of clusters (usually, $k = 2$) is proposed when attempting to identify the MDM of a data set with small dimensions and a low percentage of missing values. However, as a general rule of thumb, a third of the available number of missing CLPs can be used as the optimal number of clusters to deduce whether the MDM is possibly MAR of MCAR. If the silhouette coefficient obtained from the number of clusters equivalent to a third of the number of missing CLPs is above 0.6, it is strongly suggestive of a MAR MDM.

References

Buhi, E.R., Goodson, P., Neilands, T.B.: Out of sight, not out of mind: Strategies for handling missing data. Am. J. Health Behav. **32**(1), 83–92 (2008)

Cox, T.F., Cox, M.A.A.: Multidimensional Scaling, 2nd edn. Chapman and Hall/CRC (2001)

García-Laencina, P.J., Sancho-Gómez, J., Figueiras-Vidal, A.R.: Pattern classification with missing data: a review. Neural Comput. Appl. **19**(2), 263–282 (2010)

Gower, J.C., Hand, D.J.: Biplots. Chapman and Hall (1996)

Gower, J., Lubbe, S., Le Roux, N.: Understanding Biplots. Wiley, New York (2011)

Greenacre, M., Pardo, R.: Multiple correspondence analysis of subsets of response categories. In: Greenacre, M., Blasius, J. (eds.) Multiple Correspondence Analysis and Related Methods, pp. 197–217. Chapman and Hall/CRC, New York, NY (2006)

Greenacre, M.: Correspondence Analysis in Practice, 3rd edn. Chapman and Hall/CRC (2017)

Hendry, G., North, D., Zewotir, T., Naidoo, R.N.: The application of subset correspondence analysis to address the problem of missing data in a study on asthma severity in childhood. Stat. Med. **33**(22), 3882–3893 (2014)

Kowarik, A., Templ, M.: Imputation with R package VIM. J. Stat. Soft. **74**(7), 1–16 (2016)

Maechler, M., Rousseeuw, P., Struyf, A., Hubert, M., Hornik, K.: cluster: Cluster Analysis Basics and Extensions, Available at https://cran.r-project.org/cluster. R package (version 2.0.6) (2017)

Mitsuhiro, M., Yadohisa, H.: Reduced k-means clustering with MCA in a lowdimensional space. Comput. Stat. **30**(2), 463–475 (2015)

Nenadić, O., Greenacre, M.: Correspondence analysis in R, with two- and three-dimensional graphics: The ca package. J. Stat. Soft. **20**(3), 1–13 (2007)

Nienkemper-Swanepoel, J.: Biplot methodology for analysing and evaluating missing multivariate nominal scaled data. Ph.D. thesis, Stellenbosch University (2019). https://scholar.sun.ac.za/handle/10019.1/107027

Struyf, A., Hubert, M., Rousseeuw, P.J.: Integrating robust clustering techniques in S-plus. Comput. Stat. Data An. **26**(1), 17–37 (1997)

Templ, M., Alfons, A., Filzmoser, P.: Exploring incomplete data using visualization techniques. Adv. Data Anal. Classi. **6**(1), 29–47 (2012)

Van Buuren, S.: Flexible Imputation of Missing Data. Chapman and Hall/CRC (2012)

Van der Heijden, P.G.M., Escofier, B. Analyse de correspondances: Recherches au cœur de l'analyse des données. In: Escofier B. (ed.) Multiple Correspondence Analysis with Missing Data, pp. 152–170. Presses Universitaire de Rennes (1997)

Triplet Clustering of One-Mode Two-Way Proximities

Akinori Okada and Satoru Yokoyama

Abstract Some researchers noticed that proximities of three objects are useful to disclose relationships among objects. Sometimes it is not easy to obtain one-mode three-way proximities in contrast to obtain one-mode two-way proximities. Hence, a procedure to assemble one-mode three-way proximities from one-mode two-way proximities is introduced. And a method for hierarchical clustering of the resulting one-mode three-way proximities, where three clusters (objects) form a new cluster at each step of the clustering, is introduced. The procedure is applied to one-mode two-way dissimilarities among kinship terms, and the resulting one-mode three-way dissimilarities (dissimilarities of three kinship terms) were analyzed by the method of cluster analysis for one-mode three-way dissimilarities, which is comparable to the complete linkage. The one-mode two-way dissimilarities, from which the one-mode three-way dissimilarities were assembled, were analyzed by the complete linkage cluster analysis. The comparison of the two results shows that the present analysis revealed the aspects which cannot be disclosed by the analysis using one-mode two-way cluster analysis.

Keywords Clustering · One-mode three-way proximity · One-mode two-way proximity · Proximity of three objects · Triadic relationship

1 Introduction

Cluster analysis and multidimensional scaling have been developed to disclose proximity relationships among objects by analyzing mainly the dyadic proximity or the proximity of two objects. Some researchers focus their attention onto triadic

A. Okada (✉)
Rikkyo University 3-18-1 Ozenji Higashi Asao-ku Kawasaki-shi, Kanagawa-ken 215-0018, Japan
e-mail: okada@rikkyo.ac.jp

S. Yokoyama
Aoyama Gakuin University 4-4-25 Shibuya, Shibuya-ku, Tokyo 150-8366, Japan
e-mail: yokoyama@busi.aoyama.ac.jp

T. Chadjipadelis et al. (eds.), *Data Analysis and Rationality in a Complex World*, Studies in Classification, Data Analysis, and Knowledge Organization,
https://doi.org/10.1007/978-3-030-60104-1_24

relationships which show the proximity of three objects (e.g., Heiser and Bennani 1997). Cluster analysis and multidimensional scaling which can analyze triadic relationships have been introduced by de Rooji and Heiser (2000), Nakayama (2005), Yokoyama et al. (2009). Yokoyama et al. (2009) introduced an overlapping clustering (Arabie and Carroll 1980). de Rooji and Heiser (2000), and Nakayama (2005) developed multidimensional scaling which can deal with triadic relationships based on L_p transform (de Rooji and Heiser 2000).

These cluster analysis and multidimensional scaling methods were introduced to analyze the proximities of three objects. There are some sorts of data where proximities of three objects are not easy to obtain. The brand switching is one of the examples where the proximity of three objects (brands) are not easy to obtain. The number of consumers who purchased brand j at a period and brand k at the next period is the frequency of brand switching from brands j to k, and is regarded as the similarity of brands j and k. But it is not easy to obtain the similarity of three brands by the purchase record at two consecutive periods when the brands are durable goods (e.g., car, refrigerator, $\cdots$), because those durable goods usually would not be purchased two times or more in one period. It seems useful to disclose relationships among objects by analyzing triadic relationships not only dyadic relationships (Yokoyama et al. 2009). The purpose of the present study is twofold. One is to introduce a procedure of assembling one-mode three-way proximities from one-mode two-way proximities. The other is to develop a method for hierarchical clustering of one-mode three-way proximities which were assembled from one-mode two-way proximities.

2 Procedure for Assembling One-Mode Three-Way Proximities from One-Mode Two-Way Proximities

A procedure of assembling one-mode three-way proximities from one-mode two-way proximities is described below. In the next section, a method of clustering resulting in one-mode three-way proximities is introduced. The present and the next sections are described by using the similarity (Okada and Yokoyama 2019). But this does not mean any constraint. The parallel procedure and the method are possible for the dissimilarity. Let s_{jk} is the dyadic similarity of objects j and k, and $s_{jk\ell}$ is the triadic similarity of objects j, k, and ℓ. The objective of the present procedure is to assemble the triadic similarity from the dyadic similarity. One idea is to define

$$s_{jk\ell} = s_{jk} + s_{j\ell} + s_{k\ell}, \tag{1}$$

where larger $s_{jk\ell}$ denotes the larger similarity of three objects j, k, and ℓ ($s_{jk\ell} = s_{j\ell k} = s_{kj\ell} = s_{k\ell j} = s_{\ell jk} = s_{\ell kj}$). Equation (1) is based on the L_p transformation (de

Rooji and Heiser 2000; Heiser and Bennani 1997; Nakayama 2005). But Eq. (1), has a drawback. Let $s_{jk\ell} = s_{j'k'\ell'}$, where

$$s_{j'k'\ell'} = s_{j'k'} + s_{j'\ell'} + s_{k'\ell'}. \tag{2}$$

Suppose that

$$s_{jk} \simeq s_{j\ell} \simeq s_{k\ell}, \tag{3}$$

and

$$s_{j'k'} \simeq s_{j'\ell'} \gg s_{k'\ell'}. \tag{4}$$

While $s_{jk\ell}$ and $s_{j'k'\ell'}$ are equal, it seems reasonable to consider that the similarity of objects j, k, and ℓ is larger than the similarity of objects j', k', and ℓ' (cf. Hayashi 1989). An extremely smaller or larger similarity between two objects compared with the other two would reduce the similarity of three objects. To cope with the drawback, $(s_{jk} + s_{j\ell} + s_{k\ell})$ is multiplied with the ratio $min(s_{jk}, s_{j\ell}, s_{k\ell})/max(s_{jk}, s_{j\ell}, s_{k\ell})$,

$$s_{jk\ell} = (s_{jk} + s_{j\ell} + s_{k\ell}) \times [min(s_{jk}, s_{j\ell}, s_{k\ell})/max(s_{jk}, s_{j\ell}, s_{k\ell})]. \tag{5}$$

3 Method for Hierarchical Clustering

The algorithm of the hierarchical clustering of one-mode three-way proximities assembled from one-mode two-way proximities is an extension of the algorithm of Johnson (1967). Three kinds of algorithms, comparable to the complete linkage, the single linkage, and the average linkage, are developed. Let j, k, and ℓ denote clusters. Same as the hierarchical clustering algorithm for one-mode two-way proximities, each step of the present clustering algorithm consists of two phases; to merge clusters in order to form a new cluster, and to update the proximity between the newly formed cluster and existing clusters. The algorithm of the present clustering comparable to the complete linkage is described below.

Phase 1 Merge three clusters J, K, and L which satisfy

$$s_{JKL} = max_{j<k<\ell}(s_{jk\ell}) \tag{6}$$

to form a new cluster.

Phase 2.1 Update the one-mode two-way similarity between the newly formed cluster (JKL) and existing cluster i $s_{(JKL)i}$,

$$s_{(JKL)i} = min(s_{Ji}, s_{Ki}, s_{Li}). \tag{7}$$

Phase 2.2 Update one-mode three-way similarities by Eq. (5), using one-mode two-way similarities updated at Phase 2.1. When the number of clusters is not less than three, return to Phase 1. Iterate Phases 1 through 2.2 until the number of clusters is less than three. When the number of clusters is less than three, the algorithm stops.

Phase 2 is divided into two stages (Phases 2.1 and 2.2) to deal with one-mode three-way proximities. When two clusters remain, while it is inconsistent, two clusters are merged to form a new cluster. The value of the similarity of forming the new cluster is defined by $2 \times s_{JK}$, where s_{JK} is the one-mode two-way similarity between two remaining clusters.

For carrying out the algorithm comparable to the single linkage, Eq. (7), is replaced with

$$s_{(JKL)i} = max(s_{Ji}, s_{Ki}, s_{Li}), \tag{8}$$

and the algorithm comparable to the average linkage, Eq. (7), is replaced with

$$s_{(JKL)i} = mean(s_{Ji}, s_{Ki}, s_{Li}), \tag{9}$$

where $mean(s_{Ji}, s_{Ki}, s_{Li})$ is the mean of all similarities between objects in cluster i and objects in clusters J, K, and L.

4 An Application

The present procedure of assembling one-mode three-way dissimilarities was applied to one-mode two-way dissimilarities among 15 kinship terms (Borg and Groenen 2005, p. 83); cf Arabie et al. (1987), Carroll and Arabie (1983), Rosenberg and Kim (1975). Fifteen kinship terms are 'Aunt', 'Brother', and 13 terms shown at the first column in Table 1. The one-mode two-way dissimilarities are represented in a 15×15 table whose diagonal elements are null.

One-mode three-way dissimilarities were assembled from the one-mode two-way dissimilarities among 15 kinship terms by using Eq. (5). The resulting one-mode three-way dissimilarities $s_{jk\ell}$ are represented by a set of 15 tables of 15×15 (each table is symmetric), where the jth table corresponds to kinship term j, kth row of each table corresponds to kinship term k, and ℓth column of each table corresponds to kinship term ℓ. Table 1 shows the lower triangular part of the first table of the set where $j = 1$ ('Aunt'). Table 1 has 13 rows and 13 columns ($j < k < \ell$). The first row and the first column of the original 15×15 table were eliminated, whose elements are null because $j = k = \ell = 1$. The second row ('Brother') was also eliminated because it has only one element of null, and the 15-th column ('Uncle') was also eliminated because it has only one element of null.

The obtained one-mode three-way dissimilarities among 15 kinship terms were analyzed by the algorithm comparable to the complete linkage. The resulting dendro-

Table 1 One-mode three-way dissimilarities when $j = 1$ ('Aunt'), $j < k < \ell$

Kinship term (ℓ)													
Kinship term (k)	2	3	4	5	6	7	8	9	10	11	12	13	14
3 Cousin	296.6												
4 Daughter	267.8	259.7											
5 Father	395.0	294.9	242.1										
6 Granddaughter	292.4	256.9	207.8	289.7									
7 Grandfather	295.2	299.3	278.0	303.5	258.0								
8 Grandmother	311.3	273.7	183.6	262.8	256.5	427.5							
9 Grandson	323.7	304.1	281.2	301.4	449.5	476.5	274.3						
10 Mother	299.0	283.5	268.4	385.1	183.3	306.5	171.6	323.7					
11 Nephew	280.2	175.7	249.8	245.1	247.1	262.6	272.9	288.1	298.6				
12 Niece	461.7	228.6	263.7	470.8	249.4	464.1	262.8	456.7	268.1	238.5			
13 Sister	465.5	239.1	245.6	244.2	191.4	291.5	179.3	288.8	220.1	257.9	259.2		
14 Son	414.7	310.9	463.4	462.5	293.3	308.4	309.8	350.6	294.9	289.2	465.0	275.9	
15 Uncle	476.9	257.1	491.9	408.3	485.9	453.4	453.4	479.9	408.3	240.6	238.5	479.9	491.9

Fig. 1 Dendrogram of one-mode three-way dissimilarities (comparable to complete linkage)

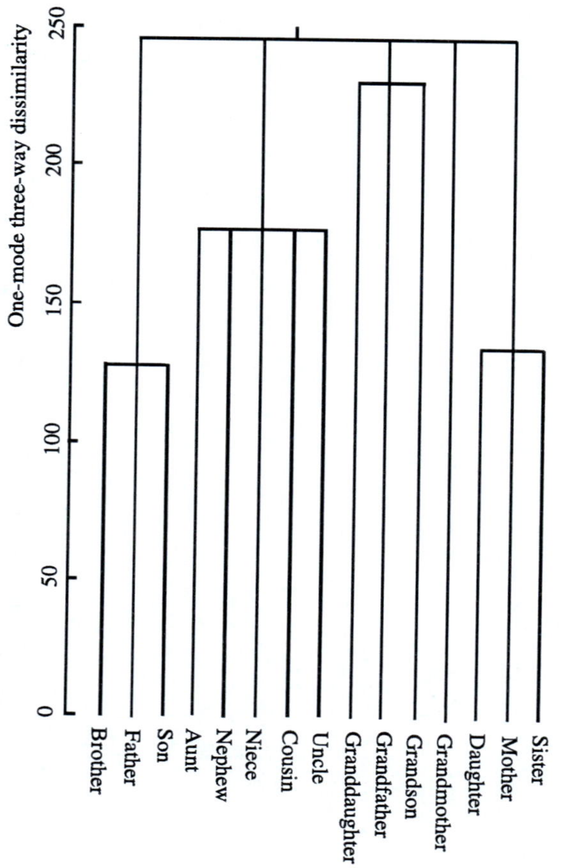

gram is shown in Fig. 1. The dendrogram tells that there are four major clusters and a singleton cluster of 'Grandmother'. The cluster, formed at the first step, is a male nuclear family ('Brother', 'Father', and 'Son'). The cluster, formed at the second step, is a female nuclear family ('Daughter', 'Mother', and 'Sister'). The difference of the two dissimilarities when two clusters were formed at steps 1 and 2, is very small. As a result, it is suggested that two clusters have much the same homogeneity within each cluster. And the homogeneity within a nuclear family of the same gender is larger than the other two clusters formed at steps 3 and 4.

In the present algorithm, three clusters (objects) form a new cluster at each step. At the third step after 'Aunt', 'Nephew', and 'Niece' formed a cluster, the newly formed cluster, 'Cousin', and 'Uncle' formed a cluster of five kinship terms. The dissimilarity when 'Aunt', 'Nephew', and 'Niece' formed a cluster and the dissimilarity when the newly formed cluster, 'Cousin' and 'Uncle' formed a cluster of five kinship terms are equal. At the fourth step, 'Granddaughter', 'Grandfather', and 'Grandson' formed a new cluster. This cluster is close to direct ancestors and descendants ± 2 generations

(Arabie et al. 1987; Carroll and Arabie 1983). At the final step, five clusters (including one singleton cluster of 'Grandmother') formed one cluster. This is caused by the same reason when the cluster was formed at the third step. There are four major clusters; Cluster 1 (Male nuclear family), Cluster 2 (Female nuclear family), Cluster 3 (Collateral relatives), and Cluster 4 ('Grandfather' and grandchild).

5 Discussion

A procedure of assembling one-mode three-way proximities from one-mode two-way proximities and a method for hierarchical clustering of the resulting one-mode three-way proximities was introduced. They were applied to one-mode two-way dissimilarities among 15 kinship terms successfully. The one-mode two-way dissimilarities, from which the present one-mode three-way dissimilarities were assembled, were cluster analyzed by the complete linkage algorithm. The resulting dendrogram is shown in Fig. 2 (cf. Rosenberg and Kim 1975, the leftmost figure of Fig. 2).

The dendrogram tells that at the earlier steps where a kinship term forms a cluster at the first time for the kinship term itself, two kinship terms which have the same position in kinship relationships but have different genders are clustered with the exception of 'Cousin'; 'Brother' and 'Sister', ..., 'Grandfather' and 'Grandmother'. This shows that the clustering is *position in kinship relationships* primary (Arabie and Hubert 1994). The present result (Fig. 1) is not position in kinship relationships primary nor gender primary. Two dendrograms illustrated in Figs. 1 and 2 are not similar, and the present result (Fig. 1) represented clusters which were not formed by the one-mode two-way analysis (cf. Yokoyama et al. 2009). But Cluster 3 (collatral relatives) was formed by both of the two analyses. Some of the clusters formed by the present analysis are somewhat similar to the clusters formed by the earlier studies (Arabie et al. 1987; Carroll and Arabie 1983; Rosenberg and Kim 1975).

The effect of the ratio $[min(s_{jk}, s_{j\ell}, s_{k\ell})/max(s_{jk}, s_{j\ell}, s_{k\ell})]$ of Eq. (5), can be adjusted by $M(\geq 0)$. When $M = 1$, Eq. (10), is equal to Eq. (5)

$$s_{jk\ell} = (s_{jk} + s_{j\ell} + s_{k\ell}) \times [min(s_{jk}, s_{j\ell}, s_{k\ell})/max(s_{jk}, s_{j\ell}, s_{k\ell})]^M. \tag{10}$$

The extension of the present procedure and the method to deal with one-mode n-way ($n > 3$) proximities is possible by replacing Eq. (5), with

$$s_{j_1 j_2 \cdots j_n} = (s_{j_1 j_2} + s_{j_1 j_3} + \cdots + s_{j_{n-1} j_n})/(sd(s_{j_1 j_2}, s_{j_1 j_3}, \ldots, s_{j_{n-1} j_n}), \tag{11}$$

replacing Eq. (6), with

$$s_{J_1 J_2 \ldots J_n} = max_{j_1 < j_2 < \cdots < j_n}(s_{j_1 j_2 \cdots j_n}), \tag{12}$$

Fig. 2 Dendrogram of one-mode two-way dissimilarities (complete linkage)

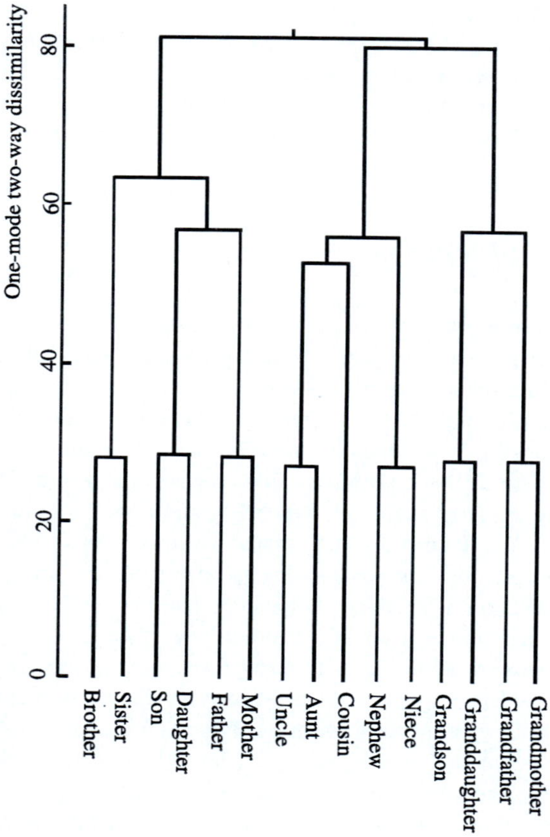

and replacing Eq. (7), with

$$s_{(J_1 J_2 \ldots J_n)i} = min(s_{j_1 i}, s_{j_2 i}, \ldots s_{j_n i}), \quad (13)$$

where the numerator of the right side of Eq. (11), is the sum of ${}_nC_2$ similarities and the denominator is the standard deviation of ${}_nC_2$ similarities. In the case of one-mode four- or more-way similarities, it seems reasonable to think the similarity among objects $j_1, j_2, \ldots, j_n$ having the smaller variability indicates the larger similarity of n objects than that having larger variability even when the sum of ${}_nC_2$ similarities are equal. The application of the present cluster analysis to the dissimilarity of three kinship terms successfully revealed clusters of the nuclear family of the same gender. The extension will help to develop cluster analysis methods which take the context of end-use into consideration (von Luxburg 2012). The extension to one-mode four-way proximities can be useful in some context, e.g., finding a nuclear family of 'Daughter', 'Father', 'Mother', and 'Son'.

Acknowledgements The authors express their gratitude to participants for their comments at the session of the IFCS-2019 conference in Thessaloniki. They also thank to anonymous reviewers for their review for the earlier version of the present manuscript.

References

Arabie, P., Hubert, L.: Cluster analysis in marketing research. In: Bagozi, R.P. (ed.) Advanced Methods in Marketing Research, pp. 160–189. Blackwell, Cambridge, MA (1994)

Arabie, P., Carroll, J.D.: A mathematical programming approach to fitting the ADCLUS model. Psychometrika **45**, 211–235 (1980)

Arabie, P., Carroll, J.D., DeSarbo, W.S.: Three-Way Scaling and Clustering. Sage Publications, Newbury Park, CA (1987)

Borg, I., Groenen, P.J.F.: Modern Multidimensional Scaling: Theory and Applications, 2nd edn. Springer, New York (2005)

Carroll, J.D., Arabie, P.: INDCLUS: an individual differences generalization of the MAPCLUS algorithm. Psychometrika **41**, 157–169 (1983)

Carroll, J.D., Chang, J.J.: Analysis of individual differences in multidimensional scaling via an N-way generalization of "Eckart-Young" decomposition. Psychometrika **35**, 283–319 (1970)

de Rooji, M., Heiser, W.J.: Triadic distance models for the analysis of asymmetric three-way proximity data. Br. J. Math. Stat. Psychol. **53**, 99–119 (2000)

Hayashi, C.: Multiway data matrix and method of quantification of qualitative data as a strategy of data analysis. In: Coppi, R., Bolasco, S. (eds.) Multiway Data Analysis, pp. 131–142. North-Holland, Amsterdam, MA (1989)

Heiser, W.J., Bennani, M.: Triadic distance models: Axiomatization and least squares representation. J. Math. Psychol. **41**, 189–206 (1997)

Johnson, S.C.: Hierarchical clustering schemes. Psychometrika **32**, 241–254 (1967)

Nakayama, A.: A multidimensional scaling model for three-way data analysis. Behaviormetika **32**, 95–110 (2005)

Okada, A., Yokoyama, S.: Triplet clustering of one-mode two-way proximities. In: Abstract book of the 16th Conference of the International Federation of Classification Societies, p. 103 (2019)

Rosenberg, E.E., Kim, M.P.: The method of sorting as a data gathering procedure in multivariate research. Multivar. Behav. Res. **10**, 489–502 (1975)

von Luxburg, U., Williamson, R.C., Guyon, I.: Clustering: Science or art?. JMLR: Wksp. Conf. Proc.: Wksp. Unsupervised Transfer Learn. **27**, 67–79 (2012)

Yokoyama, S., Nakayama, A., Okada, A.: One-mode three-way overlapping cluster analysis. Comput. Stat. **24**, 165–179 (2009)

First-Time Voters in Greece: Views and Attitudes of Youth on Europe and Democracy

Georgia Panagiotidou and Theodore Chadjipadelis

Abstract This study investigates the views, attitudes, and values of young people in Greece using a multivariate data analysis workflow. The primary objective is to investigate political mobilization and its association with various political characteristics, to perceptions toward democracy and personal moral values. The main research output is a map representing the political behavior of young first-time voters. A secondary objective is to identify the most important factors which determine their vote. The study results contribute to the voting behavior literature by revealing contemporary young voters' typologies, visualizing political competition and dynamics of political behavior, and highlighting emerging important factors which affect voting choice.

Keywords Voting behavior · Political analysis · Behavioral mapping · Conjoint analysis · Correspondence analysis · Hierarchical clustering

1 Introduction

The main objective of this study is to describe, investigate, and analyze the views, values, and attitudes of young people in Greece, who voted for the first-time in the European Elections of 2019. The aim is to identify any possible relationship between political mobilization, political attitudes, perception of democracy, and set of personal values.

For this purpose, an empirical study was designed and conducted on a sample of young voters in Greece. The first research question is whether political mobilization depends on other characteristics, such as "left-right" self-positioning, political knowledge, sources of information or simple demographics. A second research

G. Panagiotidou (✉) · T. Chadjipadelis
School of Political Sciences, Aristotle University of Thessaloniki, Thessaloniki, Greece
e-mail: gvpanag@polsci.auth.gr

T. Chadjipadelis
e-mail: chadji@polsci.auth.gr

T. Chadjipadelis et al. (eds.), *Data Analysis and Rationality in a Complex World*, Studies in Classification, Data Analysis, and Knowledge Organization,
https://doi.org/10.1007/978-3-030-60104-1_25

question is whether these emerging political behavioral patterns are also related to a more comprehensive typology of the "moral self" and the "political self" (Marangudakis and Chadjipadelis 2019), assigning a thorough set of political characteristics to each group, which explain political behavior and political competition for the case of young people in Greece. The last research question is to determine the most important factors that affect young people's voting choice.

To analyze the data, a combination of multivariate data analysis methods was employed. Specifically, Hierarchical Cluster Analysis (HCA) and Multiple Correspondence Analysis (MCA) was subsequently applied to produce a so-called representation map of young voters' political behavior. Conjoint Analysis (CA) was used to detect the most important factors which determine their vote. The proposed data analysis workflow can contribute to political research on voting attitudes and behavior by describing the political behavior of youth in this specific study, but it can also be exploited to various political environments, to analyze and comprehend political behavior dynamics of any voting population. The final research output is a two-dimensional map, which depicts the political and moral discourses of young voters in connection to their political attitudes and behavior, facilitating interpretation by the researcher. Section 2 presents the study methodology (study sample, instrumentation, and data analysis workflow), Sect. 3, presents the study results, and Sect. 4, discusses the results and concludes the paper.

2 Methodology

2.1 Sample

A sample of 3, 780 students from two universities in Greece (Aristotle University of Thessaloniki and the University of Macedonia) participated in the study. Data collection took place in April 2019, almost a month before the European elections of May 05 2019. Students were aged 17–25 years, 40% were men and 60% were women.

2.2 Study Instrument

The research instrument was a questionnaire, consisting of four sections. The first section included demographic questions, such as sex, age, region, and field of study. In the second section, respondents were asked to evaluate their level of political interest, the way they choose to mobilize when facing an issue, their intention to vote or abstain in the upcoming elections, their attitude towards EU and to their level of political knowledge. Moreover, the respondents had to position themselves on the left-right axis. In the third section, respondents were asked to choose two sources they use more often to get informed about politics and choose 3 among 12 pictures to indicate how they perceive democracy and what are their most important personal values. The two latter questions were used to analyze respondents' opinions towards constitutive goods. This approach is based on the theory for (a) the construction of the political self, that is the phenomenological process by which individuals, by analogy and homology, construct symbolic representations of democratic institutions and (b) the construction of their moral self (Marangudakis and Chadjipadelis 2019). Both sets of pictures originate from the study of (Marangudakis and Chadjipadelis 2019), and are part of a questionnaire from an ongoing research.[1] The last section of the questionnaire consisted of a set of 8 different scenarios, prioritizing differently four major factors that contribute to the selection of vote (parties, candidates, issues, and the attitude towards coalition). Respondents had to sort the scenarios in a descending order of preference.

2.3 *Data Analysis*

A three-step approach was employed using HCA (Benzecri's chi-square distance, Ward's linkage criterion) and MCA to create and interpret the representation map (Chadjipadelis 2015, 2017). In the first step, HCA was separately applied to three sets of dichotomous variables (0-not selected/1-selected), namely "information sources", "perception on democracy", and "personal values" to assign subjects into distinct groups based on their response patterns. This step produced three cluster membership variables, one for each set. The number of clusters was determined upon the empirical criterion of the change in the ratio of between-cluster inertia to total inertia, when moving from a partition with r clusters to a partition with $r - 1$ clusters (Papadimitriou and Florou 1996). In addition, the behavior typology of each cluster was investigated by detecting the contribution weight of each variable to the clusters using a series of chi-square tests (Karapistolis 2010). In the second step, the three cluster membership variables were jointly analyzed with the demographic characteristics and the variables of the second section of the questionnaire, using MCA on the Burt table (Greenacre 2007). The resulting factorial map is the representation map, which reveals behavioral patterns and abstract discourses for the variables and the subjects. In the third step, HCA was applied to the coordinates of the variable categories on the first two factorial axes to facilitate interpretation of behavioral patterns. All analyses were conducted with the software M.A.D. [Méthodes de l' Analyse des Données] (Karapistolis 2010).

The variables used to construct the representation map are the following:

- E21E (field of study) 1: Humanities, 2: Science, 3: Arts, 4: Social studies, 5: Health science
- E25 (sex) 1: Male, 2: Female
- E4N (self-positioning on left-right axis) 1: Left, 2: Center, 3: Right

---

[1] We would like to thank Prof. Marangudakis for his kind permission to use the pictures in this study.

- E5 (attitude to the elections): 1: I will vote (decided), 2: I will vote (not decided yet), 3: Invalid/Blank, 4: Abstention
- E11 (political mobilization): 1: I personally address the authorities, 2: I participate with others in collective mobilizations, 3: I take action through Social Media, 4: I let the authorities to do their job, 5: I do not know / I do not answer
- E12 (political interest): 1: very much, 2: quite, 3: a little, 4: not at all
- PK (political knowledge): 1: low, 2: moderate, 3: high
- E13N (political info source): 5 clusters
- GE14 (perception of democracy): 8 clusters
- GE15 (personal values): 9 clusters

The data obtained from the last section of the questionnaire (preferences of different scenarios) were analyzed using Conjoint Analysis. CA was applied to understand latent dynamics of each factor that contributes to the selection of vote.

3 Results

With regard to information sources about politics, HCA revealed 5 clusters of subjects, differentiating mostly those who prefer to get informed by mass media such as the TV, radio, internet, those who read the press, and those who get updated about political matters from their family circle or their friendly environment. The fifth group contained those who do not get informed at all (E13N4, E13N6, E13N8, E13N18, E13N19).

The application of HCA on data about the perception of democracy indicated 8 clusters (GE141, GE142, GE143, GE144, GE145, GE146, GE147, GE148). Figure 1 shows the pictures-concepts associated with each cluster (important concepts are shown in gray). A tick mark indicates that the respondents of this cluster select the corresponding picture, while an x mark indicates that they do not.

With regard to personal values, HCA revealed 9 groups of subjects with profiles shown in Fig. 2 (GE151, GE152, GE153, GE154, GE155, GE156, GE157, GE158, GE159). Note that to interpret personal values, we used the theoretical tool of expressivism-naturalism (Marangudakis and Chadjipadelis 2019, pp. 262–273), on how the individuals construct their moral self by choosing constitutive goods.

Overall, five behavioral profiles of young voters can be observed on the MCA representation map (Fig. 3). The first axis (horizontal) can be described by the polarization between high/low political knowledge and interest, male/female, social sciences studies/other studies, will vote/abstention, decided/undecided. Moreover, on the left of first axis, there is a small group of subjects with no political interest, knowledge, or information. The second axis (vertical) can be characterized by the polarization between right/left, naturalist/expressionist, individualistic/collective mobilization, democracy as representation, money, ancient Greece, church/ democracy as protest.

Picture-Concept	GE141	GE142	GE143	GE144	GE145	GE146	GE147	GE148
Protest								×
Ancient Greece								×✓
Direct							✓	×
e-Democracy						✓		
Representative								✓
Riot	✓							
Representation			✓					
Direct-participation								✓
Individualistic				✓				
Rebellion								×✓
Protest					✓			×
Christianity		✓						

Fig. 1 Cluster profiles about perceptions on democracy

Picture-Concept	GE151	GE152	GE153	GE154	GE155	GE156	GE157	GE158	GE159
Expressivist	✓								×
Expressivist				✓					×
Expressivist			×		✓			×	
Spirituality		✓							×
Spirituality						✓			×
Moon exploration									✓
Naturalist			✓					✓	×✓
Christian			×					×	✓
Naturalist			×					×	✓
Naturalist							✓		×
Army			✓					✓	×
Naturalist			✓					✓	×

Fig. 2 Cluster profiles about personal values

Last, based on the results of Conjoint Analysis, coalition appears to be the most important factor when selecting a party/candidate (29.6%). Party is also almost equally important (29.1%), followed by issues (23.3%) and candidates contribute with the lowest percentage (18%) to voting decision.

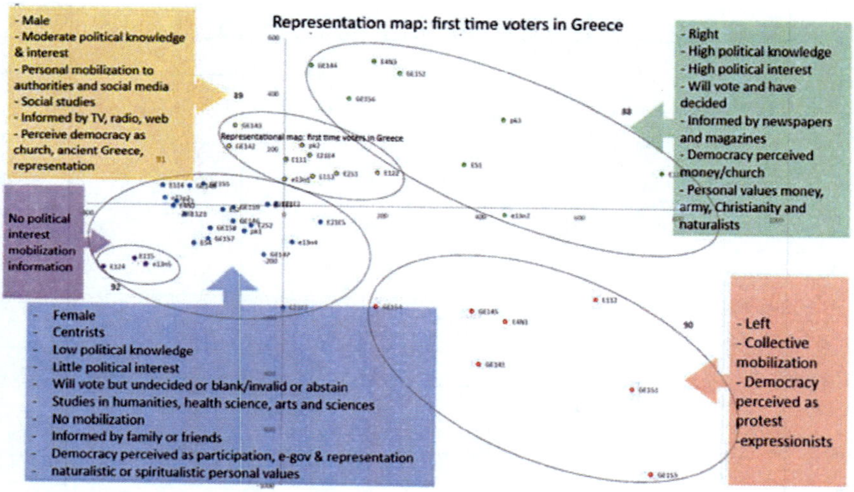

Fig. 3 Representation map of the political behavior of young voters, visualizing five distinct behavioral profiles

4 Discussion

Analysis of the political attitudes and electoral behavior of young voters in Greece, revealed two main dimensions of opposing characteristics. The first axis is characterized by the degree of political mobilization, interest, and decision to vote in a descending order from high to low, while on the second axis, the main polarization is between "left" and "right" political ideology. The map of political behavior shows five significant discourses, which include specific characteristics regarding moral and political self.

High political mobilization is characterized by strong ideological identity (left or right) and high political interest and knowledge. The type of mobilization differs between the left and the right, as the left choose collective mobilization and the right prefer individual mobilization. On this dimension, male voters lean towards the right, while female voters lean towards the center and center left. An equally important finding connects the political and moral identity to specific profiles. More specifically, collective mobilization is linked to expressionistic discourses and the perception of democracy as protest and people's fight, while the individualistic mobilization is connected with individualistic (almost autocratic) discourses which include values connected with power and past traditions (money, army, church). These two major groups of voters belong to the category of those who vote and are aware of their choice.

When interpreting the first axis, we comprehend that the polarization is between high mobilization, interest, and knowledge against those who express low interest in politics and do not necessarily vote. Also, in this step of our analysis, the relation between the discourses of political and moral self is strong. A large group includes the no mobilization voters who perceive democracy as participatory and electronic and tend to choose a more spiritualistic moral path. On the opposite side, a group of respondents of more but still low political interest, who are more engaged with the internet and the social media, perceive democracy as a representative system and include in their discourse ancient Greece and Church (which reflects the main aspects of democracy perception in contemporary Greece). This group of voters create their moral self upon naturalistic values. A small but important group exists in the "ideological center", linked with naturalistic values, including voters who abstain elections, show no interest in politics and have an absence of political knowledge. The analysis reveals five significant and distinctive profiles of the political behavior of young people in Greece and two cleavages which characterize the political competition.

The first cleavage refers to high political mobilization and low political mobilization with the relatively same difference between decided and undecided voters. The opposition follows the same pattern for the way they choose to get information about politics (traditional media vs digital media and close environment) and for their moral values (expressionists/autocrats vs naturalists/spiritualists) and their perception of democracy (protest/power vs representation/e-gov). A second cleavage is created by the opposition between left and right ideology, followed accordingly by the opposition between naturalists and expressivists, the perception of democracy as conceived in contemporary Greece or protest and finally the individual or collective mobilization. Moreover, a distinct group of apolitical behavior is separated from the above, positioning themselves in the center of the first axis.

The analysis of the criteria upon which young people decide how to vote, revealed that the attitude of parties on coalition perspective and the party identity seems to be the most important factor and candidates play a small part in the formation of their final choice.

The research contributes to the voting behavior literature, revealing some important findings on the political behavior of youth in Greece, with five prominent typologies of voters, while the representation map visualizes the dimensions of political competition and the polarization between these five types of voters. Political mobilization is strongly associated with the ideology, the moral self, and the political self. Furthermore, political interest, knowledge, field of study and source of information can affect one's level of political mobilization, leading to a higher probability of abstention in elections (or the opposite). Additionally, the results showed the emerging importance of the "parties' coalition perspective" factor in Greece's young audience.

To conclude, the methodology used in this study can be applied and transferred to other environments (countries, elections, organizations) as a comprehensive political behavior and political competition analysis tool, providing space for any adaptations needed (context, variables, characteristics) according to the features of each political environment.

References

Chadjipadelis, T.: Parties, Candidates, Issues: The Effect of Crisis. Correspondence Analysis and Related Methods. CARME 2015. Napoli, Italy (2015)

Chadjipadelis, T.: What really happened: party competition in the january and september 2015 parliamentary elections. Eur. Q. Polit. Attitudes Mentalities **6**(2), 8–39 (2017)

Greenacre, M.: Correspondence Analysis in Practice. Chapman and Hall/CRC Press, Boca Raton (2007)

Karapistolis, D.: The Software MAD (2010). http://www.pylimad.gr/

Marangudakis, M., Chadjipadelis, T.: The Greek Crisis and Its Cultural Origins. Palgrave-Macmillan, New York (2019)

Papadimitriou, G., Florou, G.: Contribution of the Euclidean and chi-square metrics to determining the most ideal clustering in ascending hierarchy (in Greek). In: Annals in Honor of Professor I. Liakis, pp. 546–581. University of Macedonia, Thessaloniki (1996)

Comparison of Hierarchical Clustering Methods for Binary Data From SSR and ISSR Molecular Markers

Emmanouil D. Pratsinakis, Lefkothea Karapetsi, Symela Ntoanidou, Angelos Markos, Panagiotis Madesis, Ilias Eleftherohorinos, and George Menexes

Abstract Data from molecular markers, which are used to construct dendrograms based on genetic distances between different plant species, are encoded as binary data. For the construction of the dendrograms, the most commonly used linkage method is the UPGMA (Unweighted Pair Group Method with Arithmetic Mean) in combination with the squared Euclidean distance. It seems that in this scientific field, this is, the "golden standard" clustering method. In this study, a comparison of 189 clustering methods (except the "golden standard"), that is seven linkage methods in the sense that this methodological scheme is used in the vast majority of the corresponding studies by 27 appropriate distances along with the Benzécri's chi-squared distance in combination with the Ward's linkage method, is attempted using data originating from molecular markers applied on pear trees species and *Sinapis arvensis* populations. Fruit trees cluster analysis was performed using SSR markers, while for *Sinapis arvensis* populations' clustering, ISSR markers were used. The results showed that the "golden standard" is not the only appropriate method for dendrogram

E. D. Pratsinakis · S. Ntoanidou · I. Eleftherohorinos · G. Menexes (✉)
Aristotle University of Thessaloniki, Thessaloniki, Greece
e-mail: gmenexes@agro.auth.gr

E. D. Pratsinakis
e-mail: epratsina@agro.auth.gr

S. Ntoanidou
e-mail: melina-nt@hotmail.com

I. Eleftherohorinos
e-mail: eleftero@agro.auth.gr

L. Karapetsi · P. Madesis
Centre for Research and Technology, Thessaloniki, Greece
e-mail: lefki8@certh.gr

P. Madesis
e-mail: pmadesis@certh.gr

A. Markos
Democritus University of Thrace, Alexandroupoli, Greece
e-mail: amarkos@eled.duth.gr

T. Chadjipadelis et al. (eds.), *Data Analysis and Rationality in a Complex World*, Studies in Classification, Data Analysis, and Knowledge Organization,
https://doi.org/10.1007/978-3-030-60104-1_26

construction based on binary data derived from molecular markers. Ten other hierarchical clustering methods could be used for the construction of dendrograms from SSR markers and thirty-seven other hierarchical clustering methods could be used for the construction of dendrograms using binary data resulted from ISSR markers.

Keywords Hierarchical clustering · Cluster validity · Benzécri's chi-squared distance · *Sinapis arvensis* · Pear

1 Introduction

Binary data are usually encoded with values of zero for the absence, and one for the presence of a characteristic or trait (Song et al. 2019). In the field of Molecular Biology, binary data are produced through an experimental workflow beginning with genomic DNA isolation from plant or animal tissue, polymerase chain reaction (PCR) with the use of molecular markers, and electrophoresis of the PCR products (Khorshidi et al. 2017). In the image of the agarose gel, the presence of a luminous band is encoded as one and the absence as zero. Molecular markers are known DNA sequences with known location at the chromosome and their length varies from one base (SNPs-single oligonucleotide) to multiple repeated bases (microsatellites). Molecular markers are used for the estimation of genetic variability, the identification of phenotypes, the exploration of the relationships among individuals, and the clustering of individuals via cluster analysis and tree-dendrogram building (Schlötterer 2004). In order to test the validity of clustering, it is important to establish validation criteria. These criteria can be based on statistical indices, on cross-validation and bootstrap methods or/and external validation criteria (Hair et al. 2010; McIntyre and Blashfield 1980; Sneath and Sokal 1973; Sharma 1996). In this study, we consider the last option.

In Molecular Biology, the "golden standard" method for clustering binary data and the construction of dendrograms is the combination of the Euclidean distance or its square with the UPGMA (Unweighted Pair Group Method with Arithmetic Mean) linkage criterion, and it is used for testing the intra-species genetic variability and the relations between individuals. After a thorough review (November 2019), we found that during the last five years, 1, 586 papers in Web of Science and 2, 175 papers in Scopus, have used the "golden standard" method.

According to Choi et al. (2010), there are at least 76 distances for binary data and seven different linkage methods (between-groups average linkage, within-groups average linkage, nearest neighbour, furthest neighbour, centroid, median, and ward's linkage criterion). The Euclidean (squared or not) distance is not considered appropriate for the construction of typologies based on binary data (Dillon and Goldstein 1984; Finch 2005; Ludwig and Reynolds 1988), based on the argument that it is suitable for continuous variables only (Hair et al. 2010). Although the Euclidean distance is not recommended for clustering binary data (Sharma 1996; Tamasauskas et al. 2012), it can be profitably used for clustering binary data with nonhierarchical clustering methods (Iodice D'Enza and Palumbo 2013).

Based on the above remarks, four questions or concerns were raised. The first is whether the Euclidean distance is suitable for binary data. The second is which one of the existing clustering methods for binary data performs better (among at least five hundred thirty-two different options). The third concern is how to assess the validity of the clustering solution. Finally, a fourth concern was raised when analyzing the proximity matrix resulting from the binary data that were produced by the SSR molecular markers. "Ties" were observed in the proximity matrix causing the following problems: the clustering solution is different when the order of individuals changed in the data set, the analysis resulted in different dendrograms, and some distances could not be calculated and the analysis was halted (van der Kloot et al. 2003; Sharma 1996; Backeljau et al. 1996).

2 Material and Methods

To answer the aforementioned questions, we compared one hundred eighty-seven clustering methods available in the IBM SPSS Statistics 23, with the "golden standard" in the field of Molecular Biology. The results of all those methods were compared to the "golden standard" dendrogram (squared Euclidean distance combined with the UPGMA linkage method), obtained via the combination of two software packages: (GenAIEx 2012), which is an add-in to EXCEL and creates the proximity matrix based on the squared Euclidean distance and MEGA7 (Kumar et al. 2016), which utilizes these distance matrices for the construction of the corresponding dendrogram.

We used the data from two plant species, *Pyrus sp.* or as widely known pear and *Sinapis arvensis* or wild mustard. In the first case, 46 pear individuals from 26 varieties and 3 areas were studied. Pear *(Pyrus sp.)* is a diploid genus ($2n = 2x = 34$) of allopolyploid origin and its species are widely spread (Fernández-Fernández et al. 2006). Seven simple sequence repeats (SSRs) molecular markers were used to study the genetic variability between these 26 Greek domestic varieties. SSRs are codominant and highly polymorphic markers, widely used in mapping, fingerprinting, and diversity studies (Fernández-Fernández et al. 2006). The identity of the samples-individuals was not known during the analyses except from two authors that were not present at the stage of the data analysis. The validation criteria were that the individuals of the same variety should be classified in the same group or cluster and the areas-origin, the leaf size, the fruit size, and the time of maturity for consumption should differentiate the groups. Note that the input order stability of clustering results was tested with the PermuCluster software, an SPSS add-in (Spaans and van der Kloot 2004; van der Kloot et al. 2003).

In the second case, we studied 5 populations of *Sinapis arvensis*. *Sinapis arvensis* is an annual winter broadleaf weed, belonging to the family of Brassicaceae and commonly found in winter cereals cultivation. It is a diploid ($2n = 18$) weed species

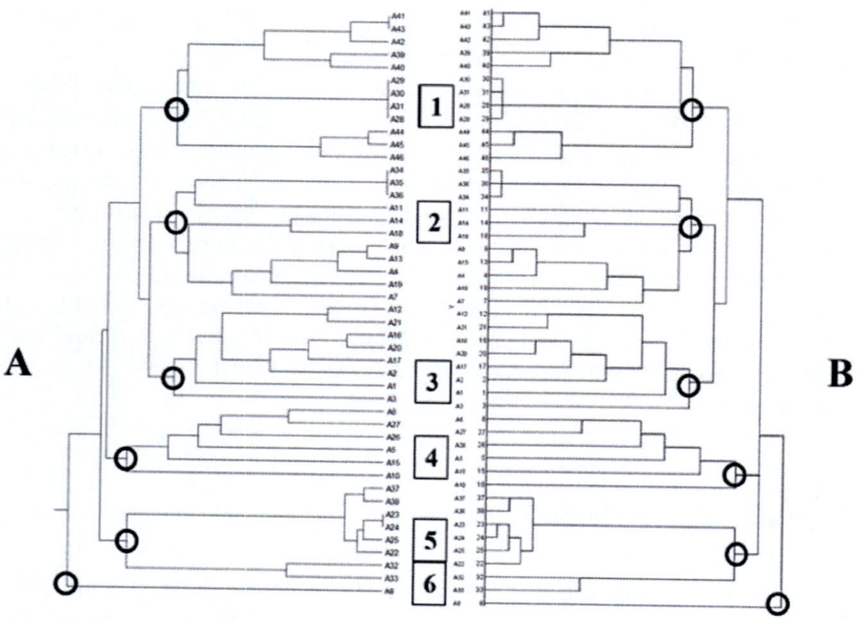

Fig. 1 Hierarchical cluster analysis dendrograms using squared Euclidean distance and the UPGMA, obtained via MEGA7 (A) and SPSS (B) software. Notes: Numbers stand for the period of maturity and the origin of the pear varieties, 1: October to November, nursery, 2: September to October, native, 3: September to November, native, 4: August to September, native, 5: August to September, nursery and 6: October to November, native

and it is self-incompatible, which means that it is pollinated by insects (Warwick et al. 2000). *Sinapis arvensis* is found in winter cereals grown in Greece, where it is reported to cause problems over the last years (Ntoanidou et al. 2017). Four out of five populations were resistant and one was susceptible to herbicides. There were 10 individuals per population collected from 5 areas and 6 ISSR molecular markers were used. ISSRs are DNA fragments, with a length of 100 – 3, 000 base pairs. ISSRs are located between microsatellite regions with opposite orientation. ISSRs were developed by Zietkiewicz et al. (1994). Note that during the analyses, the identity of the samples was not known. The validation criteria were that individuals from the same population should be classified in the same group and the areas and herbicide resistance should differentiate the groups.

3 Results

For both the data sets, the dendrograms resulting from the "golden standard" combination (GenAIEX and MEGA7) and those from SPSS were the same. For the pear data set (Fig. 1) and according to two of the validation criteria (maturity period and origin), the varieties were classified in the same subgroup. There is a large group that includes four subgroups (1–5) and an outlying individual (6). In the subgroup 2–3, varieties that are cultivated and mature for consumption from September to October and from September to November, group together. Subgroup 1 consists of varieties originating from the nursery and they are mature for consumption from October to November. Subgroup 4 consists of varieties that are native and mature for consumption from August to September. Finally, subgroup 5 contains varieties from nursery and they are mature for consumption from August to September. Subgroups 1 to 5 are merged together, following a specific seasonal pattern that is linked to the maturity period. At the base of the dendrogram, the maturity period starts in August (late summer) and as we move upwards, the maturity period follows the change of the sequence of the seasons (e.g., varieties in subgroup 2–3 and in subgroup 1, at the top of the dendrogram, their maturity period ends in November—late autumn). Also, the criterion of origin (native or nursery) differentiated the subgroups. The outlying individual belongs to a pear variety that differentiates from the others mainly for its big fruit size.

For the pear data set, 10 clustering methods satisfied the validation criteria. These are the Euclidean distance with between-groups linkage, the variance distance with between-groups linkage, the shape distance with between-groups linkage, the Hamman distance with between-groups linkage, the Sokal and Sneath 1 distance with between-groups linkage, the Euclidean distance with complete linkage, the squared Euclidean distance with complete linkage, the variance distance with complete linkage, the simple matching distance with complete linkage and Rogers and Tanimoto distance with complete linkage. A total of 34 clustering methods partially satisfied (i.e., did not satisfy all the validation criteria simultaneously) the validation criteria and for 82 clustering methods the validation criteria are not satisfied. Furthermore, for 63 clustering methods, the distances could not be calculated and the analysis halted. The SSR molecular markers produced "ties" in the proximity matrix and resulted in different partitions when the order of individuals was changed in the data set, different dendrograms, and some distances could not be calculated. This problem can be solved using the software PermuCluster, that constructs the most "unchangeable" dendrogram after permuting the order of samples in the data set in combination with bootstrap resampling (Spaans and van der Kloot 2004; van der Kloot et al. 2003).

For the wild mustard data set, 37 clustering methods satisfied the validation criteria. Some of these are the Jaccard distance with between-groups linkage, the Sokal and Sneath 3 distance with between-groups linkage, the Sokal and Sneath 5 with between-groups linkage, the Lance and Williams distance with between-groups distance, the Phi 4 point distance with between-groups linkage, Sokal and Sneath 4 between-groups linkage, Yule's Y distance with between-groups linkage, Sokal and Sneath 5 distance with furthest neighbor linkage, Yule's Q distance in combination with centroid linkage, and variance in combination with Ward's linkage. A total of 35 clustering methods partially satisfied the validation criteria and for 116 clustering methods the validation criteria were not satisfied.

Apart from the 188 methods available in SPSS (including the "golden standard" combination), the combination of Benzécri's chi-squared distance and Ward's linkage criterion (Menexes and Angelopoulos 2008), was evaluated for both data sets. To our knowledge, this combination is used the first time on binary data from molecular markers, and the results satisfy the validation criteria. These analyses were performed using the methodology proposed in Menexes (2006), and verified with the results of the CHIC analysis v.1.1 software (Markos et al. 2010).

4 Discussion

Data from molecular markers, which are used to construct dendrograms based on genetic distances (Deza and Deza 2016), are encoded as binary data. For dendrogram construction, the most commonly used linkage method is the UPGMA (or between groups average linkage method), paired with the squared (or not) Euclidean distance. This is the "golden standard" in the field of Molecular Biology. However, the Euclidean distance (squared or not) is usually not recommended for clustering binary data.

With regard to whether the Euclidean distance is suitable for binary data and which hierarchical clustering method for binary data performs better, results showed that the "golden standard" approach is appropriate for this type of data but there are also at least 10 other hierarchical clustering options (for the pear data set) and 37 options (for the *Sinapis arvensis* data set), that result in a dendrogram identical to the one obtained from the "golden standard" method and satisfy all external validation criteria.

Another issue that was addressed in the study is how to assess the validity of the clustering solution. The term validity here refers to the number of clusters in the data set, the assessment of the clustering structure and the quality and input order stability of the solution (Han et al. 2012; Mojena and Wishart 1980; Spaans and van der Kloot 2004; van der Kloot et al. 2003). In the present study, the input order stability of the resulting cluster solution was tested and verified via a bootstrapping procedure proposed by Spaans and van der Kloot (2004). According to Hair et al. (2010), the best cluster solution should not only be determined upon statistical approaches. The choice of the best clustering method for binary data is data-dependent (Wijaya et al. 2016). In this study, five external validation criteria were used for the pear data set and three external validation criteria were used for the *Sinapis arvensis* data set, in order to compare a series of hierarchical clustering methods with the "golden standard" solution. The external validation criteria focused on cluster homogeneity, cluster completeness, and the biological interpretation of the resulting clusters. These dimensions of validation represent an extrinsic method of cluster validation (Amigó et al. 2009; Han et al. 2012).

Moreover, the results showed that when there are many "ties" in the proximity matrix, special care must be taken to check the validity of the results. This is because the software often results in individuals with equal distances. As a consequence, each time the algorithm is executed on the same data set, with the same distance, and the same linkage criterion, the results (dendrograms) are different. This problem is more evident in cases where the Euclidean distance is used. One way to solve the problem is to use an approach that allows permuting the order of samples in the data set in combination with bootstrap resampling. Such an approach is implemented in PermuCluster v.1.0.

To conclude, (a) there is no need for specialized software for the construction of dendrograms using binary data from molecular markers, (b) other clustering methods can be also used and lead to clusters with the same merging pattern to the one resulted from the application of the "golden standard" method, (c) Benzécri's chi-squared distance in combination with the Ward's linkage criterion seems to be a valid approach to cluster binary data derived from molecular markers, (d) the establishment of external criteria is necessary to test the validity of dendrograms, (e) convergence results of different clustering methods enhance the results' validity, (f) the squared Euclidean distance combined with UPGMA linkage method is not the only appropriate clustering method for dendrogram construction based on binary data derived from molecular markers; ten clustering methods for the pear data set and 37 clustering methods for the *Sinapis arvensis* data set, along with the "golden standard" solution, cluster the individuals in line with the validation criteria, and finally, (g) in the case of "ties" special care must be taken to test the validity of the results and the establishment of external validation criteria is a significant aid for this kind of testing. Future research could be focused on the study of the effect of ties on the clustering of additional plant species using binary data from SSR molecular markers.

Acknowledgements Part of this research (pear data set) has been co-financed by the European Union and Greek national funds through the Operational Program Competitiveness, Entrepreneurship and Innovation, under the call RESEARCH-CREATE-INNOVATE (project code: T1EDK-05438).

References

Amigó, E., Gonzalo, J., Artiles, J., Verdejo, F.: A comparison of extrinsic clustering evaluation metrics based on formal constraints. Inf. Ret. **12**(4), 461–486 (2009)

Backeljau, T., de Bruyn, L., de Wolf, H., Jordaens, K., van Dongen, S., Winnepenninckx, B.: Multiple UPGMA and neighbor-joining trees and the performance of some computer packages. Mol. Biol. Evol. **13**(2), 309–313 (1996)

Choi, S.S., Cha S.H., Tappert, C.C.: A survey of binary similarity and distance measures. J. Syst. Cyb. Inf. **8**(1), 43–48 (2010)

Deza, M.M., Deza, E.: Encyclopedia of Distances, 4th edn. Springer, Berlin (2016)

Dillon, W.R., Goldstein, M.: Multivariate Analysis: Methods and Applications. Wiley, New York (1984)

Fernández-Fernández, F., Harvey, N.G., James, C.M.: Isolation and characterization of polymorphic microsatellite markers from European pear (*Pyrus communis L.*). Mol. Econ. Notes **6**(4), 1039–1041 (2006)

Finch, H.: Comparison of distance measures in cluster analysis with dichotomous data. J. Data Sci. **3**(1), 85–100 (2005)

GenAIEx: A comprehensive Guide to GenAIEx 6.5. Australian National University, Cambera Australia (2012)

Hair, J.F., Black, W.C., Babin, B.J., Anderson, R.E.: Multivariate Data Analysis: A Global Perspective, 7th edn. Pearson Education Inc, New Jersey (2010)

Han, J., Pei, J., Kamber, M.: Data Mining: Concepts and Techniques, 3rd edn. Elsevier, New York (2012)

Iodice D'Enza, A., Palumbo, F.: Dynamic data analysis of evolving association patterns. In: Giusti, A., et al. (eds.) Classification and Data Mining, pp. 45–53. Springer, Heidelberg (2013)

Khorshidi, S., Davarynejad, G., Samiei, L., Morhaddam, M.: Study of genetic diversity of pear genotypes and cultivars (*Pyrus communis L.*) using inter-simple sequence repeat markers (ISSR). Erwerbs-Obstbau. **59**(4), 301–308 (2017)

Kumar, S., Stecher, G., Tamura, K.: MEGA7: molecular evolutionary genetics analysis version 7.0 for bigger datasets. Mol. Biol. Evolu. **33**(7), 1870–1874 (2016)

Ludwig, J.A., Reynolds, J.F.: Statistical Ecology: A Primer in Methods and Computing. Wiley, New York (1988)

Markos, A., Menexes, G., Papadimitriou, I.: The CHIC Analysis Software v1.0. In: Locarek-Junge, H., Weihs, C. (eds.) Classification as a Tool for Research, pp. 409–416. Springer, Heidelberg (2010)

McIntyre, M., Blashfield, R.: A nearest-centroid technique for evaluating the minimum-variance clustering procedure. Mul. Beh. Res. **15**(2), 225–238 (1980)

Menexes, G.: Experimental Designs in Data Analysis. Published Ph.D. thesis, University of Macedonia, Thessaloniki, Greece (in Greek) (2006)

Menexes, G., Angelopoulos, S.: Proposals for the financing and development of Greek farms based on a clustering method for categorical data. EuroMed. J. Bus. **3**(3), 263–285 (2008)

Mojena, R., Wishart, D.: Stopping rules for Ward's clustering method. In: Proceedings of COMPSTAT, pp. 426–432. Physika–Verlag, West Germany (1980)

Ntoanidou, S., Madesis, P., Diamantidis, G., Eleftherohorinos, I.: Trp574 substitution in the acetolactate synthase of Sinapis arvensis confers cross-resistance to tribenuron and imazamox. Pest. Biochem. Phys. **142**, 9–14 (2017)

Schlötterer, C.: The evolution of molecular markers—just a matter of fashion? Nat. Rev. Gen. **5**, 63–69 (2004)

Sharma, S.: Applied Multivariate Techniques. Willey, New York (1996)

Sneath, P., Sokal, R.: Numerical Taxonomy. W. H. Freeman, San Francisco (1973)

Song, Y., Westerhuis, J.A., Aben, N., Michaut, M., Wessels, L.F., Smilde, A.K.: Principal component analysis of binary genomics data. Brief Bioinform. **20**(1), 317–329 (2019)

Spaans, A., van der Kloot, W.: Permucluster 1.0 user's guide. Department of Psychology, University of Leiden, Leiden (2004)

Tamasauskas, D., Sakalauskas, V., Kriksciuniene, D.: Evaluation framework of hierarchical clustering methods for binary data. In: 12th International Conference on Hybrid Intelligent Systems (HIS), pp. 421–426. IEEE (2012)

van der Kloot, W.A., Bouwmeester, S., Heiser, W.J.: Cluster instability as a result of data input order. In: Yanai, H., Okada, A., Shimenasu, K., Kano, Y., Meulman J. (eds.), New Developments in Psychometrics: Proceedings of the International Meeting of the Psychometric Society IMPS 2001, pp. 569–576, Springer, Tokyo (2003)

Warwick S.I., Beckie H.J., Thomas A.G., McDonald T.: The biology of Canadian weeds. 8. *Sinapis arvensis L.* (updated). Can. J. Plant Sci. **80**(4), 939–961 (2000)

Wijaya, S.H., Afendi, F.M., Batubara, I., Darusman, L.K., Altaf-Ul-Amin, M., Kanaya, S.: Finding an appropriate equation to measure similarity between binary vectors: case studies on Indonesian and Japanese herbal medicines. BMC Bioinf. **17**(520), 1–19 (2016)
Zietkiewicz, E., Rafalski, A., Labuda, D.: Genome fingerprinting by simple sequence repeat (SSR)-anchored polymerase chain reaction amplification. Genomics. **20**(2), 176–183 (1994)

One-Way Repeated Measures ANOVA for Functional Data

Łukasz Smaga

Abstract In this paper, the one-way repeated measures analysis of variance for functional data is considered. For this problem, the new test statistics are obtained by integrating and taking supremum of the constructed pointwise test statistic. To approximate the null distributions of the test statistics and construct the testing procedures, different bootstrap and permutation methods are used. The performance of the new tests and their comparisons with the known testing procedures in terms of size control and power is established in simulation studies. These studies indicate that the new tests may have different finite sample properties, but they are usually more powerful than the tests proposed in the literature.

1 Introduction

Functional data are observations treated as functions or curves. They appear when a number of subjects are observed frequently over time or space. The whole functions are usually unobservable. However, modern recording equipments can measure massive data (large number of observations for each subject at different time points recorded possibly with errors), which can be easily scanned in forms of curves or images. This is the case for many processes, for example, in chemometrics, econometrics, geophysics, medicine, and meteorology. Functional data analysis has received much attention in statistics and its applications in the last two decades. Many methods of functional data analysis for such problems as classification, clustering, estimation, hypothesis testing, regression, etc., can be found in the following monographs and the references therein: Ferraty and Vieu (2006), Horváth and Kokoszka (2012), Kokoszka and Reimherr (2017), Ramsay and Silverman (2002, 2005), and Zhang (2013).

In this paper, we consider the repeated analysis of variance (ANOVA) in the framework of functional data. The examples of repeated functional data considered

Ł. Smaga (✉)
Faculty of Mathematics and Computer Science, Adam Mickiewicz University,
Uniwersytetu Poznańskiego 4, 61-614 Poznań, Poland
e-mail: ls@amu.edu.pl

T. Chadjipadelis et al. (eds.), *Data Analysis and Rationality in a Complex World*,
Studies in Classification, Data Analysis, and Knowledge Organization,
https://doi.org/10.1007/978-3-030-60104-1_27

in the literature include the mortality rates for subsequent decades for a number of countries (Smaga 2019b), and orthosis curves (a moment of force at the knee during stepping-in-place) of a number of volunteers obtained without or with different orthoses (Smaga 2019a). The repeated ANOVA problem for functional data was considered by Martínez-Camblor and Corral (2011) and Smaga (2019a). However, the test statistic considered in these papers does not take into account the information about variance, i.e., it is constructed to consider "between group variability" only. For two groups, Smaga (2019b), proposed a modification of this test statistic to take into account also "within group variability". This resulted in more powerful tests. In this paper, we extend the tests by Smaga (2019b), to the case of the number of groups greater than two. Simulation studies show the advantages and disadvantages of the tests, as well as the differences in their finite sample properties. Moreover, the new tests are usually more powerful than the tests in Martínez-Camblor and Corral (2011).

The remainder of the paper is organized as follows: In Sect. 2, we recall the repeated ANOVA model for functional data. Then we adapt the F-type test statistic to functional data framework and construct new permutation and bootstrap tests for repeated ANOVA for functional data. In Sect. 3, the finite sample properties of the new and known testing procedures are studied in simulation experiments. In Sect. 4, we conclude the paper.

2 Repeated Measures ANOVA for Functional Data

In this section, we describe the repeated measures analysis problem for functional data and construct new tests for this problem.

2.1 *Model and Test Statistics*

Let us follow the notation of Martínez-Camblor and Corral (2011). Assume that we have n subjects, which are submitted to $l \geq 2$ different conditions. The results of the experiments are functional observations, i.e., $X_1(t), \dots, X_n(t)$ for $t \in [0, l]$ is a functional sample consisting of independent stochastic processes concerned with the subjects. In this notation, the first experimental condition is represented by $X_1(t), \dots, X_n(t)$ for $t \in [0, 1]$, the second one by $X_1(t), \dots, X_n(t)$ for $t \in [1, 2]$, and so on. This means that we ignore the possible time periods between repetitions of the experiment, but this does not mean that they do not exist. We also assume the additivity assumption $X_j(t) = m(t) + \varepsilon_j(t)$, where $j = 1, \dots, n$, $t \in [0, l]$ and $\varepsilon_j(t)$ is a random process with zero mean function and covariance function $\mathbb{C}(s, t)$, $s, t \in [0, l]$. We are interested in testing the following null hypothesis about non-significance of experimental conditions:

$$H_0 : m(t) = m(t+1) = \cdots = m(t+(l-1)) \quad \forall t \in [0,1],$$

against $H_1 : \neg H_0$. We can construct the following pointwise F-type test statistic for these hypotheses:

$$F(t) = \frac{\mathrm{SSA}(t)/(l-1)}{\mathrm{SSR}(t)/((l-1)(n-1))},$$

where $t \in [0,1]$,

$$\mathrm{SSA}(t) = n\sum_{i=1}^{l}(\bar{X}_{i\cdot}(t) - \bar{X}(t))^2$$

is the pointwise sum of squares due to hypothesis and

$$\mathrm{SSR}(t) = \sum_{i=1}^{l}\sum_{j=1}^{n}(X_j(t+(i-1)) - \bar{X}_{i\cdot}(t) - \bar{X}_{\cdot j}(t) + \bar{X}(t))^2$$

is the pointwise sum of squares due to rests or errors, and $\bar{X}_{i\cdot}(t) = n^{-1}\sum_{j=1}^{n} X_j(t+(i-1))$, $\bar{X}_{\cdot j}(t) = l^{-1}\sum_{i=1}^{l} X_j(t+(i-1))$, $\bar{X}(t) = N^{-1}\sum_{i=1}^{l}\sum_{j=1}^{n} X_j(t+(i-1))$ for $i = 1, \ldots, l$ and $j = 1, \ldots, n$.

For testing H_0, Martínez-Camblor and Corral (2011) and Smaga (2019a), constructed tests based on the following test statistic:

$$\mathscr{C}_n(l) = \int_0^1 \mathrm{SSA}(t)\mathrm{d}t.$$

Thus, the $\mathscr{C}_n(l)$ tests take into account "between group variability" (i.e., $\mathrm{SSA}(t)$) only. However, using "within group variability" (i.e., $\mathrm{SSR}(t)$) usually results in more powerful tests, which was shown in Zhang et al. (2019) and Zhang and Liang (2014), for standard ANOVA for functional data, and in Smaga (2019b), for H_0, when $l = 2$. Therefore, for testing H_0, we consider the following test statistics:

$$\mathscr{D}_n(l) = \int_0^1 F(t)\mathrm{d}t, \quad \mathscr{E}_n(l) = \sup_{t\in[0,1]} F(t).$$

In standard ANOVA for functional data, the tests based on statistic analogous to $\mathscr{D}_n(l)$ are called the globalizing pointwise F-tests Zhang and Liang (2014), while these based on statistic analogous to $\mathscr{E}_n(l)$ are named the $F_{\max}$-tests Zhang et al. (2019). In the case $l = 2$, the test statistics $\mathscr{D}_n(l)$ and $\mathscr{E}_n(l)$ are consistent with these considered in Smaga (2019b), and in this paper, we extend his results for $l \geq 2$ groups.

2.2 Testing Procedures

In the following, we construct new tests based on permutation and bootstrap methods of approximation of the null distributions of test statistics $\mathscr{D}_n(l)$ and $\mathscr{E}_n(l)$.

In the first permutation method P1 considered in Martínez-Camblor and Corral (2011) and Smaga (2019b), independent permutation samples $X_1^{\mathrm{P1},b}(t), \dots, X_n^{\mathrm{P1},b}(t)$, $t \in [0, l]$, $b = 1, \dots, P_1$ are selected in the following way: for each $j = 1, \dots, n$ separately, the observations $X_j(t), X_j(t+1), \dots, X_j(t+(l-1))$, $t \in [0, 1]$ corresponding to l experimental conditions are randomly permuted, and then $X_j^{\mathrm{P1},b}(t)$ is formed from them. The p-values of the P1 tests are as follows:

$$\frac{1}{P_1}\sum_{b=1}^{P_1} I\left(\int_0^1 F^{\mathrm{P1},b}(t)\mathrm{d}t > \mathscr{D}_n(l)\right), \quad \frac{1}{P_1}\sum_{b=1}^{P_1} I\left(\sup_{t\in[0,1]} F^{\mathrm{P1},b}(t) > \mathscr{E}_n(l)\right),$$

where

$$F^{\mathrm{P1},b}(t) = \frac{\mathrm{SSA}^{\mathrm{P1},b}(t)/(l-1)}{\mathrm{SSR}^{\mathrm{P1},b}(t)/((l-1)(n-1))},$$

$\mathrm{SSA}^{\mathrm{P1},b}(t)$ and $\mathrm{SSR}^{\mathrm{P1},b}(t)$, $t \in [0, 1]$ are the sums of squares due to hypothesis and rests computed on the permutation sample, and $I(A)$ stands for usual indicator function on a set A.

The second permutation method P2 is similar to the standard permutation procedure in ANOVA (Konietschke and Pauly 2014). Let us denote the set of all N observations by

$$A = \{X_1(t), \dots, X_n(t), X_1(t+1), \dots, X_n(t+1), \dots, X_1(t+(l-1)), \dots, X_n(t+(l-1))\}$$

for $t \in [0, 1]$. Then we draw $X_1^{\mathrm{P2},b}(t), \dots, X_n^{\mathrm{P2},b}(t)$, $t \in [0, 1]$ randomly without replacement from A; after that we draw $X_1^{\mathrm{P2},b}(t), \dots, X_n^{\mathrm{P2},b}(t)$, $t \in [1, 2]$ randomly without replacement from the remaining elements in A, and so on, independently for each $b = 1, \dots, P_2$. The p-values of the P2 tests are computed analogously to these for P1 tests.

In the first nonparametric bootstrap approach NB1 considered in Martínez-Camblor and Corral (2011) and Smaga (2019b), we first select independent bootstrap samples $X_1^{\mathrm{NB1},b}(t), \dots, X_n^{\mathrm{NB1},b}(t)$, $t \in [0, l]$, $b = 1, \dots, B_1$ drawn with replacement from the original sample $X_1(t), \dots, X_n(t)$, $t \in [0, l]$. Then the p-values of the NB1 tests are estimated by

$$\frac{1}{B_1}\sum_{b=1}^{B_1} I\left(\int_0^1 F^{\mathrm{NB1},b}(t)\mathrm{d}t > \mathscr{D}_n(l)\right), \quad \frac{1}{B_1}\sum_{b=1}^{B_1} I\left(\sup_{t\in[0,1]} F^{\mathrm{NB1},b}(t) > \mathscr{E}_n(l)\right),$$

where

$$F^{\mathrm{NB1},b}(t) = \frac{n\sum_{i=1}^{l}(\bar{X}_{i\cdot}^{\mathrm{NB1},b}(t) - \bar{X}_{i\cdot}(t) - \bar{X}^{\mathrm{NB1},b}(t) + \bar{X}(t))^2/(l-1)}{\mathrm{SSR}^{\mathrm{NB1},b}(t)/((l-1)(n-1))},$$

$\bar{X}_{i\cdot}^{\mathrm{NB1},b}(t)$, $\bar{X}^{\mathrm{NB1},b}(t)$ and $\mathrm{SSR}^{\mathrm{NB1},b}(t)$, $t \in [0, 1]$ are the appropriate sample means and the sum of squares due to rests computed on the bootstrap sample.

The second nonparametric bootstrap method NB2 is an extension of the methods considered in Konietschke and Pauly (2014) and Smaga (2019b). First, we have to center the observations, i.e.

$$X_{1,c}(t) = X_1(t) - \bar{X}_{\bullet}(t), \ldots, X_{n,c}(t) = X_n(t) - \bar{X}_{\bullet}(t)$$

for $t \in [0, l]$, where $\bar{X}_{\bullet}(t) = n^{-1}\sum_{j=1}^{n} X_j(t)$. Then the independent bootstrap samples $X_1^{\mathrm{NB2},b}(t), \ldots, X_n^{\mathrm{NB2},b}(t)$, $t \in [0, l]$, $b = 1, \ldots, B_2$ are selected as follows: $X_1^{\mathrm{NB2},b}(t), \ldots, X_n^{\mathrm{NB2},b}(t)$, $t \in [0, 1]$ are randomly drawn with replacement from $X_{1,c}(t), \ldots, X_{n,c}(t)$, $t \in [0, 1]$; independently, $X_1^{\mathrm{NB2},b}(t), \ldots, X_n^{\mathrm{NB2},b}(t)$, $t \in [1, 2]$ are randomly drawn with replacement from $X_{1,c}(t), \ldots, X_{n,c}(t)$, $t \in [1, 2]$, and so on. Finally, the p-values are estimated analogously to these for P1 tests.

The last method is the parametric bootstrap approach PB, whose idea is similar to the parametric bootstrap proposed in Konietschke et al. (2015), for nonparametric ANOVA for random vectors. In that and present papers, it is shown that although the parametric bootstrap is typically applied for parametric models, it can also be successfully used in nonparametric ones (any specific distribution of the data is not assumed). In the PB method, we first estimate the covariance function $\mathbb{C}(s, t)$ by the following unbiased estimator:

$$\hat{\mathbb{C}}(s, t) = \frac{1}{n-1}\sum_{j=1}^{n}(X_j(s) - \bar{X}_{\bullet}(s))(X_j(t) - \bar{X}_{\bullet}(t)), \quad s, t \in [0, l].$$

Then we generate the independent bootstrap samples $X_1^{\mathrm{PB},b}(t), \ldots, X_n^{\mathrm{PB},b}(t)$, $t \in [0, l]$, $b = 1, \ldots, B_3$ from the Gaussian process with zero mean function and covariance function $\hat{\mathbb{C}}(s, t)$. Such bootstrap samples satisfy the null hypothesis H_0. The p-values are estimated as these for the P1 tests.

3 Simulation Studies

In this section, the $\mathcal{D}_n(l)$ and $\mathcal{E}_n(l)$ tests constructed in Sect. 2, and the tests based on test statistic $\mathcal{C}_n(l)$ of Martínez-Camblor and Corral (2011), are compared in terms of size control and power.

3.1 Simulation Setup

We consider $l = 3$ groups and $n = 35, 50$. The simulation data were generated as $X_j(t + (i - 1)) = m(t + (i - 1)) + \varepsilon_j(t + (i - 1))$ for $t \in [0, 1]$, $i = 1, 2, 3$, $j = 1, \dots, n$, in the following two models (similar to models M4 and M6 considered in Martínez-Camblor and Corral (2011):

M1 $m(t) = m(t + 1) = m(t + 2) = (\sin(2\pi t^2))^5$, i.e., H_0 is true.
M2 $m(t) = m(t + 1) = (\sin(2\pi t^2))^5$ and $m(t + 2) = (\sin(2\pi t^2))^7$, i.e., H_1 holds.

We also considered two distributions of the observations, i.e., normal and lognormal. In normal setting, $\varepsilon_j(t) = 0.5B_{j1}(t)$, $\varepsilon_j(t + 1) = \rho\varepsilon_j(t) + 0.5(1 - \rho^2)^{1/2}B_{j2}(t)$ and $\varepsilon_j(t + 2) = \rho\varepsilon_j(t + 1) + 0.5(1 - \rho^2)^{1/2}B_{j3}(t)$ for $t \in [0, 1]$, where B_{j1}, B_{j2} and B_{j3} were independent standard Brownian Bridges and $\rho = 0, 0.25, 0.5, 0.75$, $j = 1, \dots, n$. In lognormal setting, the errors were (adequately centered) $\exp(\varepsilon_j(t))$, $j = 1, \dots, n$, $t \in [0, 3]$, where $\varepsilon_j(t)$ are the errors of normal setting.

The functional observations were discretized at design time points $t_1, \dots, t_{101}, t_1 + 1, \dots, t_{101} + 1, t_1 + 2, \dots, t_{101} + 2$, where t_k, $k = 1, \dots, 101$ were equispaced in the interval $[0, 1]$. The 1000 simulation replications were used to estimate the empirical sizes and powers of the tests. We used $P_1 = P_2 = B_1 = B_2 = B_3 = 1000$ permutation or bootstrap samples for computation of p-values. For simplicity, $\alpha = 5\%$. The simulation experiments were conducted in the R program (R Core Team 2019).

3.2 Discussion on Simulation Results

The simulation results are presented in Table 1.

In model M1, we study the type I error control of the tests. The P1, NB1, and PB $\mathscr{C}_n(3)$ tests, the all $\mathscr{D}_n(3)$ tests, and the NB1 and PB $\mathscr{E}_n(3)$ tests retain the preassigned type I error in all cases. Nevertheless, the NB1 and PB tests based on test statistics $\mathscr{D}_n(3)$ and $\mathscr{E}_n(3)$ may have slightly conservative character. The empirical sizes of the P2 and NB2 $\mathscr{C}_n(3)$ tests are very close to the significance level in the independent case ($\rho = 0$). However, when the data are correlated ($\rho > 0$), these tests are extremely conservative, and their conservativity increases with the increase of the correlation (i.e., increase of ρ), which may result in a noticeable loss of power. On the other hand, the P1, P2, and NB2 $\mathscr{E}_n(3)$ tests tend to highly over-reject the null hypothesis in case of high correlation of functional data ($\rho = 0.5, 0.75$). This means that although the permutation tests based on $\mathscr{E}_n(l)$ performed best for $l = 2$ (Smaga 2019b), they may not be such that for $l \geq 3$.

In model M2, the empirical power of the tests is investigated. The empirical powers of the tests increase with the increase of the number of observations n. They also usually increase with the increase of the correlation of functional data (i.e., increase of ρ). The two exceptions are the P2 and NB2 $\mathscr{C}_n(3)$ tests, which follows from their

Table 1 Empirical sizes (M1) and powers (M2) (as percentages) of the permutation (P1, P2) and bootstrap (NB1, NB2, PB) tests based on test statistics $\mathscr{C}_n(3)$, $\mathscr{D}_n(3)$ and $\mathscr{E}_n(3)$ obtained under normal and lognormal settings in models M1 and M2

	$\mathscr{C}_n(3)$					$\mathscr{D}_n(3)$					$\mathscr{E}_n(3)$				
ρ	P1	P2	NB1	NB2	PB	P1	P2	NB1	NB2	PB	P1	P2	NB1	NB2	PB
	M1														
	Normal														
	$n = 35$														
0.00	4.9	5.4	5.0	5.5	4.9	5.1	5.5	3.6	4.4	4.5	4.1	4.2	2.9	3.4	3.3
0.25	4.8	1.7	4.8	2.1	4.6	5.7	5.3	3.4	4.6	4.0	6.1	6.1	3.4	4.4	3.7
0.50	5.3	0.5	4.8	0.8	4.5	5.2	4.5	3.7	4.7	4.0	6.7	7.1	3.7	6.0	3.9
0.75	4.7	0.0	3.9	0.0	4.2	5.6	5.2	3.3	5.1	4.0	8.2	8.1	3.7	7.4	4.3
	$n = 50$														
0.00	4.7	4.6	4.6	5.4	4.2	5.1	5.0	3.9	4.3	4.2	4.7	4.9	3.9	4.0	3.4
0.25	5.3	2.8	5.3	2.6	4.8	5.7	5.7	4.7	5.2	4.9	5.6	5.9	4.5	5.4	4.4
0.50	5.7	0.7	5.6	0.6	5.0	6.0	5.8	4.5	5.3	4.8	6.8	6.9	4.3	6.3	4.3
0.75	5.8	0.0	5.3	0.0	5.0	5.8	5.5	4.8	5.2	5.0	9.4	9.1	5.3	8.9	5.2
	Lognormal														
	$n = 35$														
0.00	4.4	4.8	3.4	4.4	3.5	4.4	4.7	3.0	3.7	3.5	5.4	6.0	3.3	4.3	4.1
0.25	4.6	1.8	3.5	1.7	3.8	4.7	5.2	3.3	4.0	3.7	5.4	6.0	3.5	4.4	4.4
0.50	5.0	0.1	3.6	0.1	3.4	5.4	5.3	3.0	4.4	3.5	7.2	7.1	3.7	5.7	4.6
0.75	5.6	0.0	4.0	0.0	3.4	5.7	5.6	2.9	4.3	3.5	9.3	9.1	4.0	7.8	5.0
	$n = 50$														
0.00	4.8	5.0	4.4	4.6	4.1	4.8	5.3	3.6	4.0	3.9	5.1	5.2	3.7	4.1	4.4
0.25	6.0	2.1	4.3	1.9	4.8	5.4	5.5	3.8	4.4	4.3	5.9	5.9	4.1	5.0	4.5
0.50	6.5	0.3	5.1	0.4	5.0	6.3	6.0	4.4	5.1	4.6	7.9	7.4	5.0	6.5	5.9
0.75	5.8	0.0	4.2	0.0	4.0	5.6	5.7	3.2	5.4	3.9	9.6	9.0	4.2	8.6	4.9
	M2														
	Normal														
	$n = 35$														
0.00	32.5	32.1	30.6	32.0	30.0	45.0	45.1	34.8	38.4	38.5	88.7	88.3	84.4	85.8	84.5
0.25	42.9	24.7	40.6	24.5	40.2	59.5	59.0	47.6	54.0	52.1	96.1	95.9	92.1	94.8	93.3
0.50	65.6	13.7	59.3	14.8	59.6	82.1	81.9	72.3	78.8	74.0	99.9	100	98.9	99.7	99.4
0.75	97.8	5.3	96.6	5.7	96.2	99.9	99.8	98.7	99.7	99.5	100	100	100	100	100
	$n = 50$														
0.00	52.6	53.1	51.0	53.4	50.9	71.3	71.2	64.7	68.3	66.8	98.9	99.2	98.7	98.9	98.9
0.25	69.9	43.0	65.2	45.0	65.5	85.3	85.5	78.8	82.4	80.7	99.9	99.9	99.9	99.9	99.8
0.50	91.0	33.5	87.1	33.4	87.9	98.3	97.9	96.2	97.8	96.7	100	100	100	100	100
0.75	100	16.4	99.9	18.2	99.9	100	100	100	100	100	100	100	100	100	100

(continued)

Table 1 (continued)

ρ	$\mathscr{C}_n(3)$ P1	P2	NB1	NB2	PB	$\mathscr{D}_n(3)$ P1	P2	NB1	NB2	PB	$\mathscr{E}_n(3)$ P1	P2	NB1	NB2	PB
	Lognormal														
	$n = 35$														
0.00	86.8	87.3	82.9	86.4	82.3	86.7	86.7	79.5	83.4	81.7	95.5	95.7	92.4	93.6	94.4
0.25	94.5	84.1	91.3	82.5	92.0	94.2	94.1	89.2	91.8	91.4	98.7	98.8	97.7	98.6	98.5
0.50	99.1	78.3	98.7	77.1	98.6	99.1	99.1	98.3	99.1	98.5	99.9	100	99.8	100	99.8
0.75	100	68.7	100	69.6	100	100	100	100	100	100	100	100	100	100	100
	$n = 50$														
0.00	98.6	98.6	97.7	98.5	97.8	98.4	98.4	97.1	97.7	97.5	99.7	99.6	99.6	99.6	99.7
0.25	99.5	97.8	99.5	97.8	99.4	99.5	99.5	99.1	99.6	99.3	100	100	100	100	100
0.50	100	97.3	100	97.3	100	100	100	100	100	100	100	100	100	100	100
0.75	100	97.5	100	97.3	100	100	100	100	100	100	100	100	100	100	100

conservative behavior in model M1. These tests are the least powerful tests in almost all cases, except the independent case ($\rho = 0$). In both settings, all $\mathscr{E}_n(3)$ tests have similar empirical powers despite the too liberal character of some of them. Moreover, these tests are the most powerful. The empirical powers of the permutation (P1 and P2) $\mathscr{D}_n(3)$ tests are usually greater than the empirical powers of the bootstrap (NB1, NB2, and PB) $\mathscr{D}_n(3)$ tests and the $\mathscr{C}_n(3)$ tests. In normal (resp. lognormal) setting, the bootstrap $\mathscr{D}_n(3)$ tests are better than (resp. comparable with) the $\mathscr{C}_n(3)$ tests in term of power.

To sum up, the NB1 and PB tests based on test statistic $\mathscr{E}_n(3)$ seem to be the best tests in terms of size control and power. Nevertheless, the (especially permutation) $\mathscr{D}_n(3)$ tests also seem to have good finite sample properties.

4 Conclusions

We have proposed permutation and bootstrap tests for the repeated measures analysis of variance for functional data extending the results known in the literature for two groups. In simulation studies, the tests have been compared with known procedures in terms of size control and power. These studies have indicated that most of the new methods have good finite sample properties, and they are more powerful than earlier solutions. This implies that these new tests are able to detect that the null hypothesis is false with a larger possible probability than the other tests. Only the permutation tests and one bootstrap test based on the supremum of the pointwise F-type test statistic need a greater number of observations to maintain the type I error level.

References

Ferraty, F., Vieu, P.: Nonparametric Functional Data Analysis: Theory and Practice. Springer, New York (2006)

Horváth, L., Kokoszka, P.: Inference for Functional Data with Applications. Springer, New York (2012)

Kokoszka, P., Reimherr, M.: Introduction to Functional Data Analysis. Chapman and Hall/CRC (2017)

Konietschke, F., Pauly, M.: Bootstrapping and permuting paired t-test type statistics. Stat. Comput. **24**, 283–296 (2014)

Konietschke, F., Bathke, A.C., Harrar, S.W., Pauly, M.: Parametric and nonparametric bootstrap methods for general MANOVA. J. Multivar. Anal. **140**, 291–301 (2015)

Martínez-Camblor, P., Corral, N.: Repeated measures analysis for functional data. Comput. Stat. Data Anal. **55**, 3244–3256 (2011)

R Core Team: R: A language and environment for statistical computing. R Foundation for Statistical Computing, Vienna, Austria (2019). https://www.R-project.org/

Ramsay, J.O., Silverman, B.W.: Applied Functional Data Analysis. Methods and Case Studies. Springer, New York (2002)

Ramsay, J.O., Silverman, B.W.: Functional Data Analysis, 2nd edn. Springer, New York (2005)

Smaga, Ł.: Repeated measures analysis for functional data using Box-type approximation—with applications. REVSTAT **17**(4), 523–549 (2019a)

Smaga, Ł.: A note on repeated measures analysis for functional data. AStA Adv. Stat. Anal. **104**(1), 117–139 (2019b)

Zhang, J.T., Liang, X.: One-way ANOVA for functional data via globalizing the pointwise F-test. Scand. J. Stat. **41**, 51–71 (2014)

Zhang, J.T.: Analysis of Variance for Functional Data. Chapman and Hall, London (2013)

Zhang, J.T., Cheng, M.Y., Wu, H.T., Zhou, B.: A new test for functional one-way ANOVA with applications to ischemic heart screening. Comput. Stat. Data Anal. **132**, 3–17 (2019)

Flexible Clustering

Andrzej Sokołowski and Małgorzata Markowska

Abstract Flexibility of cluster analysis is sometimes understood as the robustness of final partition of objects to the changes in the list of diagnostic variables—deleting some from the list or adding some. In this paper, we propose a procedure which makes possible to calculate a distance matrix on the basis of different subsets of variables, but the selection of variables is somehow unified. The procedure starts with the classical standardization of each variable. Before the calculation of a distance between two objects, we eliminate the variables with the largest absolute value in the first object and in the second object. If by chance the same variable is pointed for elimination for both objects, the next variable with the largest absolute value (for both objects) should be eliminated. With this procedure, each element of the distance matrix is based on the same number of variables, but the variables can be different. As an example, a data set of 17 variables describing human smart society characteristics for 28 European Union countries is used.

Keywords Clustering · Distance matrix · Variable selection

1 Introduction

The word "flexible" in cluster analysis up to now was rather used with respect to the clustering method, and not to the changing list of variables. Lance and Williams (1967) discussed flexible agglomeration steered by the β parameter. New flexible aggregation concept has been proposed by Hahmann et al. (2009). Gallaugher et al. (2019) called flexible their approach to model-based clustering with generalized

A. Sokołowski (✉)
Cracow University of Economics, Rakowicka 27, 31-510 Cracow, Poland
e-mail: sokolows@uek.krakow.pl

M. Markowska
Wrocław University of Economics and Business,
Komandorska 118/120, 53-345 Wrocław, Poland
e-mail: malgorzata.markowska@ue.wroc.pl

T. Chadjipadelis et al. (eds.), *Data Analysis and Rationality in a Complex World*,
Studies in Classification, Data Analysis, and Knowledge Organization,
https://doi.org/10.1007/978-3-030-60104-1_28

hyperbolic distributions (see also Tang et al. 2018). Ni et al. (2015) consider flexibility rather as different shapes and types of network structures.

The idea of our flexible clustering has two sources. In one of the first papers on linear ordering of multivariate objects, Drewnowski (1966) suggested the flexibility criterion. In his opinion, it means that the slight change in the list of variables (by adding or by omitting) should not strongly affect the final ordering of objects. The second source is our experience in the analysis of European Union countries using data from 2008. In that year, the highest Gross Domestic Product in EU countries was observed in Luxembourg, and was equal to 80, 000 Euro. The second country was Denmark, with 42, 000 Euro per head, almost a half less than the leading country. Any type of standardization or normalization keeps this skewness and the outlier crushes all other countries to the opposite corner. This can be also considered as an unwanted, hidden weighing of the variables.

Usually, one or two variables make the object become an outlier, but these variables are not the same for all objects. The proposed procedure allows to partially solve this problem, and classification is based on the flexible list of variables.

2 Method

We define the proposed approach in the following steps:

1. Define the set of objects to be clustered
2. Set the initial list of m diagnostic variables
3. Standardize variables
4. For each object j find variable i^* with the highest absolute value of standardized value and variable i^{**} with the second highest module
5. Calculate distance matrix **D** in such a way that distance d_{lk} between objects l and k is calculated using $m-2$ standardized variables, omitting variables i_l^* and i_k^*
6. If both objects l and k point out the same variable i^*, select variable from i_l^{**} and i_k^{**} with a higher module from variables to be eliminated as the second one
7. Perform cluster analysis using distance matrix **D**

With the described procedure each element of a distance matrix can be based on slightly different set of variables. For each object, a variable with the most outlying value is skipped.

3 Example

The aim is to clusters 28 European Union countries on the basis of 2017 data characterizing human smart development. The initial set of diagnostic variables consists of the following 17 characteristics (source: https://strateg.stat.gov.pl/dashboard/#/polityka-spojnosci/1):

$X1$	- Expenditures on R&D as % in GDP
$X2$	- % of population using internet at least once a week
$X3$	- Number of inventions reported to EPO (European Patent Office) for 1 million population
$X4$	- PISA (Programme for International Student Assessment) test—% of students on high levels in reading and interpretation
$X5$	- HDI (Human Development Index)
$X6$	- PISA test—% of students on high levels in mathematics
$X7$	- PISA test—% of students on high levels in science
$X8$	- Government and higher education sector expenditures on R&D in % of GDP
$X9$	- Corporate expenditures on R&D in % of GDP
$X10$	- % of small and medium enterprises adopting product or process innovations
$X11$	- % of population with high internet skills
$X12$	- % of employees working in R&D
$X13$	- % of young people not working nor learning
$X14$	- % of teenagers who quit education
$X15$	- % of 25–64 population still active in education
$X16$	- % of 30–34 population with higher education completed
$X17$	- % of population with basic or higher computer skills

In Table 1, we report the variables which were pointed out to be eliminated, having the highest and the second highest absolute standardized value for specific EU countries. In addition, the Table indicates if the value under the module was positive or negative.

Ward's agglomerative method has been used for clustering EU countries, with the squared Euclidean distance calculated per variable, in order to make possible the comparison between the classical and the flexible approach for 17 variables. With the first important increase of the agglomerative distance as a cut-off point for constructing the dendrogram, six groups of countries have been identified, both in the classical and the flexible approach (see Fig. 1).

The composition of the groups is very similar. Both partitions are presented in Table 2.

Three countries change their group membership in the flexible approach: Slovakia and Hungary have moved to the cluster with Malta, Portugal, Italy, and Spain, while Croatia joined Romania and Bulgaria. Mean values and standard deviations of the first three clusters (only those which change in their composition) are given in Table 3.

Rather unexpectedly, in clusters 1 and 2, more variables have lower standard deviation in the flexible version as compared to the classical one. This means that flexibility generally does not spoil cluster compactness. Of course, this cannot be considered as a general rule, since it was just observed in the first empirical example. In this example, changes in cluster membership were small, mainly because only two variables were skipped out of 17. The proposed method can be used to study the stability of clusters, and for the example provided, it seems to be quite good.

Table 1 Variables with the highest and second highest modules of standardized values

Variable	Countries with the highest standarized module	Countries with the second highest standarized module
$X1$ - Expenditures on R&D		Austria (+)
		Germany (+)
$X2$ - Internet activity	Great Britain (+)	Croatia (–)
		Luxemburg (+)
$X3$ - Inventions to EPO		Denmark (+)
$X4$ - PISA - reading and interpretation	France (+)	Spain (–)
	Ireland (+)	Finland (+)
	Slovakia (–)	
$X5$ - HDI	Hungary (–)	Bulgaria (–)
		Ireland (+)
		Portugal (–)
$X6$ - PISA - mathematics	Cyprus (–)	Estonia (+)
	Czechia (+)	Slovenia (+)
$X7$ - PISA - science	Estonia (+)	Great Britain (+)
	Slovenia (+)	
$X8$ - Non-corporate expenditures on R&D		
$X9$ - Corporate expenditures on R&D	Austria (+)	Belgium (+)
	Germany (+)	Latvia (–)
		Malta (–)
$X10$ - SME with innovations	Belgium (+)	Romania (–)
	Latvia (–)	Slovakia (–)
	Poland (–)	
	Portugal (–)	
$X11$ - High internet skills	Lithuania (+)	Czechia (–)
	Sweden (+)	France (–)
$X12$ - Employent in R&D		Greece (+)
$X13$ - Young not working or learning	Greece (+)	Netherlands (–)
	Italy (+)	
$X14$ - Teenagers quited education	Spain (+)	Poland (–)
	Croatia (–)	
	Malta (+)	
	Romania (+)	
	Slovenia (–)	

(continued)

Table 1 (continued)

Variable	Countries with the highest standarized module	Countries with the second highest standarized module
*X*15 - Adult population educating	Denmark (+)	Sweden (+)
	Finland (+)	
*X*16 - 30–34 population with higher education		Cyprus (+)
		Hungary (–)
		Italy (–)
		Lithuania (+)
*X*17 - At least basic computer skills	Bulgaria (–)	
	Luxembourg (+)	
	Netherlands (+)	

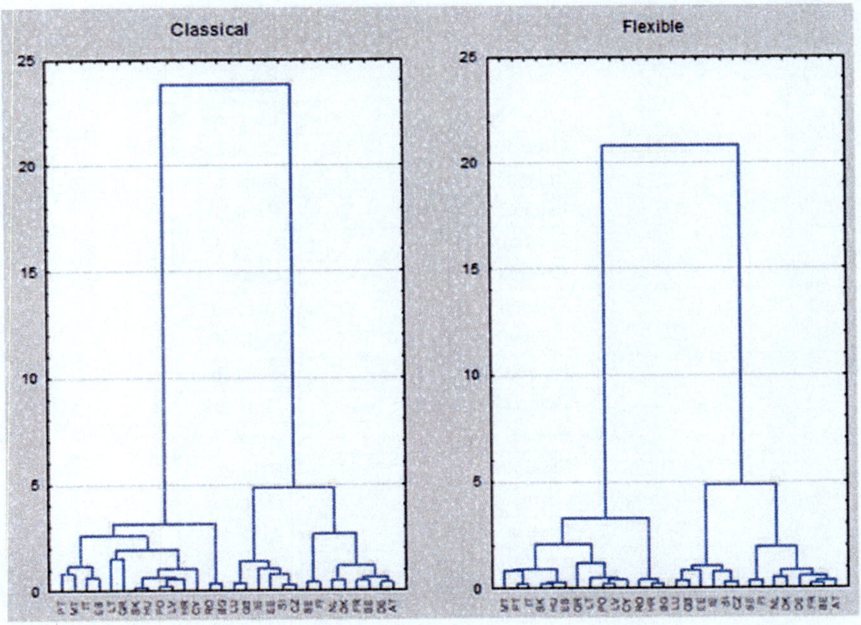

Fig. 1 Ward's dendrograms for classical and flexible approaches

Table 2 Clusters of countries

Cluster	Classical	Flexible
1	Portugal	Malta
	Malta	Portugal
	Italy	Italy
	Spain	Slovakia
		Hungary
		Spain
2	Lithuania	Lithuania
	Greece	Greece
	Slovakia	Poland
	Hungary	Latvia
	Poland	Cyprus
	Latvia	
	Croatia	
	Cyprus	
3	Romania	Romania
	Bulgaria	Bulgaria
		Croatia
4	Luxembourg	Luxembourg
	Great Britain	Great Britain
	Ireland	Ireland
	Estonia	Estonia
	Slovenia	Slovenia
	Czechia	Czechia
5	Sweden	Sweden
	Finland	Finland
6	Netherlands	Netherlands
	Denmark	Denmark
	France	France
	Belgium	Belgium
	Germany	Germany
	Austria	Austria

It would be interesting to eliminate more variables. The effect can be measured by the so-called *P-flexibility of order k*, where P is a Rand Index between the clustering with the full list of variables and flexible clustering with k variables eliminated from each of two objects. In our example, we have *0.929-flexibility of order 1*. The highest possible k is $int[m/2]$.

Table 3 Means and standard deviations in first three clusters (first row—classical; second row—flexible)

Variable	Cluster 1		Cluster 2		Cluster 3	
	Mean	SD	Mean	SD	Mean	SD
*X*1	1.11	0.38	0.90	0.28	0.63	0.18
	1.11	0.33	0.82	0.28	0.70	0.18
*X*2	75.00	5.83	74.00	5.37	61.50	0.71
	75.83	4.79	74.40	4.77	62.67	2.08
*X*3	33.06	25.68	11.39	5.20	4.60	0.66
	27.07	22.17	11.21	4.15	4.67	0.48
*X*4	4.60	1.09	4.81	1.78	2.70	1.13
	4.05	1.23	5.31	1.23	3.80	2.07
*X*5	0.87	0.02	0.85	0.01	0.81	0.00
	0.86	0.02	0.86	0.01	0.82	0.01
*X*6	10.60	2.08	6.41	1.47	4.10	0.71
	9.58	2.26	5.83	1.61	5.07	1.75
*X*7	4.98	2.13	4.03	1.68	1.55	1.34
	5.35	1.76	3.28	1.41	2.23	1.52
*X*8	0.50	0.12	0.42	0.12	0.21	0.00
	0.46	0.12	0.43	0.15	0.28	0.12
*X*9	0.54	0.37	0.47	0.27	0.41	0.17
	0.61	0.34	0.37	0.23	0.42	0.12
*X*10	32.40	12.38	22.51	10.23	8.73	8.70
	26.88	12.86	25.59	12.15	12.63	9.15
*X*11	16.41	6.87	17.38	10.62	13.50	10.61
	15.44	5.56	19.80	13.29	13.33	7.51
*X*12	1.84	0.59	1.62	0.92	0.70	0.23
	1.66	0.53	1.83	1.14	0.84	0.30
*X*13	16.25	7.02	16.79	3.97	19.10	0.57
	16.43	5.57	16.36	4.96	19.03	0.42
*X*14	15.65	2.78	7.30	2.99	15.40	3.82
	14.07	3.42	6.70	1.73	11.30	7.60
*X*15	9.55	1.16	5.09	1.82	1.70	0.85
	7.97	2.76	5.76	1.50	1.90	0.69
*X*16	33.78	5.85	42.78	10.68	29.55	4.60
	33.58	4.59	49.42	6.96	29.27	3.29
*X*17	51.50	5.80	49.43	5.57	29.00	0.00
	52.50	5.54	49.08	3.66	33.00	6.93

Acknowledgements The project is financed by the Ministry of Science and Higher Education in Poland under the programme "Regional Initiative of Excellence" 2019–2022 project number 015/RID/2018/19 total funding amount 10 721 040.00 PLN, and research fund granted to the Faculty of Management at Cracow University of Economics.

References

Drewnowski, J.: The Level of Living Index. UNSRISD, Report No 4, Geneva (1966)

Gallaugher, M.P.B., Tang, Y., McNicholas, P.D.: Flexible clustering with a sparse mixture of generalized hyperbolic distributions (2019). arXiv:1903.05054 [stat.ME]

Hahmann, M., Volk, P.B., Rosenthal, F., Habich, D., Lehner, W.: How to control clustering results? flexible clustering aggregation. In: Adams, N.M., Robardet, C., Siebes, A., Boulicaut, J.F. (eds.) Advances in Intelligent Data Analysis VIII. IDA 2009. Lecture Notes in Computer Science, vol. 5772. Springer, Berlin (2009)

Lance, G.N., Williams, W.T.: A general theory of classification sorting strategies. I. Hierarchical systems. Comput. J. **9**(4), 373–380 (1967)

Ni, J., Tong, H., Fan, W., Zhang, X.: Flexible clustering and robust multi-network clustering. In: Proceedings of the ACM SIGKDD International Conference on Knowledge Discovery and Data Mining, vol. 2015, pp. 835–844. Association for Computing Machinery (2015)

Tang, Y., Browne, R.P., McNicholas, P.D.: Flexible clustering of high-dimensional data via mixtures of joint generalized hyperbolic distributions. Stat **7**(1), e177 (2018)

Classification of Entrepreneurial Regimes: A Symbolic Polygonal Clustering Approach

Andrej Srakar and Marilena Vecco

Abstract Entrepreneurial regimes is a topic, receiving ever more research attention. Existing studies on entrepreneurial regimes mainly use common methods from multivariate analysis and some type of institutional related analysis. In our analysis, the entrepreneurial regimes are analyzed by applying a novel polygonal symbolic data cluster analysis approach. Considering the diversity of data structures in Symbolic Data Analysis (SDA), interval-valued data is the most popular. Yet, this approach requires assuming equidistribution hypothesis. We use a novel polygonal cluster analysis approach to address this limitation with additional advantages: to store more information, to significantly reduce large data sets preserving the classical variability through polygon radius and to open new possibilities in symbolic data analysis. We construct a dynamic cluster analysis algorithm for this type of data with proving main theorems and lemmata to justify its usage. In the empirical part, we use a data set of Global Entrepreneurship Monitor (GEM) for the year 2015, to construct typologies of countries based on responses to main entrepreneurial questions. The article presents a novel approach to clustering in statistical theory (with novel type of variables never accounted for) and application to a pressing issue in entrepreneurship with novel results.

Keywords Symbolic data · Polygonal cluster analysis · Entrepreneurial regimes

A. Srakar (✉)
Institute for Economic Research (IER), Ljubljana and Faculty of Economics, University of Ljubljana, Ljubljana, Slovenia
e-mail: andrej_srakar@t2.net

M. Vecco
Burgundy School of Business—Universite Bourgogne Franche-Comte, Bourgogne Franche-Comte, France
e-mail: mari.vecco@gmail.com

T. Chadjipadelis et al. (eds.), *Data Analysis and Rationality in a Complex World*, Studies in Classification, Data Analysis, and Knowledge Organization,
https://doi.org/10.1007/978-3-030-60104-1_29

1 Introduction

Several different classifications of EU countries based on welfare regimes, type of capitalism and other institutional characteristics have been proposed in the recent decades. In their formulation, Hall and Soskice (2001), identify a core distinction between two types of political economies: liberal market economies, in which firms coordinate their activities primarily via firm hierarchies and competitive market arrangements, and coordinated market economies, in which coordination relies more heavily on non-market relationships (see Dilli and Elert 2020). Esping-Andersen (1990) original classification is composed of three models: Liberal, Social-Democratic and Continental, with later studies pointing to the existence of Mediterranean and Eastern European regimes. In their analysis, Dilli and Elert (2020), derive six groups of entrepreneurial regimes, including a separate cluster for the Eastern European countries.

Entrepreneurial activities highly differ across countries. This holds true for different measures of entrepreneurship such as start-up activity, business ownership, small business share, nascent entrepreneurship and the preference and motives for entrepreneurship. Besides individual characteristics (risk tolerance, entrepreneurial culture, etc.), the level of economic development and cultural aspects are often mentioned as the principal drivers of entrepreneurial activities.

Although there are different streams and approaches of the institutional theory within the realm of entrepreneurship, the approach introduced by North (1990), is still the most used. This author defined institutions as "rules of game in the society" and constraints that structure human interaction ... made up of formal constraints (for example, rules, laws and constitutions) (North 1990, p. 360). North classified the formal and informal institutions impacting organizations and organizational actors into regulatory, normative and cognitive categories.

Furthermore, some scholars observed, relevant cross-national differences may be embedded in historical experiences, institutional heritage, norms, or cultural values. These differences can provide idiosyncratic institutional milieus for entrepreneurial activities.

As noted by scholars, the set of regulatory procedures and administrative constraints may negatively impact the entrepreneurial activity as entrepreneurs have to spend extra time and resources to fit in the administrative system instead of being devoted to develop their business content. Furthermore, quality of institutions is regarded as many scholars to have a significant impact on the modelling of entrepreneurial regimes (for more, see, e.g. Dilli and Elert 2020).

2 Polygonal Variables in Symbolic Data Analysis

Due to the nature of GEM database, which is a large scale database, including for each year more than 100,000 respondents, we utilize a symbolic data analysis, which is a special type of statistical analysis of large data sets, developed in recent decades (see, e.g. Billard and Diday 2006; Diday and Noirhomme-Fraiture 2008; Noirhomme-Fraiture and Brito 2011). In the classical data framework, one numerical value or one category is associated with each individual (microdata). However, the interest of many studies lays in groups of records gathered according to the characteristics of the individuals or classes of individuals, leading to macro data. The traditional approach for such kind of situations is to associate with each individual or class of individuals a central measure, e.g. the mean or the mode of the corresponding records. However, with this option, the variability across the records is lost. To avoid this unsatisfactory result, Symbolic Data Analysis (SDA) proposes that a distribution or an interval of the individual records' values is associated with each unit, thereby considering new variable types, named symbolic variables. One such type of symbolic variable is the histogram-valued variable, where each entity under analysis corresponds to an empirical distribution that can be represented by a histogram or a quantile function. To this purpose, it is necessary to adapt concepts and methods of classical statistics to new kinds of variables. Furthermore, our analysis derives from symbolic polygonal variable concept, differing in several aspects from polygonal spatial clustering in Joshi (2011).

A *multi-valued* symbolic variable Y is one whose possible value takes one or more values from the list of values in its domain Y. The complete list of possible values in Y is finite, and values may be well-defined categorical (qualitative, nominal) or quantitative (numerical) values. An *interval* symbolic variable Y is one whose possible value takes values in an interval, i.e. $Y = \rho = [a, b] \subset R$ with $a \leq b, a, b \in R$. The interval can be closed or open at either end. Let the random variable Y take possible values $\{\eta_k, k = 1, 2, \ldots\}$ over a domain Y. Then, a particular outcome is *modal valued* if it takes the form $Y(\omega_u) = \rho_u = \eta_k, \pi_k; k = 1, \ldots, s_u\}$ for an observation u where π_k is a non-negative measure associated with η_k and where s_u is the number of values actually taken from Y. These possible η_k can be finite or infinite in number, they may be categorical or quantitative in value. The domain Y can be finite or infinite in size.

Many methods have been developed for cluster analysis of interval data. Methods based on dissimilarities generally use adaptations of K-means. Alternative approaches propose suitable dissimilarity measures for interval data, and then use the K-means algorithm to obtain a partition that locally optimizes a criterion measuring the fit between the cluster composition and their prototypes. Such methods are, for example, de Souza and De Carvalho (2004) using City-Block L_1 distance between intervals, $d_1\left(I_i, I_j\right) = |l_i - l_j| + |u_i - u_j|$; (De Carvalho et al. 2006) using L_2 distance between intervals, $d_2\left(I_i, I_j\right) = \sqrt{\left(l_i - l_j\right)^2 + \left(u_i - u_j\right)^2}$ and Chavent et al. (2006), using various types of distances, including Hausdorff-based $d_H\left(x_1^j, x_2^j\right) = \max\left(|l_1^j - l_2^j|, |u_1^j - u_2^j|\right)$; $d_1\left(x_1, x_2\right) = \sum_{j=1}^{p} \max\left(|l_1^j - l_2^j|, |u_1^j - u_2^j|\right)$.

Fuzzy K-means methods for interval data generally result from adapting the classical fuzzy c-means algorithm, using appropriate distances, as is done for the crisp algorithms. Other extensions use adaptive distances or multiple dissimilarity matrices. Hierarchical or pyramidal clustering has been used by Brito (1994, 1995).

A monothetic clustering method using a divisive approach has been used in Chavent (1998), who uses a criterion that measures intraclass dispersion using distances appropriate to interval-valued variables. The algorithm successively splits one cluster into two sub-clusters, according to a condition expressed as a binary question on the values of one variable. Kohonen maps have been used by Bock (2002, 2008). Different dynamic algorithms have been applied as well, in particular in Chavent et al. (2006).

A novel type of symbolic variables, closely related and deriving from interval ones, have been proposed in a recent article of Silva et al. (2019). They define polytope as a convex hull of a compact non-empty finite set. In general, the face structures of a convex polytope are significantly more simple than convex hull. A polytope $P = conv\{x_1, \ldots, x_l\}$ is called a k-polytope if $dim P = k$. This means that some $(k+1)$ subfamily of $(x_1, \ldots, x_l)$ is affinely independent, but $(k+2)$ is not affinely independent. In the following we assume the same number of vertices for all variables.

Definition 1 Let Ω be polygons space (a 2-polytope is known as a polygon, i.e. a plane figure limited by a finite chain of line segment forming a closed surface) and let Z be a random variable such that $Z : \Omega \to R^2$. This random variable assumes values in polygon (P) with L vertices, then $Z = \rho = \{(a_1, b_1), \ldots, (a_L, b_L)\} \subset R^2$. It can also be rewritten by $Z = \rho = (\rho_1, \rho_2)$, where $\rho_1 = \{a_1, \ldots, a_L\}$ and $\rho_2 = \{b_1, \ldots, b_L\}$

Definition 2 Let n_z be the number of individuals in a class z. Each individual is described by a continuous variable X. A polygon P_z with l vertices for $l \leq n_z$ inscribed in circumference can be obtained by

$$P_{zl} = (a_{zl}, b_{zl}) = \left(c_z + r_z \cos\left(\frac{2\pi l}{L}\right), c_z + r_z \sin\left(\frac{2\pi l}{L}\right)\right) \tag{1}$$

where c_z is the centre of the polygon z (mean of X in class z) and $r_z = 2 \times sd\,(\chi_z)$ is the radius of the polygon (or circumference) z where $sd\,(\chi_z)$ is the standard deviation of X in class z , respectively. P_{zl} represents a vertex of the polygon P_z and $l = 1, 2, \ldots, L$ for $L \in N \geq 3$ is the number of vertices of this polygon.

Exact formulas for empirical statistics have been derived by Silva et al. (2019). The mean of the plane surface is given by centre of gravity on x and y , defined by coordinates for the centroid as

$$(C_x, C_y) = \left(\frac{\int_A x dA}{\int_A dA}, \frac{\int_A y dA}{\int_A dA}\right)$$

where dA expresses the infinitesimal area element, A is the area of the surface.

When the surface is a polygon the expression below can be rewritten and is given by Silva et al. (2019)

$$(C_x, C_y) = \frac{1}{6A}\left(\sum_{i=1}^{L}(a_i + a_{i+1})\, c_i, \sum_{i=1}^{L}(b_i + b_{i+1})\, c_i\right)$$

Empirical second moments can be expressed as

$$(I_x, I_y) = \left(\frac{1}{12}\sum_{i=1}^{L}\left(b_i^2 + b_i b_{i+1} + b_{i+1}^2\right) c_i, \frac{1}{12}\sum_{i=1}^{L}\left(a_i^2 + a_i a_{i+1} + a_{i+1}^2\right) c_i\right)$$

One of main advantages of using polygonal data is relaxation of the equidistribution hypothesis, inherent (and problematic) for interval data (for more see Silva et al. 2019). Polygonal equidistribution hypothesis (which is a generalization of interval-related equidistribution) can be expressed as

1. Each observation $u \in S$ is equiprobable, i.e. each observation is selected with probability $1/m$ where m is the cardinality of sample space (S);
2. We define Z_u in the polygon for each $u \in S$, and Z_u has uniform distribution in the polygon.

Empirical pdf in any polygon

$$fz(\rho) = \begin{cases} \frac{1}{m}\sum_{u \in S}\frac{1}{A_u} \; if\, \rho \in P \\ 0 \text{ otherwise} \end{cases}$$

where

$$A_u = \frac{1}{2}\sum_{u \in S} | \sum_{i=1}^{L} b_{u,i}\left(a_{u,i+1} - a_{u,i-1}\right) |.$$

Novel clustering algorithm for polygonal data

We propose a novel and, actually, the first clustering algorithm for data described above, polygonal data/variables. It is based on the combination of Hausdorff and City-Block distance, as defined below

Definition 3 We define the distance between two polygons p_1^z and p_2^z as a combination of Hausdorff and City-Block distance:

$$d_H\left(p_1^z, p_2^z\right) = \max_z\left(|a_1^z - a_2^z| + |b_1^z - b_2^z|\right) \tag{2}$$

where the maximization takes place over all correspondent vertices of the two polygons a_1^z and b_1^z, are coordinates of the corresponding vertex of polygon p_1^z , while a_2^z and b_2^z, are coordinates of the corresponding vertex of polygon p_2^z.

Cumulative distance (distance criterion) between set (matrix) of polygons is defined accordingly as

$$d_{TH}(p) = \sum_{z=1}^{p} \max_{z} \left(|a_1^z - a_2^z| + |b_1^z - b_2^z| \right) \tag{3}$$

where the summation runs over the full matrix of p polygons.

We also define the following:

Definition 4 The prototype $G = \left(g^1, \ldots, g^p\right)$ of a cluster C is a matrix of p polygons which minimizes the adequacy criterion

$$f_C(G) = \sum_{s \in C} d_{TH}(x_s, G) = \sum_{s \in C} \sum_{z=1}^{p} d_H\left(x_s^z, g^z\right) = \sum_{s \in C} \widetilde{f_C}\left(g^z\right) \tag{4}$$

Our main theoretical result is theorem below.

Theorem 1 *Derivation of prototype G is equivalent to solving two separate minimization problems like in Chavent et al. (2006)*

$$\min \sum_{s \in C} |\mu^z - c_s^z| \tag{5}$$

$$\min \sum_{s \in C} |\lambda^z - r_s^z| \tag{6}$$

The solutions $\hat{\mu}^z$ and $\hat{\lambda}^z$ are, respectively, the median of the set $\{c_z^s, s \in C\}$ of the polygon centres, and the median of the set $\{r_z^s, s \in C\}$ of their radiuses.

The problem of deriving the prototype in Definition 3 is equivalent to finding the polygon g^z for $(z = 1, \ldots, p)$ which minimizes

$$\widetilde{f_C}\left(g^z\right) = \sum_{z=1}^{p} d_H\left(x_s^z, g^z\right) = \sum_{z=1}^{p} \max_{z} \left(|g_a^z - a_l^z| + |g_b^z - b_l^z| \right) \tag{7}$$

If we insert the respective coordinates for prototype and each respective polygon vertex into (7), the equation for the Hausdorff/City-Block distance is transformed into

$$\begin{aligned}
d_H\left(x_s^z, g^z\right) &= \max_z \left(|\hat{\mu}^z - c_i^z + \left(\hat{\lambda}^z - r_i^z\right)\cos\frac{2\pi l}{L}| + |\hat{\mu}^z - c_i^z + \left(\hat{\lambda}^z - r_i^z\right)\sin\frac{2\pi l}{L}| \right) \\
&\leq \max_z \left(|\hat{\mu}^z - c_i^z| + |\left(\hat{\lambda}^z - r_i^z\right)\cos\frac{2\pi l}{L}| + |\hat{\mu}^z - c_i^z| + |\left(\hat{\lambda}^z - r_i^z\right)\sin\frac{2\pi l}{L}| \right) \\
&= \max_z \left(2|\hat{\mu}^z - c_i^z| + |\left(\hat{\lambda}^z - r_i^z\right)\cos\frac{2\pi l}{L}| + |\left(\hat{\lambda}^z - r_i^z\right)\sin\frac{2\pi l}{L}| \right) \\
&= \max_z \left(2|\hat{\mu}^z - c_i^z|\right) + \max_z \left(|\left(\hat{\lambda}^z - r_i^z\right)\cos\frac{2\pi l}{L}| + |\left(\hat{\lambda}^z - r_i^z\right)\sin\frac{2\pi l}{L}| \right) \\
&= \max_z \left(2|\hat{\mu}^z - c_i^z|\right) + \max_z \left(|\left(\hat{\lambda}^z - r_i^z\right)||\cos\frac{2\pi l}{L}| + |\left(\hat{\lambda}^z - r_i^z\right)||\sin\frac{2\pi l}{L}| \right) \\
&= \max_z \left(2|\hat{\mu}^z - c_i^z|\right) + \max_z \left(|\left(\hat{\lambda}^z - r_i^z\right)| \left(|\cos\frac{2\pi l}{L}| + |\sin\frac{2\pi l}{L}| \right) \right) \qquad (8)
\end{aligned}$$

where the maximization runs over all L vertices of the polygon z. The equality in the second line inequality relationship is satisfied iff $\hat{\mu}^z - c_i^z$ and $\left(\hat{\lambda}^z - r_i^z\right)\cos\frac{2\pi l}{L}$, respectively, $\hat{\mu}^z - c_i^z$ and $\left(\hat{\lambda}^z - r_i^z\right)\sin\frac{2\pi l}{L}$ are of the same sign (positive or negative).

As the maximization runs over the vertices, it is clear that expressions $\hat{\mu}^z - c_i^z$ and $\hat{\lambda}^z - r_i^z$ are fixed for each polygon. It follows from basic trigonometry that the same sign is achieved if $0 \leq \frac{2\pi l}{L} \leq \frac{\pi}{2}$ or $\pi \leq \frac{2\pi l}{L} \leq \frac{3\pi}{2}$. Furthermore, exact maximum of $\sqrt{2}$ of the expression $|\cos\frac{2\pi l}{L}| + |\sin\frac{2\pi l}{L}|$ is achieved for $\frac{2\pi l}{L} = \frac{\pi}{4} + 2k\pi, k \in Z$. It is important to repeat this maximum does not depend on expressions $\hat{\mu}^z - c_i^z$ and $\hat{\lambda}^z - r_i^z$ which are fixed for each polygon.

The expression in (7), optimizes the adequacy criterion from Definition 3 over a set of polygons in cluster C . Hence, the exact value of maxima in (8), which depends only on specific polygon (and is fixed for each polygon) is not important. But this means the optimization in (7), is dependent only upon (separate) minimization of expressions $\min \sum_{s \in C} |\mu^z - c_s^z|$ and $\min \sum_{s \in C} |\lambda^z - r_s^z|$. This is exactly equivalent to Chavent et al. (2006), with the same set of solutions $\hat{\mu}^z$ and $\hat{\lambda}^z$, respectively, the median of the set $\{c_z^s, s \in C\}$ of the polygon centres, and the median of the set $\{r_z^s, s \in C\}$ of their radiuses.

Our clustering algorithm for polygonal data can, therefore, be stated as follows (and is based on the dynamic algorithm of Chavent et al. 2006):

Initialization: Define a random partition $P = (C_1, \ldots, C_i, \ldots, C_k)$

Allocation:

test $\leftarrow 0$

a for $s = 1$ to n do:

Find the cluster C_m to which s belongs

If $card\ (C_m) \neq 1$ for $l = 1, \ldots, k$ and $l \neq m$

Perform the new prototypes G_m of $C_m \backslash \{s\}$ and G_l of $C_l \cup \{s\}$

Perform the criterion $\Delta_l = \sum_{i=1}^{k} \sum_{s' \in C_i} d_H\left(p_{s'}, G_i\right)$

Find the cluster C_{l^*} such that $l^* = arg\ \min_{l=1,\ldots,k} \Delta_l$

If $l^* \neq m$ move s to C_{l^*}

test $\leftarrow 1$

$C_{l^*} = C_{l^*} \cup \{s\}$ and $C_m = C_m \backslash \{s\}$

If test $= 0$ then stop, otherwise go to a.

Empirical application: classification of cultural entrepreneurial regimes

For the empirical application, we use data of The Global Entrepreneurship Monitor (GEM), which is a research project and annual assessment of the national level of entrepreneurial activity in multiple, diverse countries. Based in London, England, GEM is now the largest ongoing study of entrepreneurial dynamics in the world.

The data used for the GEM is collected from two large surveys, the Adult Population Survey (APS) and the National Expert Survey (NES). Each year, the GEM assembles the survey of a minimum of 2000 adults and at least 36 experts from a country of interest into an annual report. In the 2014 report, 206,000 adults from around the world anonymously participated along with 3,936 national experts. In the application, we used a set of fifteen binary variables—not shown here due to space limitations. For transformation into polygons, binary variables were transformed into their linear probabilities version (Figs. 1 and 2).

We present basic descriptives for all fifteen variables in the form of polygonal variables—we use squares (4-angles) and octagons (8-angles). All clustering procedures converged in a reasonable time. In the best clustering solution, we got four clusters

Cluster 1: Azores, Argentina, Bosnia and Herzegovina, Croatia, Germany, Greece, Hungary, Israel, Italy, Korea, Latvia, Macedonia, Portugal, Romania, Russia, Slovenia, Spain, Taiwan, Turkey, Uruguay

Cluster 2: Angola, Bolivia, Costa Rica, Ghana, Iran, Uganda, Vanuatu, Zambia

Cluster 3: Australia, Belgium, Finland, France, Iceland, Ireland, Japan, Malaysia, Netherlands, Norway, Sweden, Switzerland, United Kingdom, United States

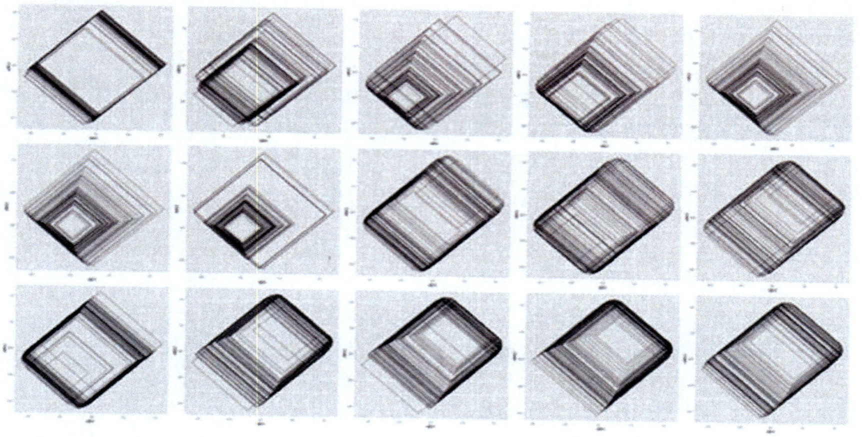

Fig. 1 Basic descriptive for 15 included variables, squares representation. *Source* Own calculations based on GEM data set for 2014

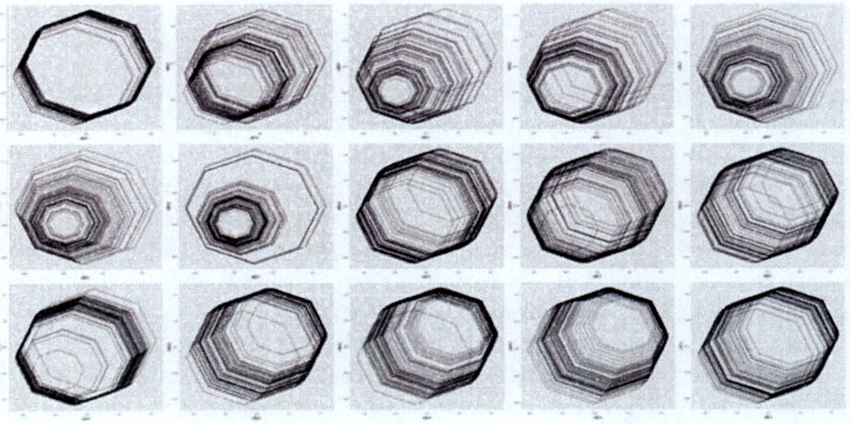

Fig. 2 Basic descriptive for 15 included variables, octogons representation. *Source* Own calculations based on GEM data set for 2014

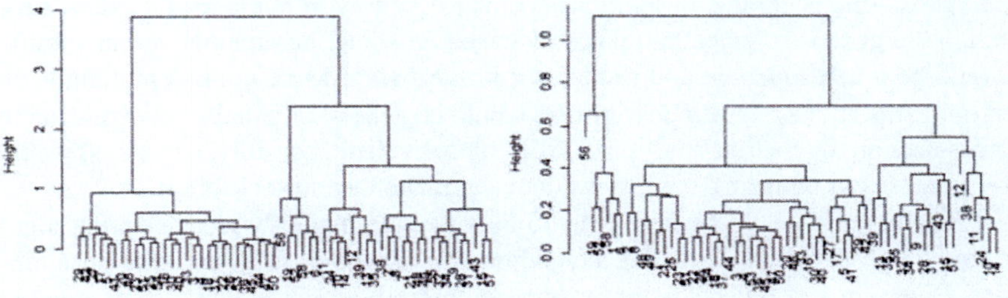

Fig. 3 Dendrograms of polygonal clustering, squares (left) and octagons (right)

Cluster 4: Brazil, Chile, China, Colombia, Ecuador, Egypt, Guatemala, Jamaica, Mexico, Montenegro, Pakistan, Peru, Saudi Arabia, Trinidad and Tobago, Tunisia, West Bank and Gaza.

It is interesting to search for the included Central and Eastern European countries, defined initially. They largely cluster in the first cluster, which includes Bosnia and Herzegovina, Croatia, Hungary, Latvia, Macedonia, Romania and Slovenia. This cluster includes also some countries from the developed Western world, like Germany, although they largely group in the second cluster. Only Montenegro clusters in the fourth cluster, including large countries of South America and the Arab World (Fig. 3).

3 Discussion and Conclusion

In our article, we provide several important novelties in the literature. Firstly, we derive novel and first clustering algorithm for polygonal data, which are representations of real-valued variables in (but could be extended to with arbitrary ones, where the representation is defined by the polygon centre and radius. This provides large methodological possibilities for work in future, as symbolic data analysis is possibly the most powerful approach to deal with big data sets, rivalled here with machine learning approaches. We also provide the only second article in the literature explicitly dealing with such polygonal type of variables.

Possibilities to extend the work in methodological terms are vast, as polygonal data analysis is only gaining ground. We could compare the results with the main competitive approach in symbolic data analysis, interval data clustering. We could extend the analysis to different combinations of variables (higher or lower in number—our algorithm did not show significant problems depending upon the number of included variables). Important and possibly key question is selection of polygons—the accuracy of solutions seems to grow with number of vertices chosen, but a criteria of selection, possibly based on some incremental variance ratio, would be much welcome and needed for further work. More consistent definitions of moments and regression possibilities would be necessary. Finally, an extension to cross-section time-series analysis of GEM data sets (and, possibly, Amadeus) would be great, in our empirical case to validate and make the clusters robust.

For future, it would be necessary to also include model-based clustering algorithms (e.g. Gaussian, Dirichlet, nonparametric and other finite mixture possibilities; Bayesian and Bayesian nonparametric possibilities). It would also be great to combine the method with extension to machine learning possibilities. Crucially, it would be interesting to relax the assumption of uniform distribution within the polygon, but the same problem occurs with interval data in general. Finally, asymptotic behaviour and simulation studies should be performed to explore the performance of the procedure proposed.

Regarding possibilities of work in cultural entrepreneurship, we would suggest more consideration over methodological novelties and more sophisticated methods. While the field now has a "history" and several key referential works, even regression possibilities have been largely unexplored in most aspects (e.g. non- and semi-parametric methods (for example quantile regression), Bayesian modelling—which seems natural for the field, machine learning regression methods). Our article tried to provide a combination of some more complex methodological work and novelties and empirical application. We hope it will stimulate more complex methodological applications in the field.

References

Billard, L., Diday, E.: Symbolic Data Analysis: Conceptual Statistics and Data Mining. Wiley, New York (2006)
Bock, H.-H.: Visualizing symbolic data by Kohonen maps. In: Diday, E., Noirhomme-Fraiture, M. (eds.) Symbolic Data Analysis and the Sodas Software, pp. 205–234. Wiley, New York (2008)
Bock, H.-H.: Clustering methods and Kohonen maps for symbolic data. Jpn. Soc. Comput. Stat. **15**(2), 217–229 (2002)
Brito, P.: Use of pyramids in symbolic data analysis. In: Diday, E., et al. (eds.) New Approaches in Classification and Data Analysis, pp. 378–386. Springer, Berlin(1994)
Brito, P.: Symbolic objects: order structure and pyramidal clustering. Ann. Oper. Res. **55**, 277–297 (1995)
Chavent, M.: A monothetic clustering method. Pattern Recognit. Lett. **19**(11), 989–996 (1998)
Chavent, M.De., Carvalho, F.A.T., Lechevallier, Y., Verde, R.: New clustering methods for interval data. Comput. Stat. **21**(2), 211–229 (2006)
De Carvalho, F.A.T., Brito, P., Bock H-H.: Dynamic clustering for interval data based on L2 distance. Comput. Stat. **21**(2), 231–250 (2006)
de Souza, R.M., De Carvalho, F.D.A.: Clustering of interval data based on city-block distances. Pattern Recognit. Lett. **25**(3), 353–365 (2004)
Diday, E., Noirhomme-Fraiture, M. (eds.): Symbolic Data Analysis and the SODAS Software. Wiley, Chichester (2008)
Dilli, S., Elert, N.: The Diversity of Entrepreneurial Regimes in Europe. IFN Working Paper No. 1118
Esping-Andersen, G.: The Three Worlds of Welfare Capitalism. Princeton University Press, Princeton (1990)
Hall, P.A.: Soskice, D.: Varieties of Capitalism: The Institutional Foundations of Comparative Advantage. Oxford University Press, Oxford (2001)
Joshi, D.: Polygonal Spatial Clustering. Dissertation, University of Nebraska-Lincoln (2011)
Noirhomme-Fraiture, M., Brito, P.: Far beyond the classical data models: symbolic data analysis. Stat. Anal. Data Min. **4**(2), 157–170 (2011)
North, D.C.: Institutions, Institutional Change and Economic Performance. Cambridge University Press, Cambridge (1990)
Silva, W.J.F., Souza, R.M.C.R., Cysneiros, F.J.A.: Polygonal data analysis: a new framework in symbolic data analysis. Knowl. Based Syst. **163**, 26–35 (2019)

Multidimensional Factor and Cluster Analysis Versus Embedding-Based Learning for Personalized Supermarket Offer Recommendations

George Stalidis, Theodosios Siomos, Pantelis I. Kaplanoglou, Alkiviadis Katsalis, Iphigenia Karaveli, Marina Delianidi, and Konstantinos Diamantaras

Abstract Multidimensional factor and cluster analysis and embedding-based machine learning were evaluated toward a knowledge-based recommendation system for supermarket e-marketing. The goal was to produce personalized notifications on special offers, optimized per individual customer's predicted response. To this purpose, we firstly applied Multiple Correspondence Analysis and Hierarchical Clustering to extract insights on the ordering behaviors and to identify customer classes associated with predictable preference patterns. Secondly, a neural network model based on embeddings was developed to predict the customers' ordering actions on a personalized level at large scale. Application of the factor and cluster analysis on the Instacart dataset resulted in the identification of typical and niche patterns with prediction value. The neural network model was successfully trained to predict with satisfactory accuracy individual customers' future orders, to be used as a basis for composing personalized recommendations.

G. Stalidis (✉) · T. Siomos · P. I. Kaplanoglou · A. Katsalis · I. Karaveli · M. Delianidi · K. Diamantaras
International Hellenic University, Thermi, Greece
e-mail: stalidgi@ihu.gr

T. Siomos
e-mail: theodosissio@gmail.com

P. I. Kaplanoglou
e-mail: pikaplanoglou@gmail.com

A. Katsalis
e-mail: alkiskatsalis@gmail.com

I. Karaveli
e-mail: i.karav@hotmail.com

M. Delianidi
e-mail: delmarin35@gmail.com

K. Diamantaras
e-mail: k.diamantaras@ihu.edu.gr

T. Chadjipadelis et al. (eds.), *Data Analysis and Rationality in a Complex World*, Studies in Classification, Data Analysis, and Knowledge Organization,
https://doi.org/10.1007/978-3-030-60104-1_30

Keywords Personalized recommendations · Factor and cluster analysis · Embeddings · Multiple correspondence analysis · Supermarket offers

1 Introduction

In addition to traditional advertising, retailers can nowadays communicate special offers with customers through a variety of channels, such as mobile apps, SMS, personalized web, and push notifications, which allow for better targeting and feedback collection. Companies specialized in mobile marketing report achievements using data-driven methods such as 800% increase in push notification engagement due to personalisation based on device, location, timing, and content (Fleit 2016). While traditional techniques such as Market Basket Analysis (Agrawal 1993) can easily fail to predict the customers' needs, machine learning techniques aim at more accurate models (Christodoulou et al. 2017) being able to learn from large amounts of data with minimum human intervention. However, they have in general limited capabilities to provide interpretable results and to incorporate prior knowledge and business rules. On the other hand, explorative statistical methods from the family of Multidimensional Data Analysis (Benzecri 1992) are promising for discovering patterns, identifying clusters of customers based on their purchasing behavior, and associating them with specific preferences (Greenacre 2007).

In previous work, (Chen and Luo 2006) built a personalized recommendation system for e-Supermarket commodities, using SVM classification, content-based recommendation based on Vector Space Model and an adaptive recommendation algorithm. Wang et al. (2018) suggested a personalized recommendation methodology applied to an Internet shopping mall, based on a variety of techniques such as web usage mining, decision tree induction, association rule mining, and product taxonomy. For the evaluation of the methodology, they produce a recommender system using intelligent agents and data warehousing technologies. Cho et al. (2002) have presented a methodology for personalized recommendations in e-commerce and developed a recommender using user preferences learned from clickstream web usage data. Wang et al. (2015) built a recommendation system to predict the next basket of a user. Using a Hierarchical Representation Model, they captured both sequential behavior and user's general taste. Furthermore, contextualisation techniques have been reported, where recommendations are adjusted to circumstances/occasions such as holidays or marriage, aiming at optimisation for event marketing (Zheng 2016).

The problem addressed in this paper was to predict the next buys of individual customers, thus identifying the product categories for which promotional actions would be relevant. The intended future use of this research was to enable an intelligent recommendation system for supermarket chains to produce well-matched push notifications for special offers. Multidimensional factor analysis and hierarchical clustering methods were applied on supermarket purchase records, to explore the customers' ordering profiles and to identify customer classes with predictable buying behavior. This classification was used to predict future buys and thus the relevance of product categories to individual customers. As an alternative method with the potential to overcome the limitation to small datasets and the need for interpretation by a human analyst, a neural network model based on product embeddings was developed to automatically learn individual customers' future buys. Both approaches were evaluated in terms of prediction performance, as potential components of a recommendation system at large scale.

2 Methods

2.1 Dataset and Data Preparation

The data used for developing and evaluating the methods was the Instacart Online Grocery Shopping Dataset 2017 (The Instacart Online Grocery Shopping Dataset 2017), an open-access dataset offered as a basis for a Kaggle machine learning competition (Kaggle 2019). The dataset includes 3.2 million online grocery orders in sets of 4–100 orders across 200.000 customers, where, apart from their order history, customers were unknown. The available input was the products included in each order, categorized per department and aisle, the order's day of week and time of day, whether each product is reordered, as well as the days passed since the previous order. Our goal was to predict the aisles (i.e., product categories) from which a customer will buy, rather than specific products, since the business problem from the perspective of the shop was not focused on brands but on identifying the product categories for which a personalized offer would be relevant. As evaluation criterion, in addition to the F1 measure, we used precision, which was considered more representative, since the goal was to correctly identify relevant offers rather than guess the entire customer's order.

Prior to the main analysis stage, the order history of each customer was aggregated in one vector including (a) the buying frequency for each aisle, expressed as the ratio of the customer's orders in which products of the aisle appear, transformed to categorical (i.e., 1. Not buyer, 2. Periodic buyer, 3. Regular buyer) and (b) for each day of week and hour zone, a binary variable was used, indicating whether there is significant association between the customer's ordering day/hour and the particular day/time zone. The customer's profile was a vector with size 147 (134 aisles + 7 days + 6 hour zones). The next pre-processing step was to study the distribution of buying frequencies per aisle over all customers, in order to segment aisles according to their profile. It was found that aisles can be segmented to three groups: (a) Frequent: they tend to be included in almost every order of a large buyer group (6 aisles of common everyday products, e.g., milk, fresh fruit) (b) Occasional: they tend to be ordered once every 3–4 orders (43 aisles, e.g., soft drinks), (c) Special: they are either ordered periodically by certain customers or not at all by others (85 categories of uncommon products, e.g., diapers).

2.2 Multiple Correspondence Analysis and Hierarchical Clustering

The dataset has been split into three subsets, one for each aisle group. The customer profile in each dataset contained the buying frequencies for the aisles of the group and the variables for Day and Hour. For each subset, MCA was applied on the generalized contingency table (Burt), in order to discover ordering trends, i.e., associations among aisles and day/time of purchase. Hierarchical clustering of customers using Benzecri's chi square distance and Ward's linkage criterion was then applied on the MCA results, i.e., on the coordinates of the four most important factors. The clusters were projected on the factorial planes and were associated with aisles, days, and hour zones, revealing customer classes with specific ordering behavior. The analysis was performed using the FactoMineR R package (Le et al. 2008). The association between clusters and properties was performed through visual interpretation, aided by the p-value of each association (provided as "description of cluster by categories" by the HCPC function). The analysis was repeated for each aisle group (Frequent, Occasional, Special), resulting in three customer classifications, which express the buying behavior of each customer in relation to each aisle group. The aim of this exploratory analysis was to enhance the precision of personalized recommendations by using the class membership variables to pre-select the target customers who best fit a specific product group, e.g., buyers of Special products.

The next step was to evaluate the ability of the Factor and Clustering approach to predict the relevance of each aisle to individual customers. On the factorial axes produced by MCA as above, we projected each individual and used the projections of customer profiles to associate them with aisles. As a criterion for such associations, we used the p-value of the dependence between individual and category (which in this case was calculated on the level of individual instead of cluster). Aisles for which the significance exceeded a threshold were labeled as relevant to the customer. The analysis was performed using the MAD data analysis software (Karapistolis 2002).

2.3 Personalized Prediction With Embedding Approach

A component for personalized prediction of a customer's next order(s) was also developed based on word2vec (Mikolov et al. 2013). Word2vec is a two-layer shallow neural network, originally used to assign embeddings to text words. The main advantage of this method is that the vectors can be directly compared to each other so that vectors in small distance have related content. Similar approaches have been used in numerous applications, including product recommendations to customers (Barkan and Koenigstein 2016).

In this approach, which we called item2vec, vectors have been assigned to customers, aisles, and weekdays. Day vectors were introduced since time was found to play an important role in the customers' purchase habits. During the preprocessing phase, triplets were created for each order, consisting of the customer id, the aisle from which the product was purchased and the day of week when the order was placed. The target output of such "true" triplets was assigned to 1, while a number of random "false" triplets were added with target value 0. The model was a shallow neural network with a single layer. The input pattern was the concatenation of the customer, aisle, and day vectors, while the output was the probability of this coexistence. In order to predict the next order, the model estimated the probability of each triplet, expressing how likely is that a customer will buy from an aisle in his next order on that day. The K higher probabilities were the predicted aisles for the customer.

3 Results

MCA and hierarchical clustering were applied on the three subsets for each aisle group. On the factorial plane 1X2 for Frequent aisles (Fig. 1a), we found a Guttman curve, indicating a scaling from intense buying of frequent products to no buying. The most common frequent buys were milk and fresh fruit, which tended to be on Sundays, while more uncommon behavior was to frequently buy yogurt and packaged vegetables. Factor 3 (not visible in this Figure) differentiated ordering in the afternoon (13.00–17.00) during weekdays from ordering at night (21.00–05.00) on Saturdays and Sundays. The clustering algorithm resulted in 4 clusters, which were associated with certain ordering behavior according to their positioning on the factorial planes, e.g., C1.1 (visible in black in Fig. 1b) represents customers who order almost every time milk and fresh fruit, usually on Sundays. The choice of the number of clusters was based on the bar plot of within-group inertia, which is provided by the FactoMineR HCPC function together with the hierarchical tree (Husson et al. 2017). In this case, 4 clusters were selected since the transition to 3 ones largely increased the within-group inertia. A similar analysis was applied on Occasional and Special aisles, resulting in additional clusters, summarized in Table 1. The 3 resulting cluster membership variables express the general behavior of each customer regarding each of the 3 aisle groups.

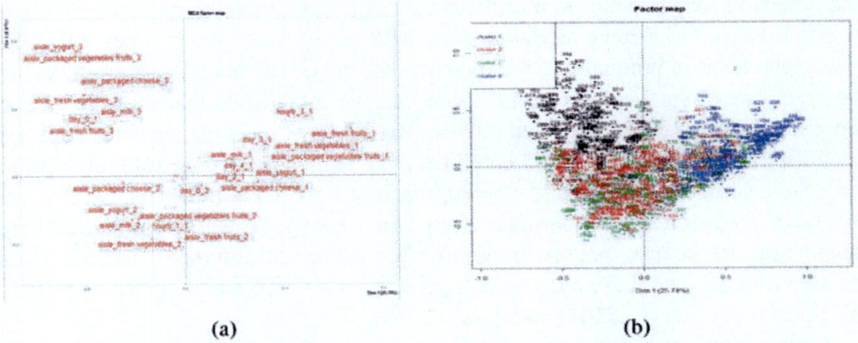

(a) (b)

Fig. 1 Factor and Cluster analysis of frequent products. **a** The factorial plane F1XF2, **b** Clusters of individuals on F1XF2

Table 1 Customer classes per type of product category

Frequent	Occasional	Special
C1.1.(21.5%) Regular buyer of milk and fresh fruit, on Sundays	C2.1.(28.2%) Buyer of various Occasional aisles	C3.1.(15.5%) Buyer of pickled goods olives, tofu meat, Asian foods, etc.
C1.2.(35.3%) Periodic buyer of frequent aisles, mainly packaged cheese and fresh fruit, in early afternoon	C2.2. (7.8%) Buyer of products for cooking (e.g., herbs, oils vinegars, spices seasonings)	C3.2.(7.8%) Buyer of meat-seafood, marinades meat preparation etc.
C1.3.(21.1%) Periodic buyer of frequent aisles, on Saturday and Sunday at night	C2.3.(13.4%) Buyer of snacks (spreads, chips pretzels, ice cream, etc.)	C3.3.(1.7%) Buyer of soap, vitamins supplements, plates bowls cups flatware, body lotions soap
C1.4.(22.1%) Not buyer of Frequent aisles	C2.4.(50.6%) Not buyer of Occasional aisles	C3.4.(74.8%) Not buyer of Special aisles

Table 2 Precision of factor and clustering methods

Aisle group	Next order	Next 3 orders	Only buyers
Overall	0.3948	0.7539	0.7651
Frequent	0.4525	0.7393	0.8162
Occasional	0.3513	0.7164	0.7567
Special	0.2059	0.5161	0.6915

In order to evaluate the Factor and Clustering approach, we applied it on a sample of 1000 customers, extracted from Instacart. The customer profiles and the exploratory analysis were based on their buying history, not including the last orders. Testing was performed by comparing the aisles predicted for each customer with the ones included in his last order. The algorithm achieved on average Precision = 0.3948, Recall = 0.6474, and F1 = 0.4585. As shown in the first column of Table 2, the precision for Frequent products was 0.4525 (higher than the overall 0.3948), it was 0.3513 for Occasional products and 0.2059 for Special ones. It is noted that the lower precision in predicting Special aisles was expected since these products are ordered rarely by a few customers. Moreover, we established that Occasional and Special products are not expected to be ordered every time, so even if they are not included in the immediately next order, they can still be labeled as truly relevant. In order to evaluate the potential of the approach in a more realistic way, we thus re-evaluated precision by considering as true positive the appearance of a predicted item in the union of the next three customer's orders. In the 2nd column of Table 2, it can be seen that the resulting precision is from this perspective much higher, especially for Special products (0.5161 instead of 0.2059).

Furthermore, we experimented with the potential of focusing the algorithm's recommendations on specific product categories by pre-selecting the most suitable customer classes. The results of the classification phase (see Table 1) were used to pre-

Table 3 Performance of item2vec

K	Precision (next order)	Recall	F1	Prec. (3 orders)	Prec. (3 orders-full dataset)
3	0.5516	0.2835	0.3316	0.7828	0.7937
5	0.4750	0.3849	0.3769	0.7160	0.7158
8	0.3950	0.4872	0.3900	0.5473	0.6232
15	0.2910	0.6294	0.3630	0.4327	0.4805

select our target customers to best fit a selected aisle group. A representative case was to focus on Special products, which were the most difficult to predict. Before applying prediction, we selected the customers classified in C3.1, C3.2, and C3.3, i.e., excluded the non-buyers of Special products. The precision in this group reached 0.6915 for Special products, which was a considerable improvement compared with the 0.5161 found in the general sample. Considering that the Special product categories were the largest portion of the offered products (85 of 134 aisles) and were the ones with the lowest a-priori probability of being found in a customer's order (mean buying frequency = 0.0038), the achieved personalized prediction accuracy of almost 70% was particularly encouraging.

The item2vec was tested both on the full dataset and, for comparison purposes, on the subset of 1000 users used by the Factor and Clustering method. Precision, recall, and F1 were measured at various fixed numbers of predicted items K, for both alternative definitions of true positive (i.e., predicted aisle found in next order or in the next 3 orders). The results are shown in Table 3. In predicting the next order, the algorithm achieved its best F1 = 0.3900, at K = 8, which was worse than the 0.4585 achieved by the factor and cluster analysis. However, the best precision was 0.5516 at K = 3, which was considerably larger than the 0.3948 achieved by the statistical methods. In the case of targeting the next 3 orders, item2vec marginally outperformed in precision the factor and clustering method, achieving 0.7828, compared to 0.7539. It is noted that the item2vec achieved its max precision when restricting the output to a small number of best predictions (K = 3) at the expense of recall, especially for customers who tend to place large orders. On the contrary, the factor and clustering method gave the best results when it was up to the algorithm to decide the number of predicted items per customer. It was also interesting that the item2vec achieved just as good or even better performance when trained and tested on the entire dataset of 200K users. Finally, it is mentioned that direct competitors dealing with the same problem were not found in the literature. The winner of the Instacart Kaggle competition achieved F1 = 0,409 in predicting the products in each customer's next order, which is, however, not directly comparable with our results, since our focus was on the precision of matching customers with relevant aisles.

4 Conclusion

The factor and cluster analysis was successful in acquiring insights in the ordering behavior of retail customers, classifying them according to their preferences and predicting the relevance of product categories to each customer. Furthermore, it was possible to segment products into groups based on their expected purchase frequency and use this information to considerably boost prediction performance in focused business scenaria, such as to promote special products which are periodically bought only by a small group of customers. The limitation of the approach was that it is not scalable and required a human interpretation step. On the other hand, the item2vec was successful in learning customers' behavior at large scale and produced encouraging results, which in the general prediction scenario outperformed the factor and clustering method. Our future steps include the consolidation of the two approaches: knowledge extraction from sampled data using factor and cluster analysis to guide a full-scale machine learning approach for personalized prediction.

Acknowledgements This research has been co-financed by the European Regional Development Fund of the European Union and Greek national funds through the Operational Program Competitiveness, Entrepreneurship and Innovation, under the call RESEARCH–CREATE–INNOVATE (project code:T1EDK-01776)

References

Agrawal, R., Imielinski, T., Swami, A.: Mining association rules between sets of items in large databases. In: Proceedings of the 1993 ACM SIGMOD International Conference on Management of Data, Washington DC, vol. 22(2) of SIGMOD Records, pp. 207–216 (1993)

Barkan, O., Koenigstein, N.: Item2vec: neural item embedding for collaborative filtering. In: IEEE 26th International Workshop on Machine Learning for Signal Processing (MLSP). IEEE (2016)

Benzecri, J.-P.: Correspondence Analysis Handbook. Marcel Dekker, New York (1992)

Chen, J., Luo, Q.: Research on adaptive recommendation algorithm in personalized E-supermarket service system. In: 8th International Conference on Signal Processing, vol. 3. IEEE (2006)

Cho, Y.H., Kim, J.K., Kim, S.H.: A personalized recommender system based on web usage mining and decision tree induction. Exp. Syst. Appl. **23**(3), 329–342 (2002)

Christodoulou, P., Christodoulou, K., Andreou, A.: A real-time targeted recommender system for supermarkets. In: 19th International Conference on Enterprise Information Systems (2017)

Fleit, B.: Breaking barriers to push notification engagement. Report by Leanplum (2016). Available at http://get.leanplum.com/push-notification-engagement/

Greenacre, M.: Correspondence Analysis in Practice. Chapman and Hall (2007)

Husson, F., Le S., Pages J.: Exploratory Multivariate Data Analysis by Example Using R, Chapman and Hall CRC Computer Science and Data Analysis (2017)

Kaggle: Instacart Market Basket Analysis (2019). Available at https://www.kaggle.com/c/instacart-market-basket-analysis

Karapistolis, D.: The software MAD (in Greek). Data Anal. Bull. **2**, 133–147 (2002)

Le, S., Josse, J., Husson, F.: FactoMineR: An R Package for Multivariate Analysis. J Stat Soft. **25**(1), 1–18 (2008)

Lee, D., Gopal, A.: When push comes to shop: on identifying the effects of push notifications on mobile retail sales. In: ICIS Proceedings, Association for Information Systems (2016)

Mikolov, T., Sutskever, I., Chen, K., Corrado, G.S., Dean, J.: Distributed representations of words and phrases and their compositionality. Adv. Neural Inf. Process Syst. **26**, 3111–3119 (2013)

The Instacart Online Grocery Shopping Dataset (2017). Available at https://www.instacart.com/datasets/grocery-shopping-2017

Wang, P., Guo, J., Lan, Y., Xu, J., Wan, S., Cheng, X.: Learning hierarchical representation model for nextbasket recommendation. In: Proceedings of the 38th International ACM SIGIR conference on Research and Development in Information Retrieval, pp. 403–412 (2015)

Wang, F., Wen, Y., Guo, T., Chen, J., Cao, B.: Personalized commodity recommendations of retail business using user feature based collaborative filtering. In 2018 IEEE International Conference on Parallel and Distributed Processing with Applications, Ubiquitous Computing and Communications, Big Data and Cloud Computing, Social Computing and Networking, Sustainable Computing and Communications (ISPA/IUCC/BDCloud/SocialCom/SustainCom), pp. 273–278 (2018)

Zheng, Y.: Context in recommender systems. In: The 31st ACM Symposium on Applied Computing. Pisa, Italy (2016). Available at https://www.slideshare.net/irecsys/tutorial-context-in-recommender-systems

Motivation for Participating in the Sharing Economy: The Case of Hungary

Roland Szilágyi and Levente Lengyel

Abstract Our research focuses on the sharing economy (SE), which has gained more and more ground in recent years and is receiving increased media coverage nowadays. The use of the sharing economy spread rapidly from around 2005 significantly. Different authors define the system differently, and they analyse the participants' motivation from different aspects. The main purpose of SE is to improve the utilisation of unused assets. Most people think that only economic factors have impact on participation, but social and environmental motivations can also be important for users. Besides, there are numerous other factors that can influence the participants of SE. The aim of our study is to analyse why people take part in sharing economy activities in Hungary. We use the Structural Equation Modelling technique to determine which are the most important motivation factors. The study employs survey data from Hungarian sharing economy users.

Keywords Structural equation modelling · Sharing economy · Motivation analysis · Collaborative consumption

1 Introduction

Nowadays it has become obvious that the leading economic powers are not able to provide remedies for global problems in every area (not even if it is in their interest), for example, regarding environmental pollution, the depletion of natural resources (Varga and Fodor 2019), starvation and overproduction.

Today people are encouraged to think globally, as can be experienced in every area of life more and more intensely. Perhaps this process is most obvious with

R. Szilágyi · L. Lengyel (✉)
Institute of Economic Theory and Methodology, University of Miskolc, Miskolc-Egyetemváros 3515, Hungary
e-mail: lengyel.levente@uni-miskolc.hu

R. Szilágyi
e-mail: roland.szilagyi@uni-miskolc.hu

T. Chadjipadelis et al. (eds.), *Data Analysis and Rationality in a Complex World*, Studies in Classification, Data Analysis, and Knowledge Organization,
https://doi.org/10.1007/978-3-030-60104-1_31

regard to thinking about IT development, financial culture and environmental problems. The development of each area can be perceived well from their appearance in the educational programmes of schools. In addition to educating of our children to develop a new way of thinking in connection with these phenomena, new activities have appeared that combine both the humanist and neoliberal ideas of globalism in the framework of the sharing economy. The development and research of the sharing economy can be an important area in today's world, enabling even everyday people to contribute to solutions of global problems by living their lives more rationally. According to the rules of scientific research, it is of key importance to define the research area. Apart from our endeavour to provide an exact definition of what the sharing economy really means, our research focuses on acquiring knowledge about how it works. Some assume that it is the opportunity of being able to contribute to solving environmental problems that encourages the actors of the sharing economy to participate in it. According to materialistic views, it is the realisation of material benefits. In addition to all these, the sharing economy may even be entertaining and it definitely widens the range of social relationships. Our research question is what really motivates the participants of the sharing economy in Hungary? To answer this, in the second section, we briefly review the properties of the sharing economy (SE) and similar research on the topic. In the third section, we attempt to show on the basis of empirical research carried out in Hungary that, based on our hypotheses derived from the literature, the most motivating factors for behavioural intention and attitudes are sustainability, economic benefit, reputation and enjoyment. In the fourth section, we briefly present our results and compare them with the conclusions of other Hungarian research, as well as Finnish, German and Hong Kong studies. Based on all this, we conclude that common features can also be observed in the results of different national studies; however, most studies reveal different motivations.

2 Sharing Economy

People have a lot of unused resources. It is sufficient to think of our own household to realise how many devices we possess that we have not used for the past half year or even longer. We can think of small objects like books, CDs, or DVDs, or bigger and more valuable household appliances and small machines.

The accumulation of countless unused resources raises the social issue of overconsumption and irresponsible management of scarce resources. At the same time, the spread of different mobile technologies, Web 2.0 and broadband internet have given rise to innovations that make our lives easier in several areas. All these, as kinds of social innovations, as Varga (2017) pointed out, may also result in the improvement of the quality of life. One such solution is the sharing economy as an innovative social and economic phenomenon.

The sharing economy has been described by several authors (Botsman and Rogers 2011; Owyang 2013; Belk 2014; Hawlitschek et al. 2016). Most of them agree that the purpose of the sharing economy is to intensify the utilisation of unused resources.

In spite of the fact that the concept is becoming more and more widely known, it often occurs that people can only identify the activity when a specific example is given, e.g. Airbnb, Uber, Kickstarter or Taskrabbit. As regards definitions, numerous variations have emerged. Some authors emphasise efficiency, while the definitions created by others are dominated by SE's peer-to-peer character. Based on several sources reviewed on the subject, the following definition was developed in an earlier study of ours:

> The sharing economy is an economic redistribution system that includes every innovative business model, platform and technology that aims to rent, exchange, let or gift underused tangible and intangible resources with extensive access and high efficiency (Lengyel 2017 p. 67).

Certainly, the creation of the sharing economy was not only induced by technological development but by economic factors as well, such as the desire to simplify transactions and increase financial flexibility. On the other hand, according to Finley (2013), the financial and economic crisis of 2008 also contributed to the spread of the sharing economy. Furthermore, with the services used via sharing economy platforms being characteristically cheaper, certain resources have been made available more widely. Besides materialistic thinking, different social and environmental awareness factors have also promoted the spread of the sharing economy. The desire for community should be highlighted from among the numerous social factors, as a kind of driving force (Owyang 2013). Another important social factor is the transparency and reliability of the system, in connection with which Botsman underlined in her TED Speach in 2010 that users' reputations can be continuously measured during their internet activities, which, according to her vision, will be one of the most dominant capital.

Sharing economy platforms can be grouped in several ways. Based on the area of activity, many groups can be formed which can be divided into further subgroups, of which the most popular categories are the following: (1) housing, (2) transportation, (3) crowdfunding, (4) sharing of workforce and/or expertise or (5) consumer products. Evidently, there are also other activities besides these, but in our view these are the most popular categories. Another basis for grouping can be the interest in profit, but the researchers of the subject have diverging views on this.

The growth of the system and the development of digital market spaces also captured the attention of the European Union, and it released the European agenda for the collaborative economy in 2016. Hence, our opinion is that it is important to investigate why people take part in the sharing economy. Reviewing the relevant studies, it appears that mainly organisational level and qualitative studies have been made on the subject. Apart from a few studies, individual level, quantitative type studies have not received significant scientific attention. So the number of studies investigating the subject from this perspective is rather limited.

Möhlmann (2015) created a research framework on the basis of a data collection in 2014 to examine the satisfaction of car2go and Airbnb users and the likelihood of the reuse of services. She set up two separate structural models for the two services with the same determinants. However, it was found that different determinants exert

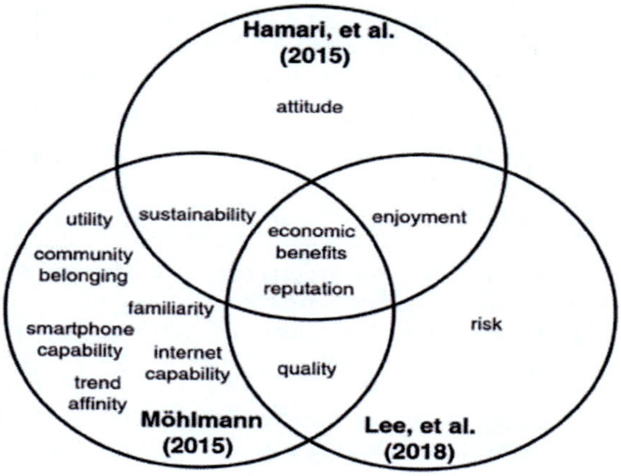

Fig. 1 Comparison of factors appearing in sources. *Source* own editing

an influence on satisfaction in the two areas. The main influencing factors in both services were familiarity, cost saving, trust and utility. In the case of car2go, quality and desire for community also appeared. The reason for such an effect could be that it is more observable in the case of motorists to have a kind of desire to belong together, even if the user takes possession of the given car only temporarily. You can also think here of the well-known motorists' fan clubs.

Lee et al. (2018) examined the motivation of Uber users. In their model, they measured each construct by three items, then they created new second-order constructs from some latent variables, thus developing a more complex structural model system. They found that the expected profit and trust in the platforms are the main influencing factors in the Uber users' intention to participate. Furthermore, the quality ensured by the platforms significantly determines the trust for the platforms, thus it indirectly influences the intention to participate.

Hamari et al. (2015) developed a much simpler structure for modelling consumer behaviour. In their model, they investigated the impact of four constructs, indirectly through the attitude and directly on the consumer's behavioural intention: (1) sustainability, (2) enjoyment, (3) reputation, (4) economic benefits. They collected their data in 2013 in Finland, and their results led them to the conclusion that attitude is directly and significantly impacted by sustainability and enjoyment. Furthermore, consumers' behavioural intention is affected by enjoyment to the greatest extent. In addition to this, they also received significant coefficients in respect of economic benefits and attitude.

Figure 1 illustrates the common constructs examined by the researchers in the aforementioned studies. In the case of Möhlmann (2015) and Hamari et al. (2015), what is common is that both studies take into account the issues of sustainability. As for Lee et al. (2018) and Hamari et al. (2015), enjoyment is the common construct. It can be observed that economic benefits, trust and reputation appear as a common intersection of all three studies. So these two constructs should be called the success factors with regard to the operation of platforms.

In the course of our study, we adapted the model of Hamari et al. (2015) and our hypotheses are also similar to those found in the cited literature.

H1a: Perceived sustainability of SE has a significant positive effect on attitude about SE.

H1b: Perceived sustainability of SE has a significant positive effect on behavioural intention to participate in SE.

H2a: Perceived enjoyment from participating in SE has a significant positive effect on attitude about SE.

H2b: Perceived enjoyment from participating in SE has a significant positive effect on behavioural intention to participate in SE.

H3a: Perceived reputation increase from participating in SE has a significant positive effect on attitude about SE.

H3b: Perceived reputation increase from participating in SE has a significant positive effect on behavioural intention to participate in SE.

H4a: Perceived economic benefits from participating in SE has a significant positive effect on attitude about SE.

H4b: Perceived economic benefits from participating in SE has a significant positive effect on behavioural intention to participate in SE.

H5: Attitude about SE has a significant positive effect on behavioural intention to participate in SE.

3 Data and Method

To investigate the subject matter, we used the set of questions created by Hamari et al. (2015). To ensure comparability we did not change the questions or the structural coherence. The statements, similarly to the study mentioned, were evaluated by the users on a 7-point Likert scale. We collected the data via electronic questionnaires in the summer of 2019 and a total of 155 responses were received. It can be observed that women are slightly overrepresented among the respondents but it can be declared that it does not cause a problem because a previous study (Lengyel 2017) showed that women outweigh men among the active users of the sharing economy.

As the method of analysis, we chose structural equation modelling (SEM). SEM can be described as a combination of regression analysis and factor analysis, the aim of which is to reveal the relationship of interdependence (Lengyel and Szilágyi 2019). Calculations were performed using SPSS Amos 24 using the maximum likelihood estimation. To verify convergent validity, we examined three indicators: average variance extracted, composite reliability and Cronbach's alpha. As can be seen from Table 1, all values met the expectations for validity. A further requirement set for the method is to have a sufficiently large sample available. Moreover, the criterion mentioned by Hamari et al. (2015) for SEM sample size requirements is satisfied.

Table 1 Measures to check convergent validity

	Average variance extracted	Composite reliability	Cronbach's alpha	N of items
Attitude	0.824	0.957	0.946	5
Behavioural intention	0.830	0.954	0.966	4
Economic benefits	0.728	0.899	0.855	4
Enjoyment	0.693	0.923	0.931	5
Reputation	0.650	0.883	0.882	4
Sustainability	0.802	0.955	0.962	5

4 Results and Conclusions

The model received as a result of the calculations can be found in Fig. 2. It can be observed that three constructs have a significant impact on the user's attitude: (1) sustainability, (2) enjoyment and (3) economic benefits. Thus, hypotheses H1a, H2a, H4a can be accepted, while hypothesis H3a can be rejected. In contrast, the consumer's behavioural intention is impacted directly only by enjoyment and attitude, and indirectly by the three other constructs. Thus, hypotheses H1b, H3b and H4b can be rejected, and hypotheses H2b and H5 can be accepted. A surprising finding was that reputation did not prove to be a significant factor, neither with respect to attitude nor to behavioural intention. This is contrary to our expectations since we assumed that the increase of the construct positively impacts the consumer's behavioural intention.

Because the previously mentioned authors and we used different methods, it is not our goal to directly compare the estimated parameters, but only to compare the similarities and differences between the observed structures. Our research reveals that the motivations of the participants of the sharing economy are not uniform. The

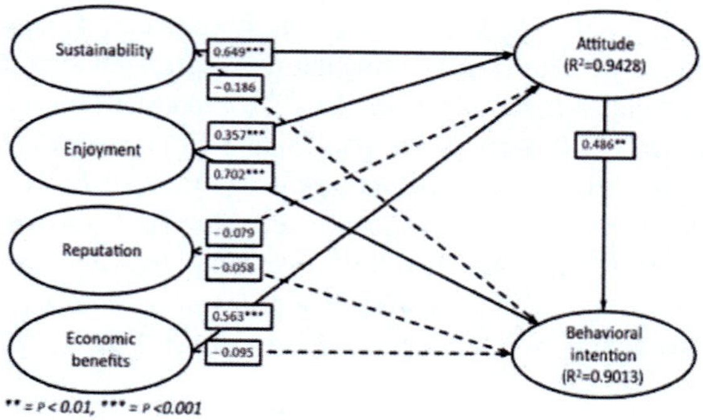

Fig. 2 Resulting model and estimated coefficients

Table 2 Comparison of effects with the Finnish experience

	Direct effect				Total effect	
	Attitude		Behavioural intention		Behavioural intention	
	Hamari et al. (2015)	Actual research	Hamari et al. (2015)	Actual research	Hamari et al. (2015)	Actual research
Attitude	n/a	n/a	Positive***	Positive***	Positive***	Positive **
Economic benefits	Negative	Positive***	Positive*	Negative	Positive*	Positive*
Enjoyment	Positive***	Positive***	Positive***	Positive***	Positive***	Positive***
Reputation	Negative	Negative	Positive	Negative	Positive	Negative*
Sustainability	Positive***	Positive***	Negative	Negative	Positive*	Positive*

Note * = $p < 0.1$, ** = $p < 0.05$, *** = $p < 0.01$

main motivations are similar but with regard to economic benefits and reputation, the direct effects are different. As can be seen from Table 2, a strongly significant direct positive effect was found in both studies for enjoyment. However, differences in effect were found between the two studies. These differences lead us to believe that there may be a national difference that also influences results. Our conclusion is that the differences should not only be searched for in the type of the shared product or service used, as it may also be possible to discover national differences among the constructs influencing behavioural intention. To examine this question, we also looked at the results of studies among German and Hong Kong SE users.

In comparison with the Finnish and Hong Kong experience, reputation exerts a negative impact on the behavioural intention of Hungarian users, contrary to expectations. As regards Hungarians, economic benefits are by far more important, which probably reflects on the materialistic mindset of the nation. Comparisons with Hong Kong results are more problematic due to the different structure of the survey, however, the positive effects of enjoyment and economic benefits on behavioural intention are the same. Compared to the German results (Möhlmann 2015), we obtained opposite results for similar factors (sustainability, economic benefits, reputation).

In order to survey partial effects more precisely, we are planning further research because the current sample does not represent the total population of the participants. Since the purpose of this study was primarily to repeat the previous study with Hungarian participants, the model was not modified in the interests of comparability. However, we suggest further development of the examined model to include further constructs. Variables must be incorporated that can capture the cultural differences between various nations. On the other hand, as Möhlmann's study (Möhlmann 2015) also reveals, differences can be detected between the platforms as well, i.e. the service areas, so with the inclusion of these factors as control variables more exact estimations could be produced. Summing up the reviewed literature and the experiences of our own research, it can be stated that the designers and developers of sharing economy solutions should keep in mind two other focuses in addition to emphasising the economic factors. One of them is the enjoyment of using the services, making them easier and more playful, which can be experienced in various areas of life, in parallel with the spread of gamification. We would mention reputation as a second focus point, since in spite of the fact that in our present study its effect did not prove to be significant, it can be called the cornerstone of the system, because if basic trust is missing, the users will leave the system.

Acknowledgements The described article was carried out as part of the EFOP-3.6.1-16-2016-00011 "Younger and Renewing University—Innovative Knowledge City—institutional development of the University of Miskolc aiming at intelligent specialisation" project implemented in the framework of the Szechenyi 2020 program. The realisation of this project is supported by the European Union, co-financed by the European Social Fund.

References

Belk, R.: You are what you can access: sharing and collaborative consumption online. J. Bus. Res. **67**(8), 1595–1600 (2014)

Botsman, R., Rogers, R.: What's Mine Is Yours: How Collaborative Consumption is Changing the Way We Live. Harper Collins Business, USA (2011)

Botsman, R.: The Case for Collaborative Consumption. TEDxSydney, Sydney (2010)

Finley, K.: Trust in the Sharing Economy: An Exploratory Study. The University of Warwick, Coventry (2013)

Hamari, J., Sjöklint, M., Ukkonen, A.: The sharing economy: why people participate in collaborative consumption. J. Assoc. Inf. Sci. Tech. **69**(9), 2047–2059 (2015)

Hawlitschek, F., Teubner, T., Gimpel, H.: Understanding the sharing economy—drivers and impediments for participation in peer-to-peer rental. In 2016 49th Hawaii International Conference on System Sciences (HICSS), pp. 4782–4791. IEEE (2016)

Lee, Z. W. Y., Chan, T. K., Balaji, M. S., Chong, A. Y.-L.: Why people participate in the sharing economy: an empirical investigation of Uber. Internet Res. **28**(3) (2018)

Lengyel, L., Szilágyi, R.: Strukturális egyenlet modellezés bemutatása és alkalmazásának lehetőségei. Debreceni Akadémiai Bizottság Műszaki Szakbizottság, Debrecen (2019)

Lengyel, L.: Új üzleti modell?—A közösségi gazdaság kihívásai Magyarországon. E-CONOM (2017) https://doi.org/10.17836/EC.2017.1.066

Möhlmann, M.: Collaborative consumption: determinants of satisfaction and the likelihood of using a sharing economy option again. J. Consum. Behav. **14**(3), 193–207 (2015)

Owyang, J., Tran, C., Silva, C.: The Collaborative Economy. Altimeter Group, San Mateo (2013)

Varga, B., Fodor, K.: Kann die logistische Regression zur Klassifizierung kritischer Rohstoffe verwendet werden? In: Karlovitz, J.T. (ed.) People and Their Values in the Society, pp. 15–25. Ausztria, Sozial und Wirtschafts Forschungsgruppe, Großpetersdorf (2019)

Varga, K.: Társadalmi innováció az önkormányzatok működésében. Társadalmi innováció és felelősségvállalás Észak-Magyarországon. **9**, 7–15 (2017)

Benchmarking Minimax Linkage in Hierarchical Clustering

Xiao Hui Tai and Kayla Frisoli

Abstract Minimax linkage was first introduced by Ao et al. (2004) in 2004, as an alternative to standard linkage methods used in hierarchical clustering. Minimax linkage relies on distances to a prototype for each cluster; this prototype can be thought of as a representative object in the cluster, hence improving the interpretability of clustering results. Bien and Tibshirani analyzed properties of this method in 2011 (Bien and Tibshirani 2011), popularizing the method within the statistics community. Additionally, they performed some comparisons of minimax linkage to standard linkage methods. In an effort to expand upon their work and evaluate minimax linkage more comprehensively, we follow the guidelines for neutral benchmark studies outlined in Van Mechelen et al. (2018), focusing on thorough method evaluation via multiple performance metrics on several well-described data sets. We also make all code and data publicly available through an R package, for full reproducibility. Similarly to Bien and Tibshirani (2011), we find that minimax linkage often produces the smallest distances to prototypes, meaning that objects in a cluster are tightly clustered around their prototype. This is true across a range of values for the total number of clusters (k). However, this is not universally true, and special attention should be paid to the case when k is the true known value. For true k, minimax linkage does not always perform the best in terms of all the evaluation metrics studied, including distance to prototype.

Keywords Minimax linkage · Hierarchical clustering · Benchmark analysis

---

X. H. Tai (✉)
University of California, Berkeley, CA, USA
e-mail: xtai@berkeley.edu

K. Frisoli
Carnegie Mellon University, Pittsburgh, PA, USA
e-mail: kfrisoli@andrew.cmu.edu

T. Chadjipadelis et al. (eds.), *Data Analysis and Rationality in a Complex World*, Studies in Classification, Data Analysis, and Knowledge Organization,
https://doi.org/10.1007/978-3-030-60104-1_32

1 Introduction

Hierarchical agglomerative clustering involves successively grouping items within a data set together, based on similarity of the items. The algorithm finishes once all items have been grouped. Given that two items are grouped together, we must determine how similar that merged group is to the remaining items (or groups of items). In other words, we have to recalculate the dissimilarity between any grouped points. This dissimilarity between groups can be defined in many ways, and these are known as linkage methods. Standard, established linkage methods include single, complete, average, and centroid linkage. Minimax linkage, which was first introduced in Ao et al. (2004) and formally analyzed in Bien and Tibshirani (2011), will be the subject of our evaluation. We describe hierarchical agglomerative clustering and the linkage methods precisely as follows.

Given a set of items, dissimilarities d_{ij} between each pair of items i and j, and dissimilarities $d(G, H)$ between groups of items $G = \{i_1, i_2, \ldots i_r\}$ and $H = \{j_1, j_2, \ldots, j_s\}$, hierarchical agglomerative clustering starts with each node (item) in a single group, and repeatedly merges groups such that $d(G, H)$ is within a threshold D. $d(G, H)$ is determined by the linkage method, defined as follows:

Single linkage $d_{\text{single}}(G, H) = \min_{i \in G, j \in H} d_{ij}$.
Complete linkage $d_{\text{complete}}(G, H) = \max_{i \in G, j \in H} d_{ij}$.
Average linkage $d_{\text{average}}(G, H) = \frac{1}{|G||H|} \sum_{i \in G, j \in H} d_{ij}$.
Centroid linkage $d_{\text{centroid}}(G, H) = d(\overline{G}, \overline{H})$, where $\overline{G} = \frac{i_1 + i_2 + \cdots + i_r}{r}$ and $\overline{H} = \frac{j_1 + j_2 + \ldots + j_s}{s}$.
Minimax linkage $d_{\text{minimax}}(G, H) = \min_{i \in G \cup H} r(i, G \cup H)$, where $r(i, G) = \max_{j \in G} d_{ij}$, the radius of a group of nodes G around i. Informally, each item i belongs to a cluster whose center c satisfies $d_{ci} \leq D$.

Bien and Tibshirani (2011) compare minimax linkage to the standard linkage methods using five data sets and two different evaluation metrics. Additionally (although not the focus of the current paper), the authors prove several theoretical properties. They also perform additional evaluations, compare prototypes to centroids, and benchmark computational speed.

The comparisons of minimax linkage to standard linkage methods in Bien and Tibshirani (2011) are summarized in Table 1. For the colon and prostate cancer data sets, distance to prototype was calculated for minimax linkage, but not for the other linkage methods, since those two data sets were used to compare prototypes to centroids, rather than compare the different linkage methods. More details on the data sets and metrics we use are in Sects. 2 and 3.

"Benchmarking in cluster analysis: A white paper" (Van Mechelen et al. 2018) makes multiple recommendations for analyses of clustering methods. We focus on the recommendations for data sets and evaluation metrics. The first recommendation with respect to choosing data sets is to "make a suitable choice of data sets and give an explicit justification of the choices made." In Bien and Tibshirani (2011), it was not explained why the particular data sets were chosen for the different evaluations, and

Table 1 Comparisons to standard linkage methods in Bien and Tibshirani (2011)

Data set	Distance to prototype	Misclassification rate
Olivetti faces	Yes	No
Grolier encyclopedia	Yes	No
Colon cancer	Not quite	No
Prostate cancer	Not quite	No
Simulations	No	Yes

features of the data sets were not fully described. In our study, we both add additional data sets and justify existing ones (which include both synthetic and empirical data) in Sect. 2.2.

With respect to evaluation metrics, Van Mechelen et al. (2018) recommend that we think carefully about criteria used, justify our choices, and consider multiple criteria if appropriate. Additionally, criteria should be applied across all data sets. This is one of our main critiques of the existing evaluation, where not all of the data sets used were evaluated on all the criteria suggested. Distance to prototype was well-justified (this is the crux of minimax linkage), but not misclassification rate (which the authors define on a pairwise basis, as whether pairs are correctly or incorrectly classified as being in the same cluster). When there are a large number of small clusters, the pairwise misclassification rate might not be the best measure of performance, as we describe in Sect. 2. Therefore, we include precision and recall as additional metrics to evaluate clustering quality.

Finally, a suggestion of Van Mechelen et al. (2018) is to fully disclose data and code. Unlike the original paper, we supply the code and data that accompanies this paper, for full reproducibility. We have also written an R package, `clusterTruster`, available on GitHub,[1] which allows the performance of additional evaluations on user-supplied data.

This paper is designed to be a neutral benchmark study of minimax linkage, and the specific contributions are

1. An evaluation of all data sets on all of the criteria in Bien and Tibshirani (2011). In other words, any instance of "Not quite" or "No" within Table 1 should be changed to "Yes."
2. A better assessment of performance with the utilization of precision and recall
3. An evaluation on additional (diverse) data sets not in Bien and Tibshirani (2011)
4. Providing publicly available code and an R package that allow for full reproducibility and transparency, while simplifying the process of making additional evaluations on user-supplied data.

[1]https://github.com/xhtai/clusterTruster.

2 Benchmark Study

2.1 Evaluation Metrics

Distance to prototype

The distance to prototype is measured by the maximum minimax radius. The radius of a group of nodes G around i was defined in Sect. 1, as $r(i, G) = \max_{j \in G} d_{ij}$. This is the distance of the farthest point in cluster G from point i. The prototype is selected to be the point in G with the minimum radius, and this radius is known as the minimax radius, $m(G) = \min_{i \in G} r(i, G)$. Now, for a clustering with k clusters, each of the k clusters is associated with a minimax radius, $m(G_k)$. The distance to the prototype is defined as the the maximum of the k minimax radii, $\max_k m(G_k)$, in other words the "worst" minimax radius across all clusters. In this sense, the maximum minimax radius can be thought of as a summary measure of the tightness of clusters around their prototype. A small value indicates that points within the clusters are close to their prototypes, meaning that the prototypes are accurate representations of points within the clusters.

Minimax linkage relies on successively merging clusters to produce the smallest maximum radius of the resulting cluster, so we would expect minimax linkage to perform the best among other linkage methods in terms of producing the smallest maximum minimax radii.

Misclassification rate

In this clustering context, the misclassification rate is defined as the proportion of misclassified *pairs* of items out of all possible pairs of items. Evaluation is on a pairwise basis: consider each of the $\binom{n}{2}$ pairs, where n is the number of individual items, and the outcome is whether the pair is predicted to be in the same cluster or not. A pair is misclassified if the clustering method predicts that the pair is in the same cluster when the true clustering says they are not, or vice versa.

A low misclassification rate typically indicates high accuracy (a good classifier). However, in cases with a class imbalance (many pairs are in different clusters and few pairs belong to the same cluster), we need to be careful with using misclassification rates because simply classifying all pairs of items as being in different clusters can produce a low misclassification rate.

Precision and recall

To take into account this potential class imbalance, we use the evaluation metrics precision and recall. These are similarly evaluated on a pairwise basis: a true positive (TP) is when two similar items are correctly assigned to the same cluster. A true negative (TN) is when two dissimilar items are correctly assigned to different clusters. A false negative (FN) is when two similar items are incorrectly assigned to different clusters and a false positive (FP) is when two dissimilar items are incorrectly assigned to the same cluster. Now, Precision $= \frac{\text{TP}}{\text{TP+FP}}$ and Recall $= \frac{\text{TP}}{\text{TP+FN}}$.

All k, best k versus true k
Again, define k as the number of clusters in the clustering. In Bien and Tibshirani (2011), evaluation for distance from prototype was conducted over all possible values of k (specifically in a data set of n items, $k \in [1, n]$). Misclassification rate, however, was reported only for the best k (the k for which misclassification rate is lowest) and the true k (the known, ground truth k).

In this paper, we evaluate all metrics using all k, and also report the metrics for true k. It is possible to derive measures for the best k, but due to the large number of data sets and evaluation metrics used, this became somewhat intractable and was not pursued further, but can be a subject of future work.

2.2 Data Sets

In terms of the data sets considered, we use all data used in Bien and Tibshirani (2011) (except for Grolier Encyclopedia which does not have information on true clusters), and introduce additional data sets (iris, NBIDE, and FBI S&W) that exhibit a wider range of data attributes. These additional data sets were included also to ensure that those used in Bien and Tibshirani (2011) were not deliberately selected to produce desired results. Brief descriptions are as follows, and more details for many of the data sets can be found in Bien and Tibshirani (2011).

Olivetti Faces
This data contains 400 images of 64×64 pixels. There are 10 images each from 40 people ($n = 400$, $p = 4096$, $k = 40$). The pairwise distance measure used is ℓ_2. We use the data from the `RnavGraphImageData` R package.

Colon Cancer
The Colon Cancer data set contains gene expression levels for 1000 genes for 62 patients, 40 with cancer and 22 healthy ($n = 62$, $p = 2000$, $k = 2$). The pairwise distance measure used is correlation. We use the data from the `HiDimDA` R package.

Prostate Cancer
The Prostate Cancer data contains gene expression levels for 6033 genes for 102 patients, 52 with cancer and 50 healthy ($n = 102$, $p = 6033$, $k = 2$). The pairwise distance measure used is correlation. There are multiple versions of the data available online and in R packages. The version we use[2] produces results that match the resulting plots produced in Bien and Tibshirani (2011).

Simulations
We repeat the simulations done in Bien and Tibshirani (2011). These involve three sets of data: spherical, elliptical, and outliers. Each data set has 3 clusters of 100 points each in $\mathbb{R}^{10}$ ($n = 300$, $p = 10$, $k = 3$ each). Both ℓ_1 and ℓ_2 distances are used as pairwise distance measures. In Bien and Tibshirani (2011), simulations were run

[2]https://stat.ethz.ch/~dettling/bagboost.html.

50 times each, but here we only ran each once. In future analyses, it is possible to perform more runs.

Iris

The iris data set (Anderson 1936; Fisher 2020) is pre-loaded in R and has been used extensively as an example data set in various applications, including clustering. It contains 50 flowers from each of 3 species ($n = 150$, $p = 4$, $k = 3$). The clusters are elliptical and well-separated. Here, we simply scale and center the features and use ℓ_2 distance as a pairwise distance measure.

NBIDE and FBI S&W

The National Institute of Standards and Technology (NIST) maintains the Ballistics Toolmark Research Database,[3] containing images of cartridge cases from test fires of various firearms. We use data from two different data sets, NIST Ballistics Imaging Database Evaluation (NBIDE) (Vorburger et al. 2007) and FBI Smith and Wesson. The former contains 12 images each from 12 different firearms ($n = 144$, $p = 144{,}000$, $k = 12$), and the latter contains 2 images each from 69 different firearms ($n = 138$, $p = 144{,}000$, $k = 69$). These are examples of high-dimensional data sets, and the latter has a large number of very small clusters. We have pre-processed and aligned these images using the R package `cartridges3D`,[4] and extracted a correlation between each pair of images. The resulting pairwise comparison data are available in the `clusterTruster` package.

3 Evaluation Results

We report our evaluation results for all data sets and evaluation metrics, using all k and true k, in an online Appendix.[5] Tables A.1 and A.2 report the results for all linkage types considered, for the true k of each data set. Figures A.1 through A.12 show the distribution of the metrics across all possible values of k. In this paper, we illustrate the results using the Olivetti Faces data set in Table 2 and Fig. 1.

3.1 Results for True k

It is important to understand how our evaluation metrics change for multiple values of k, especially because k is often unknown. However, in practice, sometimes we know a plausible range of k values and therefore results in out-of-scope regions may be irrelevant.

---

[3]https://tsapps.nist.gov/NRBTD.

[4]Available at https://github.com/xhtai/cartridges3D.

[5]https://github.com/xhtai/xhtai.github.io/blob/master/benchmarkOnlineAppendix.pdf.

Table 2 Results for Olivetti faces data set with $k = 40$ (true k)

Linkage type	Max minimax radius	Misclassification	Precision	Recall
Single	3394.93	0.40	0.04	0.78
Complete	2606.25	**0.04**	**0.31**	0.49
Average	2449.69	0.07	0.18	0.60
Centroid	3259.74	0.79	0.02	**0.83**
Minimax	**2293.45**	0.05	0.24	0.57

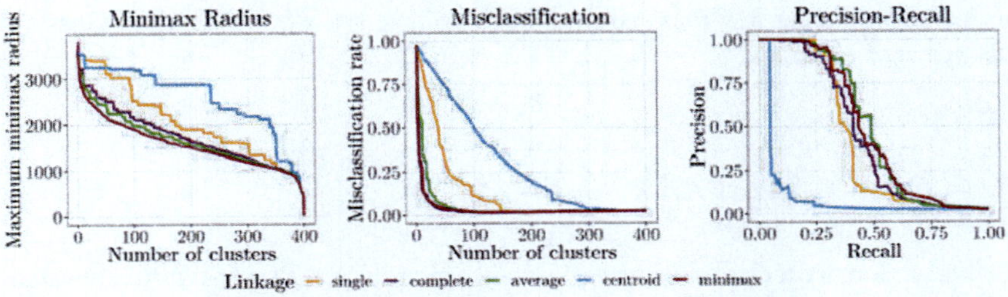

Fig. 1 Minimax radius results for Olivetti faces

In Table 2 for true k in the Olivetti Faces data set, we find that hierarchical clustering with minimax linkage produces the smallest maximum minimax radius, among all linkage types studied. This indicates that the images within each cluster are close to the cluster's prototype, and the prototype is a good representation of the cluster. Using a prototype is especially useful for interpreting cluster results in cases when averages of the data do not make practical sense (e.g., images, text). Minimax linkage does not produce the lowest misclassification rate, although the rate is comparable to both average and centroid linkages. Because average and centroid linkages result in uninterpretable cluster "representatives," minimax linkage could holistically be considered the best performer. However, minimax linkage does not produce the highest precision and recall which we argue in Sect. 2 should also be reported when determining linkage quality.

In the other data sets, minimax linkage does not always produce the best results in terms of distance to prototype. In the Colon Cancer, Prostate Cancer, Spherical-ℓ_2, Spherical-ℓ_1, and Elliptical-ℓ_2 data, other linkage methods produce the smallest maximum minimax radius, although minimax linkage does often produce similarly small radii. Bien and Tibshirani (2011) claim that "minimax linkage indeed does consistently better than the other methods in producing clusterings in which every point is close to a prototype," which we do see across multiple k values (more details in Sect. 3.2). However, when we look specifically at the true k case, which is more relevant in practice, our results dispute this claim.

In terms of the misclassification rate, minimax linkage performs well. In the Elliptical-ℓ_2, Spherical-ℓ_2, Colon Cancer, and Olivetti Faces data sets, other linkage methods produce smaller misclassification rates, but in each case the minimax linkage rate was close to (within 0.03 of) the best rate. This finding is consistent with the claims in Bien and Tibshirani (2011). For precision, minimax linkage performs worse than other methods in the Olivetti Faces, Colon Cancer, Prostate Cancer, Elliptical-ℓ_2, and FBISW data sets. For recall, minimax linkage performs worse than other methods in all data sets except NBIDE.

In summary, we find that for true k, minimax linkage does not consistently perform best in terms of smallest maximum minimax radius, highest precision, or highest recall. It does perform well in terms of misclassification. One of the core claims in Bien and Tibshirani (2011) is that minimax linkage consistently performs best in terms of producing low maximum minimax radius, but we find that this does not always hold for true k.

3.2 Results Across All k

In this section, we look at how the performance metrics change across different values of k. This could still be relevant in practice where k is unknown, or if we want an overall sense of the performance of the method across all possible clusterings.

The full results are in the online Appendix. Here, we again use the Olivetti Faces data set to illustrate the results. We find that minimax linkage performs best in terms of maximum minimax radius, but not in terms of misclassification, precision, or recall. For misclassification rate, minimax linkage performs similarly to complete linkage and both methods do better than single and centroid linkages. Average linkage performs similarly to complete and minimax linkage for large values of k.

For each value of k, we plot both precision and recall, and connect the values in order of increasing k. The area under the curve has maximum value 1, and we want this to be as large as possible. In Fig. 1, we see that for the Olivetti faces data, centroid linkage performs poorly for most values of k. Average linkage appears to perform best for most values of k although minimax and single linkage also perform well for certain values of k.

Across all data sets (see online Appendix), minimax linkage performs best across most values of k in terms of lowest maximum minimax radius. No linkage type stands out as the best for misclassification performance across the different data sets. Similarly for precision and recall, there is no best performing linkage method.

4 Discussion and Conclusion

Bien and Tibshirani (2011) analyzed minimax linkage and performed comparisons to standard linkage methods. We expand on this evaluation, taking into account guidelines recommended in Van Mechelen et al. (2018): we justify choices of data sets and evaluation metrics, apply criteria across all data sets, and fully disclose data and code. Comparing to Bien and Tibshirani (2011), we use additional data sets, include precision and recall as additional evaluation metrics, and use all metrics for all data sets. We evaluate on all possible clusterings k, as well as the true k. We highlight results for the latter case, since metrics for this value of k can be more relevant in practice.

One of the main claims of Bien and Tibshirani (2011) is that minimax linkage consistently performs best in terms of producing results with low maximum minimax radius, but we find that to not always be true (for instance, when k is the true value). We do find (similarly to Bien and Tibshirani 2011) that minimax linkage often produces the smallest maximum minimax radius across all possible values of k. As explained in Sect. 2, this result is not surprising, since minimax linkage is defined based on minimax radius, and is designed to keep this metric small. We find that minimax linkage performs well in terms of misclassification across all data sets, but it does not always produce high precision and recall. To summarize, we came to two main conclusions.

1. **For true** k: Minimax linkage does not consistently perform best in terms of smallest maximum minimax radius, highest precision, or highest recall. It does consistently produce the lowest misclassification rates.
2. **Across all** k: Minimax linkage performs best across most values of k in terms of lowest maximum minimax radius. No linkage type stands out as the best for misclassification performance, precision, or recall across all data sets.

The priority of this paper was to evaluate performance on real clustering applications, but we did include one run of the simulations that were done in Bien and Tibshirani (2011). Future work will include an increased focus on simulations. Additionally, more work can be done to properly quantify standard errors associated with the evaluation metrics. We also noted that it is possible to derive and report measures for the best k as opposed to true k, but this is left for future work. Another topic of interest that is out of scope for this paper is the analysis of data where the true number of clusters is not known. For example, we might be interested in whether it is easier to discover the number of true clusters using minimax linkage as opposed to other methods. This is an interesting question that to our knowledge has not been explored. Finally, one issue that was evaluated briefly in Bien and Tibshirani (2011), but that we have not focused on, is computational complexity.

Acknowledgements We would like to thank the Carnegie Mellon University Data Science Initiative Research Group, especially Rebecca Nugent, for her leadership and advice.

References

Anderson, E.: The species problem in iris. Ann. Mo. Bot. Gard. **23**(3), 457–509 (1936)
Ao, S.I., Yip, K., Ng, M., Cheung, D., Fong, P.-Y., Melhado, I., Sham, P.C.: CLUSTAG: hierarchical clustering and graph methods for selecting tag SNPs. Bioinformatics **21**(8), 1735–1736 (2004)

Bien, J., Tibshirani, R.: Hierarchical clustering with prototypes via minimax linkage. J. Am. Stat. Assoc. **106**(495), 1075–1084 (2011)
Fisher, R.A.: The use of multiple measurements in taxonomic problems. Ann. Eugen. **7**(2), 179–188
Van Mechelen, I., Boulesteix, A.L., Dangl, R., Dean, N., Guyon, I., Hennig, C., Leisch, F., Steinley, D.: Benchmarking in cluster analysis: a white paper (2018). arXiv:1809.10496
Vorburger, T., Yen, J., Bachrach, B., Renegar, T., Filliben, J., Ma, L., Rhee, H., Zheng, A., Song, J., Riley, M., Foreman, C., Ballou, S.: Surface topography analysis for a feasibility assessment of a National Ballistics Imaging Database. Tech. Rep. NISTIR 7362, National Institute of Standards and Technology, Gaithersburg, MD (2007)

Clustering Binary Data by Application of Combinatorial Optimization Heuristics

Javier Trejos-Zelaya, Luis Eduardo Amaya-Briceño, Alejandra Jiménez-Romero, Alex Murillo-Fernández, Eduardo Piza-Volio, and Mario Villalobos-Arias

Abstract We study clustering methods for binary data, first defining aggregation criteria that measure the compactness of clusters. Five new and original methods are introduced, using neighborhoods and population behavior combinatorial optimization metaheuristics: first ones are simulated annealing, threshold accepting and tabu search, and the others are a genetic algorithm and ant colony optimization. The methods are implemented, performing the proper calibration of parameters in the case of heuristics, to ensure good results. From a set of 16 data tables generated by a quasi-Monte Carlo experiment, a comparison is performed for one of the aggregations using L_1 dissimilarity, with hierarchical clustering, and a version of k-means: partitioning around medoids or PAM. Simulated annealing performs very well, especially compared to classical methods.

Keywords Clustering · Binary data · Simulated annealing · Threshold accepting · Tabu search · Genetic algorithm · Ant colony optimization

J. Trejos-Zelaya (✉) · E. Piza-Volio · M. Villalobos-Arias
CIMPA & School of Mathematics, Faculty of Science, University of Costa Rica, San José, Costa Rica
e-mail: javier.trejos@ucr.ac.cr

E. Piza-Volio
e-mail: eduardo.piza@ucr.ac.cr

M. Villalobos-Arias
e-mail: mario.villalobos@ucr.ac.cr

L. E. Amaya-Briceño
Guanacaste Campus, University of Costa Rica, Liberia, Costa Rica
e-mail: luis.amaya@ucr.ac.cr

A. Jiménez-Romero
School of Mathematics, Costa Rica Institute of Technology, Cartago, Costa Rica
e-mail: alejimenezr@gmail.com

A. Murillo-Fernández
Atlantic Campus, University of Costa Rica, Turrialba, Costa Rica
e-mail: alex.murillo@ucr.ac.cr

T. Chadjipadelis et al. (eds.), *Data Analysis and Rationality in a Complex World*, Studies in Classification, Data Analysis, and Knowledge Organization,
https://doi.org/10.1007/978-3-030-60104-1_33

1 Introduction

Binary data arise in several situations in research since they can encode situations where the presence (1) or absence (0) of a characteristic is studied: species present/absent, alive/dead in health sciences, yes/no in social or decision sciences, and so on. For instance, in Ecology (Salas-Eljatiba et al. 2018), it is usual to divide an area into sectors and record the presence or absence of certain vegetable species. In the study of gene expression data, several molecular techniques encode data as binary matrices (Demey et al. 2008). In pattern recognition, images may also be coded as 0/1 data indicating the presence or absence of a feature. In Sociology (Borkulo et al. 2014), Health Sciences (Zhang and Singer 1999), Economics (Jeliazkov and Rahman 2012) ...binary data are analyzed.

Several methods have been used for clustering binary data. For instance, there are partitioning methods such as dynamical clusters, which is an adaptation of Forgy's k-means (Everitt 1993), based on a representation of clusters by a kernel and iterations of two steps: allocating objects to the nearest kernel and recalculating the kernels. A variant of this k-means method is PAM (Kaufman and Roosseuw 2005), *partitioning around medoids*, where kernels are 0/1 median vectors and L1 dissimilarity is used. Another variant is for kernels selected as the object in the class that minimizes the sum of dissimilarities to the rest of objects in the class; this last version is what we will call k-means for 0/1 data in this article. These methods find local minima of the criterion to be minimized since they are based on local search procedures.

There are also hierarchical methods used for clustering binary data (Everitt 1993), using an appropriate dissimilarity for binary data (such as Jaccard, for instance). These methods find also local minima since they are based on a greedy strategy.

To overcome the local optima problem, optimization strategies have been used. We have employed combinatorial optimization metaheuristics for clustering numerical data (Trejos et al. 1998; Trejos et al. 2014). In the present article, we will use some of these heuristics for binary data, whenever it is possible. When dealing with binary data, it is necessary to adapt the criterion since Huygens theorem and other theoretical results only hold in an Euclidean context.

The article is organized as follows. Section 2 presents the clustering problem and particularly some criteria and properties for the binary case. Section 3 contains the five combinatorial optimization metaheuristics employed here and the adaptation we made for clustering binary data. In Sect. 5, the results obtained are presented, and finally, in Sect. 6, some concluding remarks are made.

2 Clustering Binary Data

Given a data set of binary vectors $\Omega = \{\mathbf{x}_1, \mathbf{x}_2, \ldots, \mathbf{x}_n\}$ with $\mathbf{x}_i \in \{0, 1\}^p$ and a number $K \in \mathbb{N}$, we seek for a partition $P = (C_1, C_2, \ldots, C_K)$ of Ω such that elements in a class C_k are more similar than elements of other classes.

We define a within inertia measure of the partition P as

$$W(P) = \sum_{k=1}^{K} \delta(C_k) \tag{1}$$

where $\delta(C_k)$ can be defined (among others) as

$$\delta_{\text{sum}}(C_k) = \sum_{i,i' \in C_k} d(\mathbf{x}_i, \mathbf{x}_{i'}),$$
$$\delta_{L_1}(C_k) = \sum_{i \in C_k} \|\mathbf{x}_i - m(C_k)\|_{L_1}.$$

d being a binary dissimilarity index in Ω (for instance, Jaccard or L_1), $m(C_k)$ the median vector of C_k. Späth (1985) and Piza et al. (2002) studied some other criteria for clustering. These indexes satisfy the following **monotonicity property**.

Proposition 1 (see Piza et al. 2002) *Let $P = (C_1, \ldots, C_K)$ and $P' = (C'_1, \ldots, C'_{K+1})$ be partitions of Ω in K and $K + 1$ non-empty classes, δ_{sum} and δ_{L_1} on 2^Ω satisfy the monotonicty property, since for all instances of the data, we have*

$$\min_{P' \in \mathcal{P}^*_{k+1}} W(P') = \sum_{j=1}^{k+1} \delta(C'_j) \leq \min_{P \in \mathcal{P}^*_k} W(P) = \sum_{j=1}^{k} \delta(C_j),$$

for every number of classes $K < n$.

Proofs can be found in Piza et al. (2002) or by request upon the authors. As a consequence of Proposition 1, the number of clusters must be predefined since best clusterings with different number of classes cannot be compared. In Piza et al. (2002), it is also proved that for δ_{sum} and δ_{L_1} all optimal partitions have non-empty classes.

Analogously to the continuous case, total inertia can be defined as follows:

$$I(\Omega) = \sum_{i=1}^{n} \sum_{i'=1}^{n} d(\mathbf{x}, \mathbf{x}'), \tag{2}$$

and, thanks to the monotonicity, the between-classes inertia is defined as follows:

$$B(\Omega) = I(P) - W(P), \tag{3}$$

and it is always positive.

3 Using Combinatorial Optimization Heuristics

We have implemented clustering algorithms using well-known combinatorial optimization heuristics. Three of them are based on neighbors, simulated annealing (SA), threshold accepting (TA), and tabu search (TS), and two based on a population of solutions, genetic algorithm (GA), and ant colony optimization (AC).

A state of the problem is a partition P of Ω in K clusters. From a current state, a neighbor is defined by the single transfer of an object from its current class to a new one, chosen according to the corresponding heuristic rules. This will be the case in SA, TA, TS, and the mutation operation in GA.

3.1 Simulated Annealing

Simulated annealing is a random search algorithm, Kirkpatrick et al. (1983), that uses an external parameter of control c_t called *temperature* that controls the random acceptance of states worse than the previous one. It employs the **Metropolis rule** for this acceptation: a new state P' generated from P is accepted if $\Delta W < 0$, where $\Delta W = W(P') - W(P)$, otherwise, it can be accepted with probability $\exp(-\Delta W)/c_t$. It is well known, Aart and Korst (1990), that under a Markov chain modeling simulated annealing has asymptotic convergence properties to the global optimum under some conditions, so it is expected that its use permits to avoid local minima. We found some simplification properties for ΔW, calculated in each iteration and useful for speeding up the algorithm.

SA parameters are χ_0 the initial acceptation rate, L length of Markov chains, γ factor reduction for c_t such that $C_{t+1} = \gamma c_t$ and ϵ stop criterion.

3.2 Threshold Accepting

TA was proposed by Dueck and Scheuer (1990) and can be seen as a particular case of SA, with a different rule for acceptation. Movements that produce an improvement for the objective function in a neighborhood are accepted and movements that worsen it are accepted if they fall into a threshold that is positive and decreases in time. Clearly, the acceptation rule is deterministic, not stochastic.

TA parameters are Th_0 the initial threshold, γ the factor reduction for threshold Th, the maximum number of iterations, and ϵ the stop criterion.

3.3 Tabu Search

TS Glover (1989) handles a tabu list T of length $|T|$ with solutions or codes of solutions that are forbidden to be attained for a certain number of iterations. In each step, current state moves to the best neighbor outside T. In our partitioning problem,

T stores a code of the transfers that define the neighbors. In this case, the indicator function of cluster contained the object that changed class during the transfer; that is, all objects that were together in the previous state are forbidden to be together for $|T|$ iterations.

TS parameters are $|T|$, maximum number of iterations, and sampling size of neighborhoods.

3.4 Genetic Algorithm

GA handles a population of solutions, which are *chromosomes* representing partitions. A chromosome is an n-vector in an alphabet of K numbers indicating the presence/absence of the ith object in the corresponding class. The *fitness* function is defined as $f(P) = \frac{B(P)}{I(\Omega)}$. In the genetic algorithm iterations, chromosomes are kept using a random wheel roulette with elitism, with a probability proportional to f. For crossover between two parents selected at random (with a uniform distribution), the dominant parent is the one with the greatest fitness; a class in it is selected uniformly at random and this class is copied in the other parent, generating a new son. Mutation is the classical one: an object is selected randomly, and it is transferred to a new class.

Parameters for GA are the number M of partitions (population size), probabilities of crossover and mutation, maximum number of iterations. Iterations stop when the fitness variance of the population is less than ϵ.

3.5 Ant Colonies

In AC (Bonabeau et al. 1999), each ant m is associated with a partition P, which is modified during the iterations. Given an object $\mathbf{x}_i$, the probability of transfering another object $\mathbf{x}_{i'}$ to the same class as $\mathbf{x}_i$ in iteration t is defined as

$$p_{ii'} = \frac{(\tau_{ii'})^{\alpha}(\eta_{ii'})^{\beta}}{\sum_{l \neq i}^{n}(\tau_{li'})^{\alpha}(\eta_{li'})^{\beta}}, \tag{4}$$

where $\eta_{ii'} = \frac{1}{d_{ii'}}$ is the visibility and the **pheromone trail** is updated with

$$\tau_{ii'}(t+1) = (1-\rho)\tau_{ii'}(t) + \rho \sum_{m=1}^{M}(\Delta^m \tau_{ii'}(t+1)), \tag{5}$$

$$\Delta^m \tau_{ii'}(t+1) = \begin{cases} \frac{B(P^m)}{I(\Omega)} & \text{if } i, i' \in \text{same class} \\ 0 & \text{otherwise.} \end{cases} \tag{6}$$

M being the number of ants, $\alpha, \beta \in \mathbb{R}$ weights, and $\rho \in \mathbb{R}$ an evaporation parameter.

Table 1 Characteristics of data tables generated, 4 factors with 2 levels each. n: number of objects, K: number of clusters, $|C_k|$: cardinality of cluster C_k, π_k probability of membership to C_k

Data table	n	K	n_k	π	$\|C_1\|$	$\|C_2\|,\ldots,\|C_K\|$	π_1	π_2	π_3	π_4	π_5
1	120	3	=	Separated	40	40	0,1	0.5	0.9		
2	120	3	=	Fuzzy	40	40	0.3	0.5	0.7		
3	120	3	≠	Separated	60	30	0.1	0.5	0.9		
4	120	3	≠	Fuzzy	60	30	0.3	0.5	0.7		
5	120	5	=	Separated	24	24	0.05	0.25	0.5	0.75	0.95
6	120	5	=	Fuzzy	24	24	0.2	0.35	0.5	0.65	0.8
7	120	5	≠	Separated	60	15	0.05	0.25	0.5	0.75	0.95
8	120	5	≠	Fuzzy	60	15	0.2	0.35	0.5	0.65	0.8
9	1200	3	=	Separated	40	40	0,1	0.5	0.9		
10	1200	3	=	Fuzzy	40	40	0.3	0.5	0.7		
11	1200	3	≠	Separated	60	30	0.1	0.5	0.9		
12	1200	3	≠	Fuzzy	60	30	0.3	0.5	0.7		
13	1200	5	=	Separated	24	24	0.05	0.25	0.5	0.75	0.95
14	1200	5	=	Fuzzy	24	24	0.2	0.35	0.5	0.65	0.8
15	1200	5	≠	Separated	60	15	0.05	0.25	0.5	0.75	0.95
16	1200	5	≠	Fuzzy	60	15	0.2	0.35	0.5	0.65	0.8

AC parameters to calibrate are α, β, ρ, size of population M, precision ϵ, and the maximum number of iterations.

4 Simulated Data

We performed a Monte Carlo-type experiment, generating 16 data tables controlling the following factors: n, the number of objects (with levels 120 and 1200); K, the number of clusters (levels 3 and 5); n_k, the cardinality of the clusters (equal cardinalities and one big cluster with 50% of the objects and the rest of objects distributed equally); and π, the probability of the Bernoulli distribution, with levels of well separated clusters ($\pi = 0.1, 0.5, 0.9$ for $K = 3$ and $\pi = 0.05, 0.25, 0.5, 0.75, 0.95$ for $K = 5$) or fuzzy separated clusters ($\pi = 0.3, 0.5, 0.7$ for $K = 3$ and $\pi = 0.2, 0.35, 0.5, 0.65, 0.8$ for $K = 5$). Table 1 presents the characteristics of each of the 16 data tables generated in the experiment, including the factors and the levels.

5 Results

We have compared the results obtained with the five metaheuristics (SA, TA, TS, GA, AC) with two classical methods: partitioning around medoids (PAM) when using the L_1 dissimilarity index, and hierarchical clustering (HC) using average linkage. For each heuristic, a parameter calibration was performed, exploring different values for each parameter. After this calibration, parameters selected for this article were

- Simulated annealing: $\chi_0 = 0.95$, $L = 50$; $\gamma = 0.91$, $\epsilon = 0.01$.
- Threshold accepting: $Th_0 = 100$, $maxiter = 50$, $\gamma = 0.9$, $\epsilon = 0.01$.
- Tabu search: $maxiter = 150$, $|T| = 5$, $s = 0.1|N(P)|$.
- Genetic algorithm: $p_m = 0.1$, $p_c = 0.8$, $M = 20$, $maxiter = 500$, $\epsilon = 0.01$.
- Ant colony: $\alpha = 0.5$, $\beta = 0.2$, $\rho = 0.5$, $M = 10$, $maxiter = 500$, $\epsilon = 0.01$.

In Table 2 we report, for a multistart of size 100, the best value of W obtained so far by any method (noted W^*), the mean value of W for each method (noted $\bar{W}$) and the attraction rate a_r or percentage of times that W^* was obtained (up to a relative error of 0.05).

Table 2 Results summary with δ_{L_1} for a multistart of size 100. W^* is the best value obtained by any method, a_r the attraction rate of W^* for each method, and $\hbar W$ the mean value for each method

Table	W^*	SA		TA		TS		GA		AC		PAM		HC
		a_r (%)	$\bar{W}$	a_r (%)	$\bar{W}$	a_r (%)	$\bar{W}$	a_r (%)	$\bar{W}$	a_r (%)	$\bar{W}$	a_r (%)	$\bar{W}$	$\bar{W}$
1	414	7	431	2	432	1	444	0	648	0	978	0	421	445
2	744	4	750	0	751	1	757	0	849	0	1017	0	780	790
3	387	0	412	0	412	0	412	0	605	0	901	100	387	412
4	367	3	387	0	429	0	430	0	611	0	901	0	400	429
5	424	2	456	5	444	0	473	0	688	0	951	0	426	451
6	587	100	587	0	607	0	620	0	797	0	963	0	600	637
7	293	0	326	0	325	0	345	0	576	0	762	100	293	305
8	513	1	525	4	522	0	559	0	720	0	853	0	542	543
9	4641	0	4868	0	5439	0	4928	0	8682	0	11350	100	4641	4983
10	7775	1	7880	0	8561	0	8210	0	11204	0	11462	0	7841	8385
11	4137	100	4137	0	4156	0	9639	0	4161	0	9636	100	4137	4379
12	4179	100	4179	0	9582	0	9582	0	4207	0	9681	100	4179	4494
13	3003	0	4304	0	4932	0	4523	0	11106	0	10642	100	3003	3277
14	6549	0	7218	0	7364	0	7272	0	11053	0	11288	100	6549	7192
15	3165	10	4512	1	8114	5	7985	0	3201	0	7883	100	3165	3337
16	5812	10	6114	10	6442	5	10558	0	5896	0	9369	100	5812	6270

6 Concluding Remarks

Generally speaking, with simulated annealing we obtain good results, although PAM obtains good results in some cases. Threshold accepting sometimes reaches the optimum and tabu search only in two cases. Population-based heuristics did not get good results with our implementation. It is worth noting that hierarchical clustering never obtained the optimum. Even if we do not report running times, SA is fast enough to be competitive. The main drawback of using heuristics is tuning the parameters though SA may have a standard choice.

Acknowledgements This research was partially supported by the University of Costa Rica (CIMPA project 821-B1-122 and the Graduate Program in Mathematics) and the Costa Rica Institute of Technology. The supports are gratefully acknowledged.

References

Aarts, E., Korst, J.: Simulated Annealing and Boltzmann Machines. Wiley, Chichester (1990)

Bonabeau, E., Dorigo, M., Therauluz, G.: Swarm Intelligence. From Natural to Artificial Systems. Oxford University Press, New York (1999)

Demey, J.R., Vicente-Villardón, J.L., Galindo-Villardón, M.P., Zambrano, A.Y.: Identifying molecular markers associated with classification of genotypes by external logistic biplots. Bioinformation **24**(24), 2832–2838 (2008)

Dueck, G., Scheuer, T.: Threshold accepting: A general purpose optimization algorithm appearing superior to simulated annealing. J. Comput. Phys. **90**(1), 161–175 (1990)

Everitt, B.S.: Cluster Analysis. Edward Arnold, London (1993)

Glover, F.: Tabu search—Part I. ORSA J. Comput. **1**, 190–206 (1989)

Goldberg, D.E.: Genetic Algorithms in Search Optimization and Machine Learning. Addison-Wesley, Reading (1989)

Jajuga, K.: A clustering method based on the L_1-norm. Comput. Stat. Data Anal. **5**(4), 357–371 (1987)

Jeliazkov, I., Rahman, M.A.: Binary and ordinal data analysis in Economics: Modeling and estimation. In: Yang, X.S. (ed.) Mathematical Modeling with Multidisciplinary Applications, pp. 1–31. Wiley, New York (2012)

Kaufman, L., Roosseuw, P.: Finding Groups in Data. An Introduction to Cluster Analysis. Wiley, New York (2005)

Kirkpatrick, S., Gelatt, D., Vecchi, M.P.: Optimization by simulated annealing. Science **220**, 671–680 (1983)

Piza, E., Trejos, J., Murillo, A.: Clustering with non-Euclidean distances using combinatorial optimisation techniques. In: Blasius, J., Hox, J., de Leeuw, E., Schmidt, P. (eds.) Social Science Methodology in the New Millennium, paper number P090504. Leske + Budrich, Darmstadt (2002)

Salas-Eljatiba, C., Fuentes-Ramirez, A., Gregoire, T.G., Altamirano, A., Yaitula, V.: A study on the effects of unbalanced data when fitting logistic regression models in ecology. Ecol. Indic. **85**, 502–508 (2018)

Späth, H.: Cluster Dissection and Analysis. Theory, Fortran Programs, Examples. Ellis Horwood, Chichester (1985)

Trejos, J., Murillo, A., Piza, E.: Global stochastic optimization techniques applied to partitioning. In: Rizzi, A., Vichi, M., Bock, H.-H. (eds.) Advance Data Analysis Classification, pp. 185–190. Springer, Berlin (1998)

Trejos, J., Villalobos, M., Murillo, A., Chavarría, J., Fallas, J.J.: Evaluation of optimization metaheuristics in clustering. In: Travieso, C.M., Arroyo, J., Ramírez, M. (eds.) IEEE International Work-Conference on Bioinspired Intelligence, pp. 154–161. IEEE, Liberia (2014)

van Borkulo, C.D., Borsboom, D., Epskamp, S., Blanken, T.F., Boschloo, L., Schoevers, R.A., Waldorp, L.J.: A new method for constructing networks from binary data. Sci. Rep. **4**, (2014)

Zhang, H., Singer, B.: Recursive Partitioning in the Health Sciences. Springer, New York (1999)

Classifying Users Through Keystroke Dynamics

Ioannis Tsimperidis, Georgios Peikos, and Avi Arampatzis

Abstract The billions of users connected to the Internet together with the anonymity that each of them can have behind a computer that is a source of many risks, such as financial fraud and seduction of minors. Most methods that have been proposed to remove this anonymity are either intrusive, or violate privacy, or expensive. We propose the recognition of certain characteristics of an unknown user through keystroke dynamics, which is the way a person is typing. The evaluation of the method consists of three stages: the acquisition of keystroke dynamics data from 110 volunteers during the daily use of their device, the extraction and selection of keystroke dynamics features based on their information gain, and the testing of user characteristics recognition by training five well-known machine learning models. Experimental results show that it is possible to identify the age group, the handedness, and the educational level of an unknown user with an accuracy of 87.6, 97.0, and 84.3, respectively.

Keywords Keystroke dynamics · User characteristic classification · Data mining · Feature selection · Information gain · Digital forensics

1 Introduction

Today there are more than 4 billion Internet users in the world who use online services in order to communicate, entertain, educate, work, etc. The way we talk over the Internet with someone else differs radically from the way we do it in person. Most of the time we do not see the face of our interlocutor, we do not hear his/her voice, and in general, the stimuli that give us information about who is and what

I. Tsimperidis (✉) · G. Peikos · A. Arampatzis
Democritus University of Thrace, 67100 Xanthi, Greece
e-mail: itsimper@ee.duth.gr

G. Peikos
e-mail: georpeik1@ee.duth.gr

A. Arampatzis
e-mail: avi@ee.duth.gr

T. Chadjipadelis et al. (eds.), *Data Analysis and Rationality in a Complex World*, Studies in Classification, Data Analysis, and Knowledge Organization,
https://doi.org/10.1007/978-3-030-60104-1_34

his/her intentions are, cease to exist. In addition, we have to consider that often a user is talking to someone completely unknown and that kids participate in these conversations, especially in social networks. It is easily understood that these lurk many dangers, such as financial fraud, seduction of minors, and anonymous threats.

One solution to this problem is to know some characteristics of the user we are talking to, such as gender, age, and so on. There are several proposals for achieving these, such as that of Cheung and She (2017) who tried to recognize the gender of users from the images generated by their mobile devices and shared in social networks. Arroju et al. (2015) try to determine the gender and age of Twitter users based on the contents of their tweets. Although in most cases gender and/or age of users are sought, there are also methods in the literature trying to discover other characteristics, such as the work of Seneviratne et al. (2014) where it is attempted to determine, among others, the religion and the spoken languages of unknown users.

All the aforementioned approaches show some limitations. For example, some of them require special equipment, such as special cameras or keyboards, or can only be applied if the target user has an account in some social network, while some others use features derived from the use of certain language, and therefore are incapable of dealing with the multilinguality of today's Internet. In contrary, methods based on keystroke dynamics features are free from such limitations. This is because the only device needed is the common QWERTY keyboard, and furthermore, these methods are language independent since the features derive mainly from how the user uses the keyboard rather than the words he/she writes in a specific language. Finally, data can be collected non-obtrusively, preserving also user privacy content-wise. The keystroke dynamics features used can be categorized into temporal and non-temporal and are described in detail in Tsimperidis et al. (2018).

The present study is not yet another work on user authentication, as is the case with most studies on keystroke dynamics, but an attempt to classify users according to some inherent or acquired characteristics of them, namely age, handedness, and educational level. The rest of the paper is organized as follows. Section 2 lists the related works in user classification through keystroke dynamics. Section 3 describes the phases of the method followed. Section 4 presents and comments on the classification results obtained by using five well-known machine learning models, i.e., the support vector machine (SVM) with polynomial kernel, the logistic regression (LR), the Bayes classifier (NB), the Bayesian network classifier (BNC), and the radial basis function network (RBFN). Finally, Sect. 5 concludes the paper.

2 Related Work

Although most research in keystroke dynamics has as its object user authentication, there are some published papers in user classification. Once again, the characteristic sought in most cases is users' gender, followed by age. For example, Buriro et al. (2016) tried to estimate user's gender, age, and handedness. They defined three age groups (teenagers, adults, and senior users) and used several machine learning models for classification. The best results were 87.9 and 95.5% accuracy in age and handedness classification, respectively. Random Forest was the most successful classifier among 7 others, in the work of Roy et al. (2017). They divided users into two classes, kids and adults, and used three fixed-text datasets. Finally, using an Ant Colony Optimization technique they achieved accuracy of 92.2%.

Studies with more age classes are those of Tsimperidis et al. (2017) who divided the users into 4 groups, and Pentel (2018) into 6 groups. The former study used 120 down-down digram latencies as features, and with a dataset of 239 logfiles presented 66.1% accuracy coming from MLP combined with a boosting algorithm, while the latter study with data from more than 7,000 users, each of which was recorded for about 320 keystrokes, and 134 keystroke dynamics features in total, reached 61.6% accuracy using Random Forest.

Handedness is a human characteristic that has been extensively researched in terms of economy, sociology, biology, criminology, etc., Goodman (2014). In the field of user classification through keystroke dynamics, Brizan et al. (2015) collected data from 329 users, and their experimental results showed an F-score of 0.223 for the left-hand class with a baseline of 0.1. Shen et al. (2016) exploited a dataset created by 51 users and extracted keystroke durations and diagram latencies as features, and achieved an accuracy of 87.75%. Another approach is that of Pentel (2017) who collected data from 504 users through an electronic questionnaire. Despite the small number of keystrokes per user in dataset, they managed to present high performance with F-score of 0.995. Similarly, in the work of Shute et al. (2017), 65 volunteers were recorded in the same laptop. The authors split the keyboard into six segments, and then fed the features, which were keystroke durations only, in 3 classifiers resulting in an accuracy of 94.5%.

User classification studies based on how users use the keyboard are quite rare. There may be no other published work in seeking age and handedness of unknown users other than those mentioned. In fact, we have not found any paper referring to user classification according to educational level, which is one of the main focuses of our work.

3 Method

Our methodology consists of three consecutive phases. In the first phase, we collected free-text data from volunteers who agreed to participate in the experiment of extracting real-life keystroke dynamics features. In the second phase, we ran a feature selection algorithm to sort the features according to their contained information. In the third phase, the age, the handedness, and the educational level of an unknown user are sought by training and hyperparameter tuning five well-known machine learning algorithms, namely SVM, LR, NB, BNC, and RBFN.

3.1 Keystroke Dynamics Dataset

Keystrokes dynamics datasets can be created by recording users either in fixed- or in free-text. The term "fixed-text" refers to the typing of a specific text usually in some closed environment, while "free-text" indicates the recording of volunteer during the typical daily use of his/her computer. In this work, the free-text approach is followed as it integrates with the subject's regular typing activities better and is less intrusive.

To create a suitable dataset, a free-text keylogger named IRecU, which can be installed on any Microsoft Windows-based devices, was designed and developed. In each of the volunteers who participated in this project, IRecU was installed on their personal computer and it was possible to record their typing at anytime, anywhere they wanted to work, and using any application, gathering data from thousands of keystrokes, in order to get the best possible approximation of the actual use of their computer.

After two recording periods totaling 18.5 months, 110 users were recorded forming a dataset of 362 log files. In each file, there are some metadata with the characteristics of the user being recorded, while keystrokes were written in the form:

```
78,2018-03-19,45743645,"dn"
79,2018-03-19,45743769,"dn"
78,2018-03-19,45743785,"up"
79,2018-03-19,45743879,"up"
```

In each record, which is a user's action on the keyboard, there are four fields separated by commas. The first field represents the virtual key code of the key used, the second indicates the date the action took place in the yyyy-mm-dd format, the third is the elapsed time since the beginning of that day (12:00 am) in milliseconds, and the forth is the action, "dn" for key-press and "up" for key-release.

Each log file contains data from about 3,500 keystrokes. Demographics of the dataset that are of interest to this research are shown in Table 1. As it can be seen, the dataset is unbalanced in each of the characteristics being studied. However, it is evident that with regard to age and educational level each class is adequately represented, while with regard to handedness the dataset is as unbalanced as it should be, since the ratio of right- to left-handers is approximately 9:1 (Dragovic and Hammond 2007).

3.2 Feature Extraction and Feature Selection

In order to keep the complexity low, among the hundreds of thousands of available keystroke dynamics features we considered the most frequently-used, namely the keystroke durations and down-down digram latencies. For the feature extraction, we developed a software application, named ISqueezeU, which reads the files created by IRecU and extracts the desired features.

Table 1 Number of volunteers and logfiles per age, dominant hand, and educational level

Characteristic	Class	Volunteers		Log files	
		#	%	#	%
Age	18–25	23	20.9	71	19.6
	26–35	37	33.6	129	35.6
	36–45	37	33.6	117	32.3
	46+	13	11.9	45	12.5
Handedness	Right-handers	98	89.1	322	88.9
	Left-handers	9	8.2	31	8.6
	Ambidextrous	3	2.7	9	2.5
Educational Level (According UNESCO)	ISCED-3	21	19.1	62	17.1
	ISCED-4	7	6.4	23	6.4
	ISCED-5	13	11.8	49	13.5
	ISCED-6	36	32.7	120	33.2
	ISCED-7-8	33	30.0	108	29.8

Although we chose to extract a small part of the available features, their number is n^2+n, with n being the number of keyboard keys, which is a large number that can lead to systems with high time complexity. Therefore, a feature selection procedure is needed.

Of the thousands of features, there must be selected those which are most capable of distinguishing users according to the studied characteristics. A method to do this is by calculating the information gain (*IG*) of each feature *f*, which is the measure that illustrates the ability of that feature to reduce the entropy of a system *x*. It is expressed as follows:

$$IG(x, f) = H(x) - H(x|f) = -\sum_{i=1}^{m} P(x_i) \cdot ln P(x_i) - \frac{1}{N}\sum_{j=1}^{k} n_j \cdot H(x_j) \,. \quad (1)$$

In Eq. 1, $H(x)$ is the entropy of the system x, and $H(x|f)$ is calculated by splitting the dataset into groups according to the value of the particular feature f. These two terms are analyzed into sums of products, as shown in the equation, where m is the length of vector x, which in the classification problem is the number of classes, and $P(x_i)$ is the probability of class x_i. Also, N is the number of instances of the initial dataset, k is the number of groups that the initial dataset was split, n_j is the number of instances of the j-th group, and $H(x_j)$ is the entropy of the j-th group.

This procedure is also described in the work of Menzies and Greenwald (2007) and if applied to every extracted feature in our classification problems, then a list with the amount of information that every feature carries will emerge. In Table 2, the

Table 2 Keystroke dynamics features with the highest IG in the three classification problems

Age				Handedness				Educational level			
#	Feature	Key(s)	IG	#	Feature	Key(s)	IG	#	Feature	Key(s)	IG
1	69	E	0.1519	1	69	E	0.0832	1	76	L	0.1801
2	69-82	E-R	0.1016	2	65	A	0.0782	2	32	(space)	0.1727
3	80	P	0.0868	3	79	O	0.0723	3	80	P	0.1354
4	32	(space)	0.0867	4	82	R	0.0618	4	77	M	0.1319
5	68	D	0.0862	5	82-65	R-A	0.0602	5	65-32	A-(space)	0.1294

first five features are ranked with the highest *IG* for age, handedness, and educational level classification problems, where each of them is represented by the virtual key code of the keys that compose it.

3.3 Experimental Procedure and Validation of Models

The feature selection procedure indicated 715, 230, and 727 features with non-zero information gain for the age, handedness, and educational level classification problems, respectively. Since we try to predict user characteristics with high precision, we decided to take advantage of any feature that carries some information, and thus all those with non-zero *IG* were used.

Several machine learning models were tested which presented low accuracy and/or too long training times. The five models which presented the best performance in terms of accuracy and time complexity were SVM, LR, NB, BNC, RBFN. Therefore, the results of these models will be presented.

The purpose of model validation is to ensure that the implementations of the models are correct and work as they should. There are many techniques that can be utilized to verify a model and several of them were adopted to validate the five models used in this work.

Firstly, to assess the performance of the five models fairly, we use the well-known 10-folds cross-validation, which divides the data into 10 disjoint parts, uses 9 of them for training and the remaining one for testing, in a round-robin fashion. Secondly, to evaluate the effectiveness of the feature selection procedure, we additionally use F-score, as a combined measurement of precision and recall, because accuracy alone cannot fully give the picture of the overall performance of a model when classes are imbalanced. Finally, to assess the ranking ability of the classifiers, we use the area under the ROC curve (AUC) or ROC index.

Table 3 Performance of the five models in the classification problems

Model	Age				Handedness				Educational Level			
	Acc. (%)	TBM	F1	AUC	Acc. (%)	TBM	F1	AUC	Acc. (%)	TBM	F1	AUC
SVM	77.1	0.22	0.767	0.868	94.8	0.05	0.943	0.824	77.9	0.22	0.778	0.897
LR	73.5	4.87	0.734	0.871	95.6	0.80	0.952	0.970	72.9	6.89	0.729	0.896
NB	69.9	0.02	0.694	0.842	89.0	0.01	0.879	0.621	68.0	0.02	0.670	0.850
BNC	71.8	2.47	0.717	0.903	96.1	0.03	0.959	0.969	63.8	0.09	0.640	0.861
RBFN	87.6	5.39	0.875	0.940	97.0	0.56	0.971	0.961	84.3	6.98	0.842	0.893

4 Experiments and Results

For each user characteristic contemplated in this paper and for each of the five mentioned models, a large number of experiments of multiclass classification were conducted in Weka to find the values of classifiers' hyperparameters that lead to the best performance, in terms of accuracy (Acc.), time complexity (TBM—Time to Build Model), F-score (F1), and ROC index (AUC).

The results after hyperparameter tuning are shown in Table 3.

From Table 3, it can be seen that RBFN outperforms in terms of accuracy and F-score all other models in every classification problem examined in this work. In regard to AUC, RBFN has the best performance in age classification and similar value with LR and BNC in handedness classification, and with SVM and LR in educational level classification. However, the fastest model in each experiment proved to be NB followed by BNC and SVM, on the contrary, it has disadvantage in accuracy, F-score, and AUC in almost every case. RBFN and LR have the longest training time, but they are not prohibitive for their use as they are (without condensation method, reducing the dimensionality, etc.). In conclusion, it seems that the RBFN model is the most suitable for user classification according, mainly because it correctly predicts the age group, handedness, and educational level of an unknown user with 87.6%, 97.0%, and 84.3%, respectively.

5 Conclusion

Often, full anonymity on the Internet can make it difficult for users to access useful services, or even worse, be the advantage of malicious users. Existing methods that achieve user characteristics recognition require specific data, or are intrusive, or violate privacy. On the contrary, keystroke dynamics provide a non-intrusive low-cost method using data coming only from the way users use the keyboard.

This study presents a process in which the most suitable keystroke dynamics features are selected to identify age, handedness, and educational level of an unknown user. To accomplish the objective, a new keystroke dynamic dataset was created from recording users during the daily usage of their devices. Then, a feature selection procedure was followed and several machine learning models were tested to show that it is possible to recognize the aforementioned three characteristics of an unknown Internet user with accuracy of 87.6%, 97.0%, and 84.3%, respectively, using only a few hundred features and in a short time of model training.

Having the ability to recognize some characteristics of an unknown user who types a certain piece of text has significant value in digital forensics, targeted advertisement, and facilitating users.

Possible extensions of this research are, firstly, the combination of the results from the various models and their fusion using the theory of Dempster-Shafer. Secondly, the extension of the existing dataset and its balancing with undersampling, in order to re-conduct experiments and re-confirm our conclusions. Thirdly, the assessment of a model whether it does better or worse between consecutive classes or classes that are far apart (such as age groups) may provide useful directions on how to improve effectiveness. Finally, the implementation of an adaptive system that will utilize data related to the way users type and predict their characteristics. The system will modify its parameters according to new data, with possible changes in the way user types, so as to improve its ability to predict.

References

Arroju, M., Hassan, A., Farnadi, G. (2015). Age, gender and personality recognition using tweets in a multilingual setting. In: Proceedings of 6th Conference and Labs of the Evaluation Forum: Experimental IR Meets Multilinguality, Multimodality, and Interaction, pp. 23–31. Toulouse, France

Brizan, D.G., Goodkind, A., Koch, P., Balagani, K., Phoha, V.V., Rosenberg, A.: Utilizing linguistically enhanced keystroke dynamics to predict typist cognition and demographics. Int. J. Hum.-Comput. Stud. **82**, 57–68 (2015)

Buriro, A., Akhtar, Z., Crispo, B., Del Frari, F. (2016). Age, gender and operating-hand estimation on smart mobile devices. In: Proceedings of 2016 International Conference of the Biometrics Special Interest Group, pp. 273–280. Darmstadt, Germany

Cheung, M., She, J.: An analytic system for user gender identification through user shared images. ACM Trans. Multimed. Comput. Commun. Appl. **13**(3), 30:1–30:20 (2017)

Dragovic, M., Hammond, G.: A classification of handedness using the Annett Hand Preference Questionnaire. Br. J. Psychol. **98**(3), 375–387 (2007)

Goodman, J.: The wages of sinistrality: Handedness, brain structure, and human capital accumulation. J. Econ. Perspect. **28**(4), 193–212 (2014)

Menzies, T., Greenwald, J., Frank, A.: Data mining static code attributes to learn defect predictors. IEEE Trans. Softw. Eng. **33**(1), 2–13 (2007)

Pentel, A. (2017). High precision handedness detection based on short input keystroke dynamics. In: Proceedings of 8th International Conference on Information, Intelligence, Systems and Applications, pp. 1–5. Larnaca, Cyprus

Pentel, A.: Predicting user age by keystroke dynamics. In: Silhavy, R. (ed.) Artificial Intelligence and Algorithms in Intelligent Systems, pp. 336–343. Springer International Publishing, Switzerland (2018)

Roy, S., Roy, R., Sinha, D.D.: ACO-Random forest approach to protect the kids from Internet threats through keystroke. Int. J. Eng. Technol. **9**(3S), 279–285 (2017)

Seneviratne, S., Seneviratne, A., Mohapatra, P., Mahanti, A.: Predicting user traits from a snapshot of apps installed on a smartphone. ACM SIGMOBILE Mob. Comput. Commun. Rev. **18**(2), 1–8 (2014)

Shen, C., Xu, H., Wang, H., Guan, X. (2016). Handedness recognition through keystroke-typing behavior in computer forensics analysis. In: Proceedings of 2016 IEEE Trustcom/BigDataSE/ISPA, pp. 1054–1060. Tianjin, China

Shute, S., Ko, R.K.L., Chaisiri, S. (2017). Attribution using keyboard row based behavioural biometrics for handedness recognition. In: Proceedings of 2017 IEEE Trustcom/BigDataSE/ICESS, pp. 1131–1138. Sydney, NSW, Australia

Tsimperidis, I., Rostami, S., Katos, V.: Age detection through keystroke dynamics from user authentication failures. Int. J. Digit. Crime Forensics **9**(1), 1–16 (2017)

Tsimperidis, I., Arampatzis, A., Karakos, A.: Keystroke dynamics features for gender recognition. Digit. Investig. **24**, 4–10 (2018)

Technological Innovation and the Critical Raw Material Stock

Beatrix Varga and Kitti Fodor

Abstract We live in a dynamically changing world. There have been so many innovations in the last few years that raw materials have become really indispensable. The European Commission collected information in separate studies for Critical Raw Materials (CRMs) and for non-critical raw materials. This paper is based on that data. In 2011 there were 14 materials on the critical raw material list, but in 2017 this list had grown to contain 27 materials. Critical raw materials play a key role in technological innovation; they are the necessary raw materials for many innovations. In this paper, our aim is to identify groups using hierarchical cluster analysis and to identify which clusters are important for innovation. We selected three variables for cluster analysis: Economic Importance (EI), Supply Risk (SR), and End of Life recycling input rate (EoL), and we identified five homogenous groups. There is one group that seems particularly important because it includes only critical raw materials.

Keywords Critical raw materials · Clusters · Ward's method

1 Introduction

At the beginning of the third millennium, we will have to face serious economic and environmental issues. In this paper, we explore the question of what kind of effect today's technical innovations will have on the stocks of raw material.

We live in a dynamically changing world. There have been periods in our history when the use of cars was unimaginable, and today, by contrast, self-driving cars are available, and electric cars are becoming increasingly popular. If we think about communication, telephones were not invented a century and a half ago, and today

B. Varga (✉) · K. Fodor
University of Miskolc, Institute of Economic Theory and Methodology,
H-3515 Miskolc-Egyetemváros, Hungary
e-mail: varga.beatrix@uni-miskolc.hu

K. Fodor
e-mail: fodor.kitti@uni-miskolc.hu

T. Chadjipadelis et al. (eds.), *Data Analysis and Rationality in a Complex World*,
Studies in Classification, Data Analysis, and Knowledge Organization,
https://doi.org/10.1007/978-3-030-60104-1_35

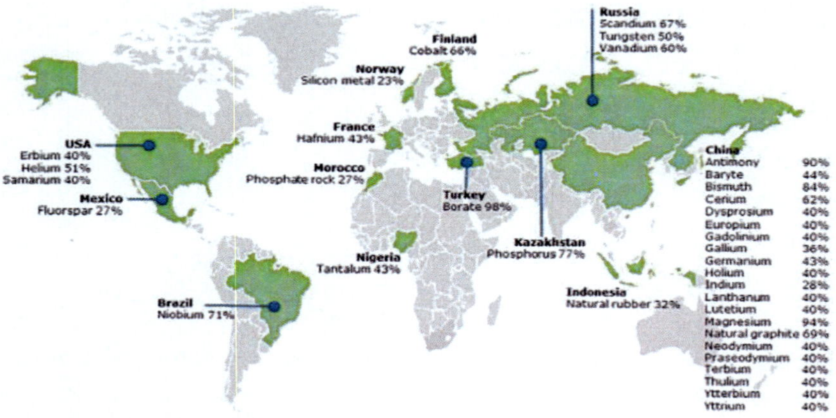

Fig. 1 Countries accounting for largest share of global supply of CRMs. *Source* European Commission https://ec.europa.eu/growth/sectors/raw-materials/specific-interest/critical_hu

we carry most of the entire world within our pockets. Therefore, finding the answer to what the future may bring is not easy.

Countless numbers of innovations have taken place in the last years and decades, in which raw materials have played an important role. The European Commission has collected information on critical and non-critical raw materials, and these raw materials form the focus of our research.

Technical innovation regularly rewrites the list of important raw materials, since these raw materials are an indispensable part of production, as their substitution with other raw materials is in many cases unsolved. A good example is semiconductors, whose specific resistance is between conductors and insulators. At room temperature, they have poor current conductivity, but as the temperature increases, their resistance decreases. Their use is diverse, as these elements are needed in LEDs, solar panels, and integrated circuits. Such materials are silicon and germanium. Silicon was classified as a critical raw material by the European Commission in 2014 and 2017 and germanium in all three years under review. The survey also reveals that there are countries that have a strong influence on the raw material market. Figure 1 illustrates these countries.

It can be said that China has a strong monopoly, which has had negative effects over the past few years. China has repeatedly invoked this situation as a weapon in recent years. China imposed several export restrictions that restricted access by companies outside China to the products concerned. These measures distorted competition and put the Chinese industry at an advantage, to the detriment of EU businesses and consumers. Therefore, the European Union has repeatedly initiated proceedings against restrictions on Chinese exports of raw materials important to European industries.

One such move was for China to restrict exports of rare earth elements in the early 2010s, and even to temporarily halt export to Japan. China has the chance to do this because the production of China is a significant part of the world production of these raw materials and a significant part of the world's stock of rare earth elements is in China. However, the mining and refining of these raw materials is extremely damaging to both human health and the environment, which, in addition to its natural potential, has also contributed to China's leadership in the field. In recent years, the rest of the world has also awakened from its sleep and is trying to boost extraction in other countries.

In 2010, the U.S. Government Accountability Office (GAO) issued a report to protect the rare earth supply chain. A GAO investigation had already alerted the U.S. Senate to the dangers of rare earth dependence. Several mineral-rich countries, such as Mongolia, Angola, and Nigeria, were visited by the German Chancellor in 2011, with the undisguised intention of providing German industry with essential critical raw materials. The news in 2019 was that the Romanian government is preparing to reopen hundreds of mines.

2 Critical Raw Materials in the World

Individual countries and organizations have recognized the need to pay attention to strategically important raw materials, so many have made a list, analysis, or study of raw materials that are important to them and have defined strategies for raw materials. Developments in recent years have also been followed up by the European Commission and surveys of relevant raw materials have been carried out. In the first year under review, in 2011, only 14 items were classified as critical, but in six years the number of these raw materials nearly doubled. Critical elements are those which are economically and strategically important, whose supply risk is high, and whose substitution is difficult. The European Commission categorizes raw materials according to two main parameters, which are the supply risks and economic importance.

> In the EC methodology, the Economic Importance is calculated based on the importance of a given material in the EU end-use applications and performance of available substitutes in the applications. (European Commission 2017b, p. 7)

> In the EC methodology, the Supply Risk is calculated based on factors that measure the risk of disruption in supply of a specific material. (European Commission 2017b, p. 9)

Such critical raw materials are germanium, helium and magnesium, or cobalt and graphite, which are required for lithium-ion batteries of electric cars.

In the case of cobalt, more than half of its yearly stock is also supplied by the politically unstable Democratic Republic of the Congo (European Commission 2017c). In the past few years, the demand for cobalt has risen, which is clearly reflected by its price. At the beginning of 2016, the price for a ton of cobalt was USD 22,000, which then more than quadrupled within two years. Only after this growth did the price begin to decrease; at this point (Q4 2019) its price is the twice of the starting price. However, the list of critical raw materials is constantly rewritten as technology evolves and new raw materials become more important.

It can be stated that the same raw materials are not necessarily strategically important for each country. Comparing the critical raw materials of the European Commission (2017) and the USA (2018), we can see that the raw materials on the list are not exactly the same. There is a list of raw materials that the European Commission did not classify as critical, but the USA does such as aluminum (bauxite), chromium, lithium, manganese, potash, rhenium, tellurium, tin, titanium, and zirconium. However, there are raw materials that were not included in the European Commission's survey but US experts believe are critical. These materials are arsenic, caesium, rubidium, strontium, and uranium.

Furthermore, it is important to look at China's strategy, as China plays a key role in the market for strategically important raw materials. China has been part of the rare earth market since the mid-twentieth century, and it took only a few decades for its leaders to realize that this is the key to the economic growth in China. The Chinese labor market and regulatory background helped the country to become an absolute monopoly by the end of the 2000s. More than two decades ago, China decided to designate rare earths as "protected and strategic materials". This strategy included the retention of rare earth metals for domestic use.

3 Database and Methodology

For our study, we have created a database which is based on the research of the European Commission. 73 raw materials were included in the database, and 54.8% of these elements were categorized as critical in 2017. The database includes 15 variables, for example, the name of the raw material, supply risk, economic importance, EU import reliance, end-of-life recycling input rate, major world producers, and critical classification in 2011, 2014, and 2017.

In our study, our goal is the identification of groups with the help of hierarchical cluster analysis, and the identification of clusters that are important from the perspective of innovation.

Cluster analysis means the grouping of items that are similar to each other, and its goal is to categorize the observed units into homogeneous groups. Basically, this is a technique for exploration, and there is no such thing as a best result. As in all analyses, it is important to examine the conditions. This includes managing outliers, scaling appropriately, and examining the correlation of variables. There are several methods of cluster analysis to choose from, for example, hierarchical and partitioning methods, model-based methods, etc. Each method has its own advantages and disadvantages, and we strove to perform our analysis keeping this in mind. We chose hierarchical cluster analysis because there was no a priori information about the number of the clusters, and the hierarchical method can help in defining the number of clusters. Further, information regarding our analysis will be addressed in the following sections of the study.

4 Empirical Research

Our goal was to assign the raw materials into homogeneous groups. Presumably, this can help with the identification of important raw materials in regard to innovation. The clusters were based on the Economic Importance (EI), Supply Risk (SR) and End of Life recycling input rate (EoL). Since the scaling of these three variables is different, standardization was used. The SPSS software package offers several options when it comes to standardization, and we chose z-score from these. We paid extra attention to the management of outliers. The identification of outliers was done with the "nearest neighbor" method. We identified two outliers, which were magnesium and titanium. Based on a dendrogram, we reached the conclusion that the creation of 5 clusters would be ideal. We used Ward's Method and Squared Euclidean distance. As the three-dimensional scatter plot (Fig. 2) shows, the groups are nicely separated from each other.

For the validation of our analysis, we used K-means clustering, and the number of iterations was 5.

We then verified whether the groups are similar. As Fig. 3 shows, similar groups were generated as in the case of Ward's method.

For the validation, we used another method as well, which is the Rand Index. The Rand Index is a measure of the similarity between two clusterings.

Based on the Table 1, the value of the Rand Index is 0.961. This means that similar groups were created using the two methods. For testing the internal validity, we used the Average Silhouette Width (ASW). The mean of the values was 0.5 and the maximum was 0.772. We did not get a negative value. In the case of silhouette calculations, a value close to 1 indicates that the clustering can be considered good for the given cluster number, so we can conclude that clustering can be considered moderately good. The highest values were in the first and second groups, which are extremely important for innovation. The lowest value was in the third group. This

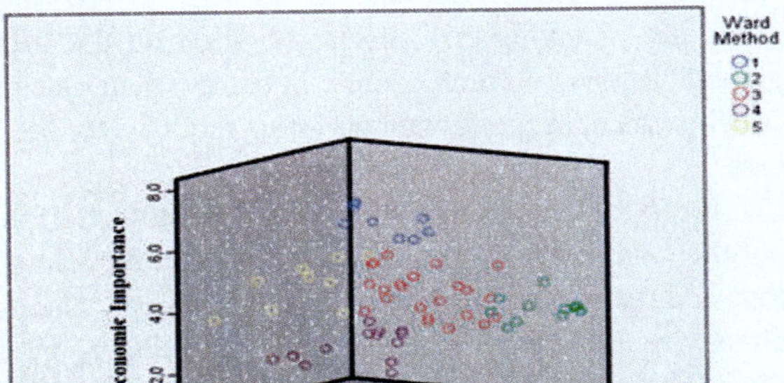

Fig. 2 Raw material clusters in 3D scatter. *Source* Own editing, SPSS output

Ward Method * Cluster Number of Case Crosstabulation

Count

		Cluster Number of Case					Total
		1	2	3	4	5	
Ward Method	1	9	0	0	0	0	9
	2	0	0	16	0	0	16
	3	1	0	1	21	1	24
	4	0	0	0	0	13	13
	5	2	7	0	0	0	9
Total		12	7	17	21	14	71

Fig. 3 Crosstabulation table for validation. *Source* Own editing, SPSS output

Table 1 Rand Index results Source: Own editing, SPSS output

			Cluster method 1		Total
			Pair assigned to		
			The same cluster	Different cluster	
Cluster method 2	Pair assigned to	The same cluster	490	56	46
		Different cluster	42	1897	1939
Total			532	1953	2485

may be because the classification of materials in this group is constantly changing. More information about this will be included in the characterization of the clusters.

Characterization of the Clusters

The first cluster contains 9 elements with high economic importance and medium supply risk. Only 1% of the world's production takes place in Europe, so import dependence can also be considered high. A growing number of elements in this group are receiving critical classification on the lists of the EU. This group was named "Innovation dependent". Such raw materials are, for example, tungsten and cobalt, which, as mentioned above, is required for the lithium-ion battery of electric cars.

The 16 raw materials in the second cluster are of medium economic importance, but their supply risk is extremely high. The recycling rate of these raw materials is low, and their production in Europe is minimal. The majority of these raw materials already received a critical classification in 2011, but all of them became critical in 2017. The main producer, with the exception of ruthenium, is located in Asia, so this cluster is named "Treasures of Asia". Such raw materials are, for example, samarium and antimony.

The 24 raw materials in the third cluster are considered to be of medium economic importance as well as supply risk. Europe yields 10% of the world's total production, so the dependence on import is also moderate. However, the analysis of the

EU confirms that these raw materials are becoming increasingly important, since a growing number of them are classified as critical. This cluster was named "Emerging Criticals". Such element are natural graphite, germanium, and helium.

The 13 raw materials of the fourth cluster have neither a significant supply risk nor high economic importance. One third of global production comes from Europe. In the survey of 2014, 8 of these materials received a critical classification, but this did not occur in the previous or in the following survey. Most of the major producers are located outside of Asia. Based on the above characteristics, the cluster deserves to be named "Harmless". For example, gypsum and lithium are in this cluster.

The 9 elements of the last, fifth cluster are characterized by a moderate-high economic importance and the lowest supply risk. Their recycling rates are high, at over 41%, which mitigates the fact that Europe accounts for only 1% of world production. In 2017, 2 of these materials were classified critical. The major producers come from Asia. Based on previous characteristics, the group was named "Recyclables". Such raw materials are lead, silver, and copper.

As mentioned above, 9 raw materials have been identified as critical for the Americans but not identified as critical by the European Commission. One of these raw materials, titanium was identified as an outlier. Of the remaining 8 raw materials, 5 are members of a cluster called Emerging Criticals or Innovation Dependent. These raw materials are aluminum (bauxite), chromium, manganese, potash, and tellurium. We believe that these raw materials deserve special attention and should be closely monitored.

5 Summary

Can we say that from the five clusters there is one which significantly more important from the perspective of innovation than the others? There is indeed a cluster that includes only critical raw materials, but it cannot be stated that it is more important than the other clusters. Let's take the batteries of electric cars, for example, which require manganese, lithium, cobalt, nickel, graphite, aluminum, and copper. This is only a single product, but it affects four out of the five groups. However, the analysis can be of help with the adjustment of attitude toward each group and raw materials.

However, we must not lose sight of the fact that our world is constantly changing and new technologies are emerging; these factors make it difficult to predict in advance what raw materials will be needed. At any time, technologies that are still new today may become history tomorrow, and rewrite the list of strategically important raw materials.

Acknowledgements The described work was carried out as part of the "Sustainable Raw Material Management Thematic Network—RING 2017", EFOP-3.6.2-16-2017-00010 project in the framework of the Széchenyi 2020 Program. The realization of this project is supported by the European Union, co-financed by the European Social Fund.

References

Balázs, Á.: Félvezetők: Az Alkalmazott fizika I. eloadás alapján (29 Jan 2016), [online], http://ispanovity.web.elte.hu/teaching/2016_2017/kor_musz_alk_i/jegyzet/T.pdf [05 Dec 2019]

CriticEl – Kritikus Elemek. http://kritikuselemek.uni-miskolc.hu/

CRM Alliance. http://criticalrawmaterials.org/

European Commission (a) (2017): Methodology for establishing the EU list of critical raw materials (Publications Office of the EU, Luxembourg). https://publications.europa.eu/en/publication-detail/-/publication/2d43b7e2-66ac-11e7-b2f2-01aa75ed71a1

European Commission (b): Study on the review of the list of Critical Raw Materials—Criticality Assessment (Publications Office of the EU, Luxembourg) (2017). https://publications.europa.eu/en/publication-detail/-/publication/08fdab5f-9766-11e7-b92d-01aa75ed71a1/language-en

European Commission (c): Study on the review of the list of Critical Raw Materials—Critical Raw Materials Factsheets (Publications Office of the EU, Luxembourg) (2017). https://publications.europa.eu/en/publication-detail/-/publication/7345e3e8-98fc-11e7-b92d-01aa75ed71a1/language-en

European Commission (d): Study on the review of the list of Critical Raw Materials—Non-critical Raw Materials Factsheets (Publications Office of the EU, Luxembourg) (2017). https://publications.europa.eu/en/publication-detail/-/publication/6f1e28a7-98fb-11e7-b92d-01aa75ed71a1/language-en

European Commission: Critical Raw Materials. https://ec.europa.eu/growth/sectors/raw-materials/specific-interest/critical_en

Eurostat [online], https://ec.europa.eu/eurostat/home [10 Oct 2019]

EU-USA-Japán összefogás a kínai ritkaföldfém-kontroll ellen [online], https://www.portfolio.hu/befektetes/20120313/eu-usa-japan-osszefogas-a-kinai-ritkafoldfem-kontroll-ellen-164263 [10 Dec 2019]

Gere László: A Föld stratégiai fontosságú nyersanyagai (19. April 2018). http://www.geopolitika.hu/hu/2018/04/19/a-fold-strategiai-fontossagu-nyersanyagai/#_edn8

Gombkötő Imre – Magyar Tamás: Mik azok a kritikus nyersanyagok, mi lenne velünk nélkülük? https://matarka.hu/koz/ISSN_2062-204X/4_evf_2_sz_2013/ISSN_2062-204X_4_evf_2_sz_2013_058-078.pdf

U.S. Government Accountability Office (GAO): Rare Earth Materials in Defense Supply Chain [online], https://www.gao.gov/new.items/d10617r.pdf [14. May 2020]

Hajdu, O.: Többváltozós statisztikai számítások. Központi Statisztikai Hivatal, Budapest (2003)

Így tartja markában Kína a hitech ipart (2. August 2011). https://index.hu/tudomany/2011/08/02/igy_tartja_markaban_kina_a_hitech_ipart/

Imre, K.: Világmodellek - A Római Klub jelentéseitol az ENSZ kezdeményezéséig. Közgazdasági és Jogi Könyvkiadó, Budapest (1980)

Interior Releases 2018's Final List of 35 Minerals Deemed Critical to U.S. National Security and the Economy (18 May 2018) [online], https://www.usgs.gov/news/interior-releases-2018-s-final-list-35-minerals-deemed-critical-us-national-security-and [30 Nov 2019]

László, E. e. a.: *Goals for Mankind.* The New American Library of Canada Limited, Scarborough (1978)

Malhotra, N.K.: Marketingkutatás. Akadémiai Kiadó, Budapest (2008)

Meadows, D.H., Meadows, D.L., Randers, J.: Beyond the Limits. Earthscan publications Limited, London (1992)

Nyersanyag, ami elöbb elfogy, mint az olaj [online], https://index.hu/tudomany/2011/04/03/ot_nyersanyag_amibol_elobb_kifogyunk_mint_az_olajbol/ [3 Apr 2011]

Nyersanyagstop: Itt a kínai bosszú a Huaweiért [online], https://infostart.hu/tudositoink/2019/05/29/nyersanyagstop-itt-a-kinai-bosszu-a-huaweiert [12 Dec 2019]

Sajtos, L., Mitev, A.: SPSS kutatási és adatelemzési kézikönyv. Alinea Kiadó, Budapest (2007)

Silberglitt, R. et al.: Critical Materials – Present Danger to U.S. Manufacturing [online], https://www.rand.org/content/dam/rand/pubs/research_reports/RR100/RR133/RAND_RR133.pdf [20 August 2019]

Spiegel: Merkel joins the Global Hunt for Natural Resources [online], https://www.spiegel.de/international/world/help-for-german-industry-merkel-joins-the-global-hunt-for-natural-resources-a-795256.html [10. May 2020]

Szüle, B.: *Klaszterszám-meghatározása módszerek összehasonlítása*, Statisztikai Szemle, (2019) 97. evf., 5. sz. pp. 421–438

The International Raw Materials Observatory [online], https://intraw.eu/reports-factsheets/ [23 Oct 2019]

Több száz bánya újranyitására készül a kormány Romániában[online],https://uzletem.hu/regio/tobb-szaz-banya-ujranyitasara-keszul-a-kormany-romaniaban [05. May 2020]

Trading Economics [online], https://tradingeconomics.com/ [17 Nov 2019]

Újabb kínai ritkaföldfém-kvóták (27 May 2011) [online], https://kitekinto.hu/kelet-azsia/2011/05/27/ujabb_kinai_ritkafoldfem-kvotak [20 Dec 2019]

Varga B., Szilágyi R.: Kvantitatív információképzési technikák https://www.tankonyvtar.hu/hu/tartalom/tamop425/0049_08_kvantitativ_informaciokepzesi_technikak/3268/index.html (2011)

Visnovitz, P.: Hadat üzent Kína a jövő technikáit kereső országoknak (02.01.2011) [online], https://www.origo.hu/nagyvilag/20101230-ritkafoldfem-katonai-fuggesbe-taszithatjak-a-vilagot-kina-nelkulozhetetlen-ritkafoldfemei.html [15 Dec 2019]

WKO: Neue Lister der kritische Rohstoffe der EU und Implikationen für Wirtschaft [online], https://news.wko.at/news/oesterreich/2017-21-Kritische-Rohstoffe.pdf [29 Sept 2017]

Redundancy Analysis for Binary Data Based on Logistic Responses

Jose L. Vicente-Villardon and Laura Vicente-Gonzalez

Abstract Redundancy Analysis (RDA) is one of the many possible methods to extract and summarize the variation in a set of response variables that can be explained by a set of explanatory variables. The main idea is to use multivariate linear regression to explain the responses as a linear function of the explanatory and then use Principal Component Analysis (PCA) or a biplot to visualize the result. When response variables are categorical (binary, nominal, or ordinal), classical linear techniques are not adequate. Some alternatives such as Distance-Based RDA have been proposed in the literature. In this paper, we propose versions of RDA based on generalized linear models with logistic responses. The natural visualization methods for our techniques are the *Logistic Biplots*, recently proposed. The procedures are illustrated with an application to real data.

Keywords Logistic biplot · Redundancy analysis · Binary data

1 Introduction

Redundancy Analysis (RDA) is an extension of Multiple Regression to cope with several response variables, so it tries to explain a set of responses from a set of predictors. It was proposed by Rao (1964) and later rediscovered by Van der Wollenberg (1977) as an alternative to Canonical Correlation Analysis (CCA). *Redundancy* means *explained variance* in this context. RDA is essentially a Principal Components Analysis of the responses in which the resulting components are linear combinations of the predictors. It is also called *Constrained PCA* or *PCA with external information* (Takane and Tadashi 1991).

J. L. Vicente-Villardon (✉) · L. Vicente-Gonzalez
Departamento de Estadística, Universidad de Salamanca, Salamanca, Spain
e-mail: villardon@usal.es

L. Vicente-Gonzalez
e-mail: laura20vg@usal.es

T. Chadjipadelis et al. (eds.), *Data Analysis and Rationality in a Complex World*,
Studies in Classification, Data Analysis, and Knowledge Organization,
https://doi.org/10.1007/978-3-030-60104-1_36

Israels (1984) extends the analysis to include qualitative variables using quantifications as in the Gifi system. The procedure has not been extensively used in practice, and it is practically forgotten in the most recent literature. Then, there is still a need for techniques that can deal with qualitative data. Our approach will be based on logistic or generalized linear models rather than quantification.

Other approaches (Legendre and Anderson 1999) treat the qualitative variables using distances. The method used the principal coordinates based on distances calculated with the response variables.

In this paper, we extend RDA to binary variables using a binary logistic biplot (Vicente-Villardon et al. 2006) as a visualization technique. The generalization is straightforward although there are some details that we have taken into count to obtain a suitable generalization.

2 RDA for Quantitative Variables

Let $\mathbf{Y}_{n\times p}$ and $\mathbf{X}_{n\times q}$ denote two data matrices containing the observations of p response variables and q predictors for n individuals. The two data matrices are centered and probably standardized by columns if the variables are not dimensionally homogeneous. In RDA, we search for a linear combination $\mathbf{X}\beta$ that maximizes the explained variances of *all* p response variables. Then, the sample redundancy analysis is obtained from the eigenvalues and eigenvectors of $(\mathbf{X}^T\mathbf{X})^{-1}\mathbf{X}^T\mathbf{Y}\mathbf{Y}^T\mathbf{X}$ or equivalently from the eigenvalues and eigenvectors of $\mathbf{Y}^T\mathbf{X}(\mathbf{X}^T\mathbf{X})^{-1}\mathbf{X}^T\mathbf{Y}$.

In Legendre and Legendre (2012), RDA is described as a two step procedure that includes Linear Regression (LR) and PCA. The procedure is as follows:

- Regress each column of $\mathbf{Y}$ on $\mathbf{X}$. The whole matrix of regression coefficients is $\mathbf{B} = (\mathbf{X}^T\mathbf{X})^{-1}\mathbf{X}^T\mathbf{Y}$. This is also called a *multivariate regression*. Because both sets are centered no intercepts have to be used.
- Calculate the fitted values of the regression $\hat{\mathbf{Y}} = \mathbf{XB}$. $\hat{\mathbf{Y}}$ contains the part of $\mathbf{Y}$ explained by $\mathbf{X}$. $\mathbf{E} = \mathbf{Y} - \hat{\mathbf{Y}}$ contains the residuals, i.e., the non explained part.
- Perform a PCA of the fitted values from its Singular Value Decomposition (SVD).

$$\hat{\mathbf{Y}} = \mathbf{U}\Lambda\mathbf{V}^T \tag{1}$$

where $\mathbf{U}$ are the eigenvectors of $\hat{\mathbf{Y}}\hat{\mathbf{Y}}^T$, $\mathbf{V}$ the eigenvectors of $\hat{\mathbf{Y}}^T\hat{\mathbf{Y}}$, and Λ the square roots of the non-negative eigenvalues of both matrices that are the same. There are only $min(p, q, n-1)$ non-zero eigenvalues, and normally a smaller subset is necessary to describe the explained variation. $\mathbf{V}$ are also interpreted as a set of principal components but constrained to be linear combinations of the predictors. This means that $\mathbf{V}$ is a set of orthogonal vectors which define a subspace in the same way as in PCA.

- A biplot $\hat{\mathbf{Y}} = \mathbf{P}\mathbf{Q}^T$ can obtained from the SVD in (1) taking $\mathbf{P} = \mathbf{U}\Lambda$ and $\mathbf{Q} = \mathbf{V}$ (or $\mathbf{P} = \mathbf{U}$ and $\mathbf{Q} = \mathbf{V}\Lambda$). **P** are called the *fitted site scores* in ecological applications and are also $\mathbf{P} = \hat{\mathbf{Y}}\mathbf{V} = \mathbf{X}\mathbf{B}\mathbf{V}$.
- As in PCA, the values in **Y** can also be projected onto the constrained components $\mathbf{T} = \mathbf{Y}\mathbf{V}$ to approximate the observed rather than the fitted values. This defines a biplot $\mathbf{T} = \mathbf{Y}\mathbf{V}^T$ for the responses. **T** are called the *site scores* in ecological applications.
- The matrix $\mathbf{C} = \mathbf{B}\mathbf{V}$ gives the coefficients of the explanatory variables **X** in the calculations of the fitted site scores, thus can serve as the vectors in an interpolation biplot to project new observations. Given a new set of individuals with values $\mathbf{X}_h$ on the predictors, the fitted scores are $\mathbf{P}_h = \mathbf{X}_h\mathbf{C} = \mathbf{X}_h\mathbf{B}\mathbf{V}$, and the predictions of the responses $\hat{\mathbf{Y}} = \mathbf{P}_h\mathbf{V}^T$.
- There are many possible biplot scalings depending on what we can represent better. We will not describe here all the details.
- A prediction biplot for **X** can also be obtained by regressing the predictors on the fitted site scores. The markers on the biplot would be $\mathbf{H} = (\mathbf{P}^T\mathbf{P})^{-1}\mathbf{P}^T\mathbf{X}$ and the biplot for the predictors $\mathbf{X} \simeq \mathbf{P}\mathbf{H}^T$. A typical triplot would be the combination of **P**, **Q** and **H**.

In summary, from our point of view, RDA is a technique to extracting the part of a set of response variables explained by a set of predictors and visualizing it in a reduced dimension representation. The advantage over separated linear regressions is that we have a joint picture of the responses and the predictors.

3 RDA for Binary Variables

When response variables are binary, linear models are not suitable and generalized linear models must be used instead.

Let $\mathbf{Y}_{n\times p}$ and $\mathbf{X}_{n\times q}$ denote two data matrices containing the observations of p binary responses and p explanatory variables (probably quantitative) for n individuals. The observed values of the response are either 0 or 1 for presence or absence. The matrix of predictors is centered and probably standardized by columns if the variables are not dimensionally homogeneous. The responses can not be centered in the usual way.

Let $\Pi_{n\times p}$ be the matrix of expected probabilities whose columns are obtained fitting logistic regressions to each column of **Y** on the whole set of predictors and $\mathbf{Z}_{n\times p}$ the matrix containing the expected logits.

The expected probabilities given by a *logistic regression*,

$$\text{logit}(\pi_{ij}) = \ln(\frac{\pi_{ij}}{1-\pi_{ij}}) = z_{ij} = \beta_{j0} + \beta_{j1}x_{i1} + \ldots + \beta_{jq}x_{iq}$$

with $i = 1, \ldots, n$ and $j = 1, \ldots, p$.

We describe the generalization of the algorithm for quantitative data with the necessary adaptations to binary responses.

- Regress each column of $\mathbf{Y}$ on $\mathbf{X}$ using standard logistic regression. Because the responses can not be centered, we have to keep the intercept in the models. The whole matrix of regression coefficients for the predictors is $\mathbf{B}_{p\times q} = (\beta_{jk}$ and the intercepts $\mathbf{b_0} = (\beta_{j0})$. Some penalization, as Ridge, may be used to avoid overfitting when the number of predictors is high or a separation is present.
- Calculate the fitted logits of the regression

$$\mathbf{Z} = \mathbf{1b}_0^T + \mathbf{XB} \tag{2}$$

- Perform a PCA of the fitted logits from the Singular Value Decomposition (SVD) of the second term in (2).

$$\mathbf{Z} = \mathbf{1b}_0^T + \mathbf{U}\Lambda\mathbf{V}^T \tag{3}$$

- A biplot $\mathbf{Z} = \mathbf{1b}_0^T + \mathbf{PQ}^T$ can obtained from the SVD in (3) taking $\mathbf{P} = \mathbf{U}\Lambda$ and $\mathbf{Q} = \mathbf{V}$. $\mathbf{P}$ are the *fitted site scores*. This is a logistic biplot in logistic scale as defined by Vicente-Villardon et al. (2006). That is, the logit is approximated by

$$\text{logit}(\pi_{ij}) = b_{j0} + \sum_{k=1}^{q} q_{jk}p_{ik} = b_{j0} + \mathbf{p}'_i\mathbf{q}_j$$

  In the original scale, the expected probability is approximated by

$$\pi_{ij} = \frac{e^{\left(b_{j0}+\sum_k p_{jk}q_{ik}\right)}}{1 + e^{\left(b_{j0}+\sum_k p_{jk}q_{ik}\right)}} \tag{4}$$

  The geometry of the representation is studied by Vicente-Villardon et al. (2006) and Demey et al. (2008). The general idea is that the directions on the representation defined by the regression coefficients are the directions that better predict the expected probabilities. The biplot directions are completed with graded scales to predict probabilities.
- A prediction biplot for $\mathbf{X}$ can also be obtained by regressing the predictors on the fitted site scores in the same way we did for quantitative variables. The markers on the biplot would be $\mathbf{H} = (\mathbf{P}^T\mathbf{P})^{-1}\mathbf{P}^T\mathbf{X}$ and the biplot for the predictors $\mathbf{X} \simeq \mathbf{PH}^T$. A typical triplot would be the combination of $\mathbf{P}$, $\mathbf{Q}$ and $\mathbf{H}$.

Then, we have the part of the binary variables explained by the predictors visualized in a biplot. An alternative way to proceed is center $\mathbf{Z}$, calculate its principal components an fit a logistic regression to each column of $\mathbf{Y}$ using the component scores as independent variables, in order to recalculate de logistic biplot. Then, we have obtained a logistic biplot in which the responses have a logistic relation to the dimensions, and the dimensions are linear combinations of the predictors.

The calculation of the site scores using the original values rather than the expected probabilities is more difficult and needs some special attention. We describe it in more detail. To estimate the parameters of the model in (4), we maximize the cost function

$$L = \sum_{i=1}^{n}\sum_{j=1}^{p}\left[-y_{ij}\log(\pi_{ij}) - (1-y_{ij})\log(1-\pi_{ij})\right] \tag{5}$$

using the gradient descent method rather than Newton–Raphson method traditionally used to obtain maximum likelihood estimates. The iterative algorithm will update the values of the parameters at each iteration as

$$p_{ik} := p_{ik} - \alpha\frac{\partial L}{\partial p_{ik}}; \quad q_{jk} = q_{jk} - \alpha\frac{\partial L}{\partial q_{jk}}; \quad b_{j0} = b_{j0} - \alpha\frac{\partial L}{\partial b_{j0}}$$

where α is a constant and

$$\frac{\partial L}{\partial p_{ik}} = \sum_{j=1}^{p} q_{jk}(\pi_{ij} - y_{ij}) \quad (k = 1, \ldots, r)$$

$$\frac{\partial L}{\partial q_{jk}} = \sum_{i=1}^{n} p_{ik}(\pi_{ij} - y_{ij}) \quad (k = 1, \ldots, r)$$

$$\frac{\partial L}{\partial b_{j0}} = \sum_{i=1}^{n}(\pi_{ij} - y_{ij})$$

are the gradients and r the dimension of the solution. The equations can be used, together with some initial guessing, in an optimization routine to obtain the solution of the problem. If some of the parameters are known in advance, just the needed equations will be used.

4 Examples

4.1 Spiders Data

For the illustration of the procedure, we have selected the spiders data originally published in Smeenk-Enserink and Van der Aart (1974), because these data have been used in various papers to illustrate ordination techniques, in particular by Ter Braak (1986) in its original paper about Canonical Correspondence Analysis.

The data set concerns the distribution of wolf spiders in a dune area. The original data consist of counted abundances of twelve species captured at 100 sites (pitfall

Table 1 Spiders data including species and environmental variables

Site	Arctl	Prdl	Zrsp	Prdn	Prdp	Allb	Trct	Alpcn	Prdm	Alpcc	Alpf	Arctp	Watcont	Barsand	Covmoss	Ligrefl	Falltwi	Coverher
s01	0	0	1	1	1	1	1	1	1	1	0	0	5	0	7	8	0	9
s02	0	1	1	1	1	1	1	1	1	0	0	0	8	0	2	3	3	9
s03	1	1	1	1	1	1	1	1	1	1	1	0	6	0	5	8	0	9
s04	1	1	1	1	1	1	1	1	1	1	0	0	6	0	5	6	0	9
s05	1	1	1	1	1	1	1	1	1	1	0	0	8	0	0	5	0	9
s06	1	0	1	1	1	1	1	1	1	0	0	0	9	5	5	1	7	6
s07	1	1	1	1	1	1	1	1	1	1	0	0	8	0	1	5	0	9
s08	0	1	1	1	1	1	1	1	1	0	0	0	6	0	2	1	9	6
s09	0	0	0	1	1	0	1	1	1	1	0	0	5	0	9	7	0	6
s10	0	0	0	0	0	0	1	0	1	1	1	0	4	8	7	8	0	5
s11	0	0	0	0	1	1	1	1	1	1	1	0	4	0	9	8	0	7
s12	0	0	0	1	1	0	1	1	1	1	0	0	5	0	8	8	0	8
s13	1	1	1	1	1	1	1	1	1	1	1	0	9	3	1	7	3	9
s14	1	1	1	1	1	1	1	1	1	0	0	0	8	0	4	2	0	9
s15	0	1	1	0	0	0	1	0	0	0	0	0	9	0	1	1	9	5
s16	0	1	1	1	0	0	1	1	0	0	0	0	8	0	1	0	9	0
s17	0	1	1	0	0	0	1	0	0	0	0	0	9	0	1	2	9	5
s18	0	1	0	0	0	0	1	1	0	0	0	0	8	0	0	2	9	5
s19	0	1	1	1	0	0	1	1	0	0	0	0	7	0	3	0	9	2
s20	0	1	1	0	0	0	1	1	0	0	0	0	8	0	1	0	9	0
s21	0	1	1	0	1	0	1	1	1	0	0	0	7	0	1	0	9	2
s22	0	0	0	0	0	0	1	0	1	1	1	1	1	7	9	8	0	0
s23	0	1	0	0	0	0	1	0	1	1	1	1	0	6	9	9	0	6
s24	0	0	0	0	0	0	0	0	1	1	1	1	2	7	9	9	0	5
s25	0	1	1	1	0	1	1	1	1	1	1	0	3	7	2	5	0	8
s26	0	0	0	0	0	0	1	0	0	1	1	1	0	9	4	9	0	2
s27	0	0	0	0	0	0	0	0	1	1	1	1	0	5	8	8	0	6
s28	0	0	0	0	0	0	1	0	1	1	1	1	0	7	8	8	0	6

traps) in a dune area in the Netherlands. We use the observed abundances at the 28 sites where environmental variables were monitored. The species are *Arctosa lutetiana, Trochosa terricola, Pardosa lugubris, Alopecosa cuneata, Zora spinimana, Pardosa monticola, Pardosa nigriceps, Alopecosa accentuata, Pardosa pullata, Alopecosa fabrilis, Aulonia albimana, Arctosa perita.* Previous to the processing of the data, it has been converted in presences/absences to fit the type of data for the proposed analysis. The presence and absence of each spider species will be our response variables. Our predictors will be some environmental variables that may explain the presence or absence of the species. Six environmental variables were measured at 28 sites. These were *water content, bare sand, moss cover, light reflection, fallen twigs,* and *cover herbs*. The complete set of data is shown in Table 1.

First, we applied the logistic biplot without constraints to compare it with the constrained version. The unconstrained version obtains sites and species scores without explicitly using the environmental information while in the constrained one the site scores are calculated to be linear combinations of the environmental variables. As measures of the goodness of fit for each species, we use the Nagelkerke pseudo R-squared and the percent of correctly classified individuals for each column of the binary matrix. The same measures can also be used for the constrained version. The comparison of both is shown in Table 2. Observe that the measures of the goodness of fit for the binary variables are slightly better for the unconstrained version, i.e., it captures better the structure of the binary matrix.

Table 3 shows the percent of variance of the environmental variables explained by the ordination dimensions of the species. Observe that now the constrained version captures much better the relation of the species presence with the environmental variables, so it contains the part of the species structure captured by the environment.

Table 2 Columns fit in the unconstrained and constrained models. This measures the relation among the observed variables as function of the site scores for the first two dimensions

	Nagelkerke		% Correct	
Species	Unconstrained	Constrained	Unconstrained	Constrained
Arctl	1.00	0.77	1.00	0.86
Prdl	0.61	0.64	0.82	0.89
Zrsp	0.98	0.74	1.00	0.93
Prdn	0.96	0.57	0.96	0.68
Prdp	0.86	0.72	0.93	0.89
Allb	1.00	0.74	1.00	0.82
Trct	0.99	0.66	1.00	0.89
Alpcn	0.99	0.63	1.00	0.93
Prdm	0.99	0.75	1.00	0.89
Alpcc	0.98	0.91	1.00	0.93
Alpf	0.96	0.74	1.00	0.86
Arctp	0.99	0.99	1.00	1.00

Table 3 Percent of variance of the environmental variables explained by the ordination

	Watcont	Barsand	Covmoss	Ligrefl	Falltwi	Coverher
Unconstrained	76.38	52.72	54.13	75.12	65.07	54.53
Constrained	90.11	69.05	73.90	89.92	90.30	93.27

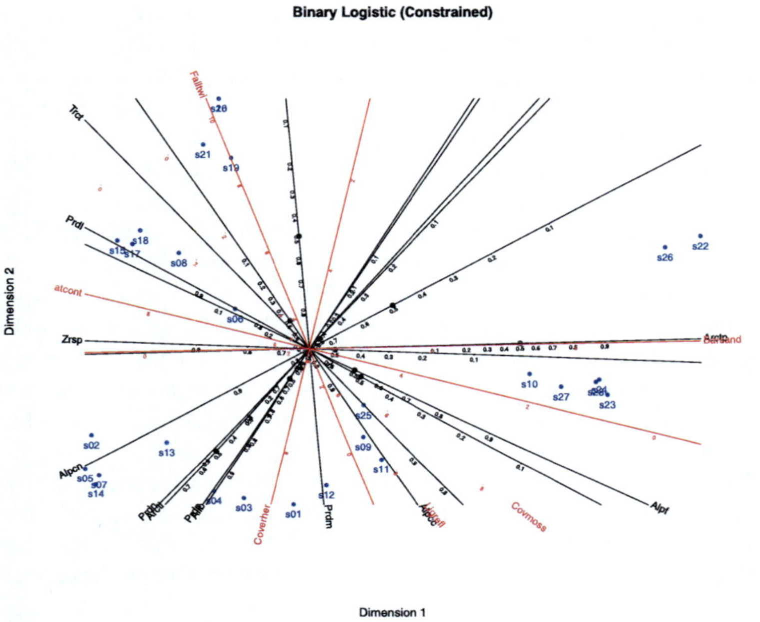

Fig. 1 Constrained Logistic Biplot for the binary species data

Figure 1 contains the triplot for the constrained version. The sites are represented with points, species, and environmental variables with straight lines (black and red, respectively). Projecting site points onto species lines, we obtain the expected probabilities of presence of the species. Projecting onto environmental lines, we obtain expected values for the variables.

A close view of the plot allows for the interpretations of the relations among sites, species and environmental variables. For example, on the right of the plot, we have a cluster of sites (s10, s27, s23, s24, s28, 26, 22). Those sites have similar species presences and environmental characteristics. Species *Arctp*, *Alpf*, *Alpcc*, or *Prdm* have high expected probabilities of being present in those sites that are also characterized by high content of *bare sand*, *moss cover*, or *light reflection* and much smaller values of the rest. We can also conclude that those species are highly related to the listed environmental variables, i.e., angles among variables are interpreted in terms of relation: acute means high direct relation, flat means high inverse relation and right, no relation.

5 Software Note

All the procedures explained in this paper can be calculated using the R (R Core Team 2019) package: *MultbiplotR* (Vicente-Villardon 2019).

Acknowledgements This research was supported by grant RTI2018-093611-B-I00 from the Ministerio de Ciencia Innovación y Universidades of Spain.

References

Demey, J.R., Vicente-Villardon, J.L., Galindo-Villardon, M.P., Zambrano, A.Y.: Identifying molecular markers associated with classification of genotypes by External Logistic Biplots. Bioinformatics **24**(24), 2832–2838 (2008)

Gabriel, K.R.: The biplot graphic display of matrices with application to principal component analysis. Biometrika **58**(3), 453–467 (1971)

Israels, A.Z.: Redundancy analysis for qualitative variables. Psychometrika **49**(3), 331–346 (1984)

Legendre, P., Anderson, M.J.: Distance-based redundancy analysis: testing multispecies responses in multifactorial ecological experiments. Ecol. Monogr. **69**(1), 1–24 (1999)

Legendre, P., Legendre, L.F.: Numerical Ecology. Elsevier, Amsterdam (2012)

R Core Team R: A language and environment for statistical computing. R Foundation for Statistical Computing, Vienna, Austria (2019). Available via. https://www.R-project.org/

Rao, C.R.: The use and interpretation of principal component analysis in applied research. Sankhyā: Indian J. Stat. Ser. A **26**(4), 329–358 (1964)

Smeenk-Enserink, N., Van der Aart, P.J.M.: Correlations between distributions of hunting spiders (Lycosidae, Ctenidae) and environmental characteristics in a dune area. Neth. J. Zool. **25**(1), 1–45 (1974)

Takane, Y., Tadashi, S.: Principal component analysis with external information on both subjects and variables. Psychometrika **56**(1), 97–120 (1991)

Ter Braak, C.J.F.: Canonical correspondence analysis: a new eigenvector technique for multivariate direct gradient analysis. Ecology **67**(5), 1167–1179 (1986)

Van der Wollenberg, A.L.: Redundancy analysis an alternative for canonical correlation analysis. Psychometrika **42**(2), 207–219 (1977)

Vicente-Villardon, J.L., Galindo-Villardon, M.P., Blázquez-Zaballos, A.: Logistic biplots. In: Greenacre, M., Blasius, J. (eds.) Multiple Correspondence Analysis and Related Methods, pp. 503–521. Chapman and Hall/CRC, London (2006)

Vicente-Villardon, J. L.: MultBiplotR: Multivariate Analysis using Biplots. R package version 19.11.19 (2019). Available via. http://biplot.usal.es/multbiplot/multbiplot-in-r/

Predictive Power of School Motivation Clusters in Secondary Education

Matthijs J. Warrens and W. Miro Ebert

Abstract In many applications of cluster analysis in educational research, the solutions found have very limited predictive power for relevant outcomes. In this paper, we explore whether the clusterings found have more predictive power (in terms of explained variance) if relevant outcomes are included in the estimation procedure, using a real-world data set on school motivation. We compare various normal mixture models with different distal outcomes involved such as no outcome variable, a single outcome, all outcomes. All models were estimated using the simultaneous estimation (one-step) procedure for distal outcomes in Latent GOLD. Partial eta squared (η_p^2) was used to assess predictive power. Including relevant outcomes will in most cases increase the predictive power of the models. Furthermore, the increase in power is more substantial, in the absolute sense, when the correlation between the outcome variable and input variables is higher.

Keywords Mastery motivation · Performance motivation · Social motivation · Extrinsic motivation · Normal mixture models

1 Introduction

Motivation for school is an important research topic in educational sciences (Korpershoek et al. 2015; McInerney and Ali 2006; Wilson et al. 2016). Students can exhibit different levels and orientations of motivation. A first distinction can be made between students that want to master a new skill, and students that aim to receive

---

M. J. Warrens (✉)
University of Groningen, Faculty of Behavioural and Social Sciences, Groningen Institute for Educational Research, Grote Rozenstraat 3, 9712 Groningen, TG, The Netherlands
e-mail: m.j.warrens@rug.nl

W. M. Ebert
University of Groningen, Faculty of Behavioural and Social Sciences, Grote Kruisstraat 2/1, 9712 Groningen, TS, The Netherlands
e-mail: w.m.ebert@student.rug.nl

T. Chadjipadelis et al. (eds.), *Data Analysis and Rationality in a Complex World*, Studies in Classification, Data Analysis, and Knowledge Organization,
https://doi.org/10.1007/978-3-030-60104-1_37

high grades. A mastery goal orientation is often related to students' engagement with school tasks (Gonida et al. 2009), high levels of metacognitive activity (Schmidt and Ford 2003), and their self-efficacy beliefs (Coutinho and Neuman 2008; Pajares et al. 2000), whereas performance goal orientation generally is related to high levels of effort (Elliott and Church 1997) and school performance (Giota 2010). McInerney and Ali (2006) considered two additional goal orientations, namely a social goal orientation and an extrinsic goal orientation. Students with the former goal orientation focus on the social gains of academic achievements (McInerney and Ali 2006; Urdan and Maehr 1995). Students with the latter goal orientation aim to gain rewards or praise through their studying (Ryan and Deci 2000).

We may differentiate between avoidance and approach versions of mastery and performance goal orientation: students either aim for mastery or high performance (approach), or are motivated to avoid negative consequences, i.e., grades (Elliott and Church 1997; Elliot and McGregor 2001). In this study, we focus on the four approach tendencies of mastery, performance, social, and extrinsic motivation. Students tend to show a mixed profile of the different orientations that motivate them to achieve in school (Elliot and McGregor 2001; Giota 2010; Van Yperen 2006).

To find clusters (groups, profiles) in data, researchers commonly use cluster analysis (Hennig et al. 2015). There are many different clustering methods, which can be divided into several approaches (Hennig et al. 2015). Examples of approaches are partitioning methods (e.g., K-means; Lloyd 1982), hierarchical methods (Murtagh and Legendre 2014), spectral clustering (Von Luxburg 2006), density-based methods (Louhichi et al. 2017), and model-based methods (Fraley and Raftery 2002; McLachlan and Peel 2004). Examples of meaningful clusters pertain to the reading behavior of pupils (Karlsson et al. 2018) and to conceptual and procedural knowledge of fractions of pupils (Hallett et al. 2012).

A sample of students can of course be clustered (partitioned) in many different ways, and different clusterings of the same data may be interesting for different reasons (Van Mechelen et al. 2018). Therefore, an important topic in cluster analysis is cluster validation, which is the process of evaluating the 'goodness' of a clustering. A useful criterion for evaluating a grouping of students is to assess whether the grouping can predict (explain) relevant or distal outcomes, which are variables that were not used to find the groups. A clustering can be considered very useful if it has high predictive power, e.g., in terms of explained variance, for some distal outcomes.

It is striking that in many applications of cluster analysis in educational research, the solutions found have very limited predictive power for distal outcomes. In a random sample of 90 articles that assessed the predictive value of profiles of students in terms of variance explained, we found 18 (20%) with no predictive power, 64 (71%) with very limited, and 8 (9%) with good predictive power. The lack of predictive power of these studies is perhaps no surprise: since the distal outcomes are not used to find the groups, clustering methods do not optimize the classification of students with respect to the relevant outcomes.

In this paper, we explore, using a real-world data set on school motivation, whether the clusterings (models) found have more predictive power if distal outcomes are involved in the estimation procedure. In the context of model-based clustering, there are several ways of including external variables (Asparouhov and Muthén 2014; Bakk and Vermunt 2016; Lanza et al. 2013). The approaches differ in how the clusters and distal outcomes are estimated (simultaneous, two-step, or three-step estimation). The approaches have been compared and validated in simulation studies, but applications and comparisons in the context of real-world-data are lacking.

The paper is organized as follows. In the next section, we describe the sample, the measurements, and the statistical analysis plan. We estimated various normal mixture models. The models differed in what distal outcomes are involved in the estimation of the clusters: no outcome variable, a single outcome, all outcomes. We compare the performance of the models in terms of the predictive power (variance explained with partial eta squared) regarding the outcome variables. Next, Sect. 3 contains the results, and Sect. 4 describes a short discussion.

2 Method

2.1 Sample and Population

The data originated from the large-scale longitudinal research study COOL5-18 performed in the Netherlands between 2007 and 2016 (www.cool5-18.nl; Timmermans et al. 2017). This research study followed students from age five to eighteen in their school career. In school year 2007/2008, 80 secondary schools with in total 21,384 9th graders participated in COOL5-18. For 9,544 students, one or more scores on the variables were missing. We did not impute these scores, but we ignored all students with at least one score missing. Although this was not investigated, we did not expect that listwise deletion of students with missing data had a major impact on representativeness of the sample and the results. The final sample consists of $N = 11{,}840$ students. There are 5,769 boys (49%) and 5,971 girls (50%), and for 100 students their gender was not recorded (1%). The research study COOL5-18 provided a representative reference sample for students aged five to eighteen in the Netherlands by using different criteria of sampling (e.g., direction, province, degree of urbanization of the school, and the school performance relative to other Dutch schools; Timmermans et al. 2017). The population of interest consists of all 9th graders in the Netherlands.

2.2 Measurements

The Dutch version of the Inventory of School Motivation (ISM) (McInerney and Ali 2006) was used to assess the four school motivation constructs. This inventory consists of 32 items with a 5-point Likert scale ($1 =$ strongly disagree to $5 =$ strongly agree). A high reliability in terms of Cronbach's α (Cronbach 1951) was found for all subscales of the inventory: mastery motivation (9 items, $\alpha = 0.77$), performance motivation (7 items, $\alpha = 0.84$), social motivation (7 items, $\alpha = 0.74$), and extrinsic motivation (9 items, $\alpha = 0.86$). We considered four external variables, namely, reading comprehension, mathematics, academic self-efficacy, and school commitment. The first two constructs were assessed with standardized achievement tests (Korpershoek et al. 2015; Timmermans et al. 2017). Academic self-efficacy was assessed with an adapted version of the self-efficacy scale by Midgley et al. (2000). Students' commitment to school was assessed with an adapted version of the Utrecht-Groningen Identity Development Scale (U-GIDS) (Crocetti et al. 2008). A high reliability in terms of Cronbach's α was found for all distal outcomes: reading comprehension (46 items, $\alpha = 0.92$), mathematics (50 items, $\alpha = 0.94$), self-efficacy (6 items, $\alpha = 0.82$), and school commitment (13 items, $\alpha = 0.80$).

2.3 *Statistical Analysis Plan*

In this study, we used model-based clustering. It may be the case that other clustering approaches are more appropriate for these data, but this has not been thoroughly investigated in the literature (Blom et al. 2020; Van Mechelen et al. 2018). In model-based clustering, it is assumed that the data are generated from a mixture of underlying probability distributions (Fraley and Raftery 2002; Hennig et al. 2015). In this study, we assumed that these probability distributions are mixtures of normal distributions. These distributions are commonly used, although it also has not been thoroughly investigated whether this assumption is well founded.

In total, we estimated 54 normal mixture models. The models differed in what distal outcomes were involved in the estimation of the clusters (none, single outcome, all outcomes) and the number of clusters (2–10). All models were estimated using the maximum likelihood estimator in Latent GOLD, using the default maximum posterior estimation for estimating the variance-covariance matrix, and the simultaneous estimation (one-step) procedure for distal outcomes (Vermunt and Magidson 2013). The probabilistic clustering method produces posterior probabilities. These probabilities specify for each object the degree of membership to the clusters. We obtained partitions by assigning each student to the cluster with the highest associated posterior probability. To assess the predictive power of the clustering found, we then performed for each model one-way analyses of variance using cluster membership as a factor and the separate distal outcomes as dependent variables. To quantify the predictive power, we calculated the partial eta squared statistic (η_p^2; see, e.g., Field 2013) associated with each analysis of variance.

To explore whether the models found have more predictive power if distal outcomes are involved in the estimation procedure or not, we compared models with the same number of clusters. A priori, it was unclear whether including distal outcomes will increase the predictive power of the models. Furthermore, it was unclear whether including one or all variables would make a difference. Note that we did not focus on finding the optimal number of clusters and the substantive meaning of the clusters. For these data, the latter issues are addressed in Blom et al. (2020) and Korpershoek et al. (2015).

3 Results

Table 1 presents the Pearson correlations between the eight variables. The values between the motivation constructs are moderate or substantial. The outcome variable that exhibits the largest correlations with the motivation dimensions is school commitment. Furthermore, the correlation values between academic self-efficacy and the motivation variables are small to moderate. Moreover, the values between both reading comprehension and mathematics, on the one hand, and the motivation constructs, on the other hand, are very small.

Table 2 presents the η_p^2 values corresponding to the four external variables for all 54 normal mixture models. The first nine rows (labeled 'None') correspond to the models that did not use any of the distal outcomes in the estimation procedure, whereas the bottom nine rows (labeled 'All') correspond to models that included all four distal outcomes. The middle nine rows (labeled 'Single') present the η_p^2 values for a total of 36 models. For each of the 36 models, including one outcome variable, Table 2 only contains the η_p^2 value associated with the external variable that was included in the model. The η_p^2 values of the outcomes that were not used in the estimation procedure are not reported in Table 2.

There are substantial differences between the η_p^2 values of the four outcome variables. For each model, the η_p^2 value associated with school commitment is always the highest, followed by the values associated with self-efficacy. Overall, the η_p^2 values of school commitment are large to very large, the values of self-efficacy are moderate to large, and the values corresponding to reading comprehension and mathematics are small and sometimes even negligible. There seems to be a relationship between these patterns and the Pearson correlations between the motivation dimensions, on

Table 1 Pearson correlation coefficients between the eight variables

Variables	1	2	3	4	5	6	7
1. Mastery motivation	1						
2. Performance motivation	0.378	1					
3. Social motivation	0.465	0.159	1				
4. Extrinsic motivation	0.525	0.567	0.353	1			
5. Reading comprehension	0.080	−0.045	0.062	−0.061	1		
6. Mathematics	0.012	0.096	0.023	−0.018	0.344	1	
7. Academic self-efficacy	0.334	0.258	0.144	0.165	0.079	0.17	1
8. School commitment	0.661	0.241	0.317	0.329	0.106	0.051	0.363

Table 2 Values of partial eta squared of four outcome variables for various normal mixture models

Outcomes	# Clusters	Reading comp.	Mathematics	Self-efficacy	School com.
None	2	0.001	0.001	0.038	0.107
None	3	0.000	0.002	0.065	0.165
None	4	0.004	0.003	0.073	0.199
None	5	0.007	0.009	0.068	0.196
None	6	0.006	0.008	0.081	0.231
None	7	0.014	0.010	0.076	0.208
None	8	0.011	0.010	0.085	0.239
None	9	0.015	0.012	0.086	0.239
None	10	0.018	0.012	0.091	0.242
Single	2	0.001	0.003	0.068	0.235
Single	3	0.001	0.006	0.117	0.398
Single	4	0.039	0.008	0.140	0.482
Single	5	0.035	0.018	0.142	0.535
Single	6	0.024	0.017	0.154	0.532
Single	7	0.053	0.015	0.136	0.556
Single	8	0.039	0.017	0.154	0.570
Single	9	0.040	0.026	0.223	0.591
Single	10	0.076	0.017	0.250	0.597
All	2	0.000	0.001	0.073	0.231
All	3	0.007	0.003	0.122	0.391
All	4	0.019	0.007	0.119	0.464
All	5	0.021	0.008	0.144	0.517
All	6	0.022	0.008	0.146	0.519
All	7	0.029	0.010	0.151	0.546
All	8	0.042	0.009	0.167	0.563
All	9	0.048	0.024	0.219	0.564
All	10	0.053	0.022	0.217	0.536

the one hand, and the external variables, on the other hand: larger Pearson correlations go along with larger η_p^2 values. Furthermore, the η_p^2 values seem to increase, although in a non-monotonic way, with the number of clusters. The latter may be an artifact since the variance of the factor in the analysis of variance will increase with the number of clusters.

For a fixed number of clusters, the model that includes a single distal outcome always has a larger η_p^2 value than the model that did not use any of the distal outcomes in the estimation. Furthermore, except for a few cases with relatively small η_p^2 values, the model that includes all outcomes has a larger η_p^2 value than the model that included none of the distal outcomes. Thus, overall we may conclude that including the distal outcomes will increase the predictive power of the models. However, there are substantial differences between the outcome variables in this respect: the increase in effect size is larger, in the absolute sense, when the correlation between the outcome variables and input variables is larger. The η_p^2 values of school commitment were already large for the models that did not use the outcome variables in the estimation, but they skyrocket if school commitment is used in the estimation procedure.

Finally, there are notable differences between the outcome variables with regard to including a single variable or all variables in the estimation. For a fixed number of clusters, the model that includes school commitment only always has a larger η_p^2 value than the model that includes all outcomes. It should be noted that the differences are relatively small compared to the effect sizes. Nevertheless, school commitment does not seem to benefit from including the other variables. For the other three variables, the results are highly mixed, and it depends on the number of clusters which model produces the largest η_p^2.

4 Discussion

In this study, we explored whether clustering models have more predictive power in terms of variance explained if relevant outcomes are involved in the estimation procedure. We compared various normal mixture models using motivation data. Including relevant outcomes will in most cases increase the predictive power of the models. The increase in power is more substantial, in the absolute sense, when the correlation between the outcome variable and input variables is higher. If the clustering aim (Van Mechelen et al. 2018) is to produce a solution that has maximal predictive power with respect to one or several distal outcomes, we recommend that the distal outcomes are involved in the estimation procedure.

This study has several limitations. We considered only one data set and one way of including distal outcomes in the estimation of the clusters. More comprehensive benchmarking studies (Van Mechelen et al. 2018) are required that compare the various methods (Asparouhov and Muthén 2014; Bakk and Vermunt 2016; Lanza et al. 2013), and these studies should explore more real-world data sets for possible systematic differences. Furthermore, future benchmarking studies could take into account the hierarchical structure of the data (students nested in classes nested in schools). Moreover, in this study, we only focused on one criterion, the predictive power in terms of partial eta squared, ignoring any substantive interpretation of the clusters. However, the inclusion of distal outcomes may influence the normal mixture measurement model, potentially changing the substantive meaning of the clusters, which is something that should also be explored in future studies.

References

Asparouhov, T., Muthén, B.: Auxiliary variables in mixture modeling: Three-step approaches using Mplus. Struct. Equ. Model. **21**, 329–341 (2014)

Bakk, Z., Vermunt, J.K.: Robustness of stepwise latent class modeling with continuous distal outcomes. Struct. Equ. Model. **23**, 20–31 (2016)

Blom, D.M., Warrens, M.J., Faber, M.: School motivation profiles of students in secondary education: a cluster benchmarking study. Manuscript under review

Coutinho, S.A., Neuman, G.: A model of metacognition, achievement goal orientation, learning style and self-efficacy. Learn Environ. Res. **11**, 131–151 (2008)

Crocetti, E., Rubini, M., Meeus, W.: Capturing the dynamics of identity formation in various ethnic groups. Development and validation of a three-dimensional model. J. Adolesc. **31**, 207–222 (2008)

Cronbach, L.J.: Coefficient alpha and the internal structure of tests. Psychometrika **16**, 297–334 (1951)

Elliot, A.J., McGregor, H.A.: A 2 × 2 achievement goal framework. J. Pers. Soc. Psychol. **80**, 501–519 (2001)

Elliott, A.J., Church, M.A.: A hierarchical model of approach and avoidance achievement motivation. J. Pers. Soc. Psychol. **72**, 218–232 (1997)

Field, A.: Discovering Statistics Using IBM SPSS Statistics, 4th edn. Sage, London (2013)

Fraley, C., Raftery, A.E.: Model-based clustering, discriminant analysis, and density estimation. J. Am. Stat. Assoc. **97**, 611–631 (2002)

Giota, J.: Multidimensional and hierarchical assessment of adolescents' motivation in school. Scand. J. Educ. Res. **54**, 83–97 (2010)

Gonida, E.N., Voulala, K., Kiosseoglou, G.: Students' achievement goal orientations and their behavioral and emotional engagement: Co-examining the role of perceived school goal structures and parent goals during adolescence. Learn Individ. Differ. **19**, 53–60 (2009)

Hallett, D., Nunes, T., Bryant, P., Thorpe, C.M.: Individual differences in conceptual and procedural fraction understanding: the role of abilities and school experience. J. Exp. Child Psychol. **113**, 469–486 (2012)

Hennig, C., Meilă, M., Murtagh, F., Rocci, R.: Handbook of Cluster Analysis. Chapman and Hall/CRC, Boca Raton (2015)

Karlsson, J., van den Broek, P., Helder, A., Hickendorff, M., Koornneef, A., Van Leijenhorst, L.: Profiles of young readers: evidence from thinking aloud while reading narrative and expository texts. Learn Individ. Differ. **67**, 105–116 (2018)

Korpershoek, H., Kuyper, H., Van der Werf, G.: Differences in students' school motivation: a latent class modelling approach. Soc. Psychol. Educ. **18**, 137–163 (2015)

Lanza, S.T., Tan, X., Bray, B.C.: Latent class analysis with distal outcomes: a flexible model-based approach. Struct. Equ. Model. **20**, 1–26 (2013)

Lloyd, S.P.: Least squares quantization in PCM. IEEE T Inform. Theory **28**, 129–137 (1982)

Louhichi, S., Gzara, M., Ben-Abdallah, H.: Unsupervised varied density based clustering algorithm using spline. Pattern Recognit. Lett. **93**, 48–57 (2017)

McInerney, D.M., Ali, J.: Multidimensional and hierarchical assessment of school motivation: Cross-cultural validation. Educ. Psychol. **26**, 595–612 (2006)

McLachlan, G., Peel, D.: Finite Mixture Models. Wiley, New York (2004)

Midgley, C., Maehr, M.L., Hruda, L.Z., Anderman, E., Anderman, L., Freeman, K.E., et al.: Manual for the Patterns of Adaptive Learning Scales (PALS). University of Michigan, Ann Arbor, MI (2000)

Murtagh, F., Legendre, P.: Ward's hierarchical agglomerative clustering method: which algorithms implement Ward's criterion? J. Classif. **31**, 274–295 (2014)

Pajares, F., Britner, S.L., Valiante, G.: Relation between achievement goals and self-beliefs of middle school students in writing and science. Contemp. Educ. Psychol. **25**, 406–422 (2000)

Ryan, R.M., Deci, E.L.: Self-determination theory and the facilitation of intrinsic motivation, social development, and well-being. Am. Psychol. **55**, 68–78 (2000)

Schmidt, A.M., Ford, J.K.: Learning with a learner control training environment: The interactive effects of goal orientation and metacognitive instruction on learning outcomes. Pers. Psychol. **56**, 405–429 (2003)

Timmermans, A. C., Naayer, H. M., Keuning, J., Zijsling, D. H.: Cohortonderzoek COOL5-18. Basisrapport VO-3 in 2014. GION Onderwijs/Onderzoek< Groningen (2017)

Urdan, T.C., Maehr, M.L.: Beyond a two-goal theory of motivation and achievement: a case for social goals. Educ. Res. Rev. **65**, 213–243 (1995)

Van Mechelen, I., Boulesteix, A.L., Dangl, R., Dean, N., Guyon, I., Hennig, C., Leisch, F., Steinley, D.: Benchmarking in cluster analysis: A white paper (2018). ArXiv:1809.10496

Van Yperen, N.W.: A novel approach to assessing achievement goals in the context of the 2 × 2 framework: Identifying distinct profiles of individuals with different dominant achievement goals. Pers. Soc. Psychol. Bull. **32**, 432–445 (2006)

Vermunt, J.K., Magidson, J.: Technical Guide for Latent GOLD 5.0: Basic, Advanced, and Syntax. Belmont Massachussetts, Statistical Innovations Inc. (2013)

Von Luxburg, U.: A tutorial on spectral clustering. Stat. Comput. **17**, 395–416 (2006)

Wilson, T.M., Zheng, C., Lemoine, K.A., Martin, C.P., Tang, Y.: Achievement goals during middle childhood: Individual differences in motivation and social adjustment. J. Exp. Educ. **84**, 723–743 (2016)